The IMA Volumes in Mathematics and Its Applications

Volume 3

Institute for Mathematics and Its Applications
IMA

The **Institute for Mathematics and Its Applications** was established by a grant from the National Science Foundation to the University of Minnesota in 1982. The IMA seeks to encourage the development and study of fresh mathematical concepts and questions of concern to the other sciences by bringing together mathematicians and scientists from diverse fields in an atmosphere that will stimulate discussion and collaboration.

The IMA Volumes are intended to involve the broader scientific community in this process.

Hans Weinberger, Director
George R. Sell, Associate Director

IMA Programs

1982–1983 Statistical and Continuum Approaches to Phase Transition

1983–1984 Mathematical Models for the Economics of Decentralized Resource Allocation

1984–1985 Continuum Physics and Partial Differential Equations

1985–1986 Stochastic Differential Equations and Their Applications

1986–1987 Scientific Computation

1987–1988 Applied Combinatorics

1988–1989 Nonlinear Waves

Springer Lecture Notes from the IMA

The Mathematics and Physics of Disordered Media
Editors: Barry Hughes and Barry Ninham
(Lecture Notes in Mathematics, Volume 1035, 1983)

Orienting Polymers
Editor: J. L. Ericksen
(Lecture Notes in Mathematics, Volume 1063, 1984)

New Perspectives in Thermodynamics
Editor: James Serrin
(Springer-Verlag, 1986)

Models of Economic Dynamics
Editor: Hugo Sonnenschein
(Lecture Notes in Economics, Volume 264, 1986)

Metastability and Incompletely Posed Problems

Edited by
Stuart S. Antman, J.L. Ericksen,
David Kinderlehrer, and Ingo Müller

With 46 Illustrations

Springer-Verlag
New York Berlin Heidelberg
London Paris Tokyo

Stuart S. Antman
Department of Mathematics, University of Maryland, College Park, MD 20742, U.S.A.

J.L. Ericksen
School of Mathematics and Department of Aerospace Engineering and Mechanics, University of Minnesota, Minneapolis, MN 55455, U.S.A.

David Kinderlehrer
School of Mathematics, University of Minnesota, Minneapolis, MN 55455, U.S.A.

Ingo Müller
FB9-Hermann Föttinger Institut, Technical University, Berlin, F.R.G.

AMS Classification: 35-06, 35R25, 35B35

Library of Congress Cataloging-in-Publication Data
Metastability and incompletely posed problems.
(The IMA volumes in mathematics and its applications ; v. 3)
Proceedings of a workshop held during the IMA 1984–85 program on continuum physics and partial differential equations.
1. Differential equations, Partial—Improperly posed problems—Congresses. 2. Field theory (Physics)—Congresses. 3. Stability—Congresses. I. Antman, S. S. (Stuart S.) II. University of Minnesota. Institute for Mathematics and its Applications. III. Title.
QA377.M47 1986 515.3′53 86-28033

Softcover reprint of the hardcover 1st edition 1987

9 8 7 6 5 4 3 2 1
ISBN-13: 978-1-4613-8706-0 e-ISBN-13: 978-1-4613-8704-6
DOI: 10.1007/978-1-4613-8704-6

The IMA Volumes in Mathematics and Its Applications

Current Volumes:

Volume 1: Homogenization and Effective Moduli of Materials and Media
Editors: J.L. Ericksen, David Kinderlehrer, Robert Kohn, and J.-L. Lions

Volume 2: Oscillation Theory, Computation, and Methods of Compensated Compactness
Editors: Constantine Dafermos, J.L. Ericksen, David Kinderlehrer, and Marshall Slemrod

Volume 3: Metastability and Incompletely Posed Problems
Editors: Stuart S. Antman, J.L. Ericksen, David Kinderlehrer, and Ingo Müller

Forthcoming Volumes:

1984–1985: Continuum Physics and Partial Differential Equations

Theory and Applications of Liquid Crystals
Amorphous Polymers and Non-Newtonian Fluids
Dynamical Problems in Continuum Physics

1985–1986: Stochastic Differential Equations and Their Applications

Random Media
Percolation Theory and Ergodic Theory of Infinite Particle Systems
Hydrodynamic Behavior and Interacting Particle Systems and Applications
Stochastic Differential Systems, Stochastic Control Theory and Applications

D
A
B
C
E

CONTENTS

FOREWORD

This IMA Volume in Mathematics and its Applications

Metastability and Incompletely Posed Problems

represents the proceedings of a workshop which was an integral part of the 1984-85 IMA program on CONTINUUM PHYSICS AND PARTIAL DIFFERENTIAL EQUATIONS. We are grateful to the Scientific Committee:

J.L. Ericksen

D. Kinderlehrer

H. Brezis

C. Dafermos

for their dedication and hard work in developing an imaginative, stimulating, and productive year-long program.

George R. Sell

Hans Weinberger

Preface

Most equilibrium events in nature do not realize configurations of minimum energy. They are only metastable. Available knowledge of constitutive relations and environmental interactions may be limited. As a result, many configurations may be compatible with the data. Such questions are incompletely posed. The papers in this volume address a wide variety of these issues as they are perceived by the material scientist and the mathematician. They represent a portion of the significant activity which has been underway in recent years, from the experimental arena and physical theory to the analysis of differential equations and computation.

One may ask what circumstances tend to promote stability or how does metastability manifest itself in nature. Traditionally, knowledge of a material or process is achieved by observation and analysis of small departures from an equilibrium configuration. With increasing frequency, however, phenomena encountered display not only larger but more dramatic changes, such as occur in phase transitions. By grappling with specific problems, the authors provide a little experience from which we might aspire to abstract viable methods for the analysis and prediction of metastable behavior. There is a speculative character to much of this work, partly because of problems stated but not resolved.

These issues were among the principal themes of the 1984-1985 I.M.A. Program, Continuum Physics and Partial Differential Equations. Additional contributions to this subject may be found in the other volumes of this series listed at the end of this book.

The workshop brought together researchers in a number of areas of engineering, mathematics, and physics. The conference committee greatly appreciates the concerted efforts of the speakers and discussants to make their presentations intelligible to a mixed audience. Included in this volume are papers connected to the topic of the workshop which were presented at other times during the year.

The conference committee would like to take this opportunity to thank the staff of the I.M.A., Professors Weinberger and Sell, Ms. Susan Anderson, Ms. Pat Kurth, and Mr. Robert Copeland Ms. Mary Saunders, Ms. Ceil McAree and Ms. Julie Hicks for their assistance in arranging the workshop. Special thanks are due to Ms. Debbie Bradley and Ms. Kaye Smith for their preparation of manuscripts. We gratefully acknowledge the support of the National Science Foundation and the Army Research Office.

S.S. Antman

J.L. Ericksen

D. Kinderlehrer

I. Müller

Conference Committee

DISSIPATIVE MECHANISMS

Stuart S. Antman

Department of Mathematics
University of Maryland
College Park, MD 20742

and

Reza Malek-Madani

Department of Mathematics
U.S. Naval Academy
Annapolis, MD 21402

1. Introduction

A compressible heat-conducting homogeneous Newtonian fluid is described by the following system of equations for the velocity $\mathbf{v}$, density ρ, and temperature θ at position $\mathbf{x}$ and time t:

$$(1.1)\qquad \rho(\mathbf{x},t)\mathbf{v}_t(\mathbf{x},t) = -\rho(\mathbf{x},t)\mathbf{v}(\mathbf{x},t)\cdot\nabla\mathbf{v}(\mathbf{x},t) - \nabla\pi(\rho(\mathbf{x},t),\theta(\mathbf{x},t)) + \nabla\cdot[\mu(\rho(\mathbf{x},t),\theta(\mathbf{x},t))\nabla\mathbf{v}(\mathbf{x},t)],$$

$$(1.2)\qquad \rho_t(\mathbf{x},t) = -\mathbf{v}(\mathbf{x},t)\cdot\nabla\rho(\mathbf{x},t) - \rho(\mathbf{x},t)\nabla\cdot\mathbf{v}(\mathbf{x},t),$$

$$(1.3)\qquad \frac{\partial}{\partial t}\varepsilon(\rho(\mathbf{x},t),\theta(\mathbf{x},t)) = \nabla\cdot[\kappa(\rho(\mathbf{x},t),\theta(\mathbf{x},t))\nabla\theta(\mathbf{x},t)].$$

Here and below, we denote partial derivatives with subscripts. The functions $[0,\infty)\times[0,\infty)\ni(\rho,\theta)\mapsto\pi(\rho,\theta),\mu(\rho,\theta),\varepsilon(\rho,\theta),\kappa(\rho,\theta)$ are prescribed constitutive functions. π represents a modified pressure, μ the viscosity, ε the internal energy, and κ the thermal conductivity. These constitutive functions are usually required to satisfy

$$(1.4)\qquad \pi_\rho > 0,\ \mu > 0,\ \varepsilon_\theta > 0,\ \kappa > 0.$$

In many circumstances it is reasonable to assume that

(1.5) $$\pi_\theta = 0, \quad \mu = \text{const}, \quad \varepsilon = (\text{const})\theta, \; \kappa = \text{const}.$$

Equations (1.1), (1.2), (1.3) are respectively the Navier-Stokes equations, the continuity equation, and the heat equation. If ρ and θ are regarded as given positive-valued functions, then (1.1) is a semilinear parabolic system for $\mathbf{v}$ when μ is everywhere positive. If ρ alone is given and (1.4) is satisfied, then (1.3) is a quasilinear parabolic equation for θ.

If $\partial\pi/\partial\theta = 0$ and if $\mu = 0$, then (1.1) reduces to the Euler equations for an inviscid compressible fluid. The Euler equations and (1.2) form a quasilinear hyperbolic system for $\mathbf{v}$ and ρ. In fluid dynamics, these Euler equations are used to approximate the Navier-Stokes equations. Since the Euler equations and (1.2) admit shocks, their analysis presents serious mathematical difficulties. To treat these, it is often convenient to adopt the paradoxical position that the Navier-Stokes equations are to be used to approximate the Euler equations. Indeed, the question of how solutions of (1.1), (1.2) (and possibly (1.3)) behave as the parameter μ approaches 0 is still open. But for analogous model problems, this limit process can be analyzed rather completely (cf. Hopf (1950)) and represents one of the best tools for probing the structure of shocks for the resulting hyperbolic equations.

Shock structure can also be approached from an alternative route in which conditions across shocks are determined by an entropy principle reflecting an interpretation of the Second Law of Thermodynamics (cf. Dafermos (1984)). The most popular version of this law for nonlinear continuum physics is the Clausius-Duhem Inequality. One of the major pieces of evidence supporting its adoption is that it delivers the second and fourth inequalities of (1.4) in a very natural way; in fact it was designed to do so. (Cf. Truesdell (1984).)

Thus the mechanism of viscosity for Newtonian fluids, characterized by the operator $\nabla \cdot (\mu\nabla\mathbf{v})$ of (1.1), has influenced greatly (and perhaps unduly) the treatment of evolution equations from all of continuum physics. In this article we examine some puzzling issues that arise in the study of simple problems for materials admitting more varied dissipative mechanisms. Now one kind of more interesting mechanism is obtained by replacing the semilinear operator

$\nabla \cdot [\mu(\rho,\theta)\nabla v]$ with a quasilinear operator $\nabla \cdot [f(\rho,\theta,\nabla v)]$, where $f(\rho,\theta,\cdot)$ is nonlinear. Within fluid dynamics, however, there is no experimental warrant for such generality. The only kinds of nonlinear dissipative mechanisms that seem reasonable for fluids are associated with memory effects and lead to functional-differential equations. No such prohibitions arise in solid mechanics. Below we study nonlinear viscous effects in solids that do not involve memory. The resulting equations are quasilinear partial differential equations.

2. Longitudinal Motion of a Viscoelastic Rod

Let the axis of a uniform viscoelastic rod in its straight unstressed reference configuration lie along the segment (0,1) of the x-axis. A typical material point of this axis (or equivalently, a typical material cross-section of the rod) is identified with the distance x of the point from 0 in this configuration. We consider longitudinal motions of this rod in which the axis of the rod is confined to the x-axis. Let $u(x,t)$ be the distance from the origin (along the x-axis) of material point x at time t. We require that $u_x > 0$ so that the local ratio of deformed to reference length never be reduced to 0. Then the longitudinal motion of this rod may be described by

$$u_{tt}(x,t) = \frac{\partial}{\partial x}\sigma(u_x(x,t),u_{xt}(x,t)), \tag{2.1}$$

where $(\xi,\eta) \mapsto \sigma(\xi,\eta)$ is the prescribed constitutive function giving the contact force in terms of the stretch u_x and the rate of stretch u_{xt}.

Existence, uniqueness, and regularity of solutions for various initial-boundary value problems for (2.1) have been studied by several authors for special forms of σ. In particular, Greenberg, MacCamy, & Mizel (1968), Greenberg (1969), Andrews (1980) treated problems for which

$$\sigma(\xi,\eta) = \tau(\xi) + \lambda\eta \tag{2.2}$$

where λ is a positive constant. Note that (2.2) corresponds to the dissipative mechanism of (1.1) with μ = const. Kanel' (1968) and MacCamy (1970) treated problems for which

$$\sigma(\xi,\eta) = \tau(\xi) + \lambda(\xi)\eta \tag{2.3}$$

when there is a number $K > 0$ such that $\lambda(\xi) \geq K$. The choice of (2.3) corresponds to the dissipative mechanism of (1.1) with μ depending on ρ. Dafermos (1969) treated problems for which there are numbers $K > 0$, $N > 0$ such that

$$\sigma_\eta \geq K, \ (\sigma_\xi)^2 \leq N^2\sigma_\eta. \tag{2.4a,b}$$

Note that in each case (2.1) is parabolic. In each case, the analysis of (2.1) relies on the derivation of a suitable number of a priori estimates. A crucial a priori estimate for (2.3) (and a fortiori for (2.2)) depends on the identity

$$\lambda(u_x(x,t))u_{xt}(x,t) = \frac{\partial}{\partial t}\int_1^{u_x(x,t)} \lambda(y)dy. \tag{2.5}$$

The corresponding estimate for (2.4) is based on the combined use of the Cauchy-Bunyakovskii-Schwarz inequality and inequality (2.4b).

If σ is to describe the longitudinal response of a rod, then it is reasonable to require that

$$\sigma(\xi,\eta) \to -\infty \quad \text{as} \quad \xi \to 0 \tag{2.6}$$

for each fixed η. Condition (2.6) says that an infinite compressive force must accompany a total compression. We now show that for a large class of materials conditions (2.4b) and (2.6) are incompatible. Thus the ostensible generality of assumptions (2.4) is illusory for the description of the longitudinal motion of rods.

For arbitrary σ, define $\tau(\xi) \equiv \sigma(\xi,0)$ so that σ has the form

$$\sigma(\xi,\eta) = \tau(\xi) + f(\xi,\eta) \tag{2.7}$$

where $f(\xi,0) = 0$. We study the family of σ's of the form (2.7) for which

$$\lambda(\xi)g'(\eta) \geq f_\eta(\xi,\eta) \geq \mu(\xi)h'(\eta), \tag{2.8}$$

$$g(0) = 0 = h(0),\ g'(\eta) > 0\ \forall\eta,\ h'(\eta) > 0\ \forall\eta,\ \mu(\xi) > 0\ \forall\xi > 0.$$

2.9 Proposition. If σ satisfies (2.4), (2.7), (2.8) with

$$\mu(\xi) \geq \sqrt{\lambda(\xi)} + o(\sqrt{\lambda(\xi)}) \text{ as } \lambda(\xi) \to \infty, \tag{2.10}$$

then σ cannot satisfy (2.6).

Proof. Conditions (2.4b) and (2.8) imply that

$$N^2\lambda(\xi)g'(\eta) \geq N^2 f_\eta(\xi,\eta) \geq [\tau'(\xi) + f_\xi(\xi,\eta)]^2, \tag{2.11}$$

whence

$$N^2\lambda(\xi)g'(0) \geq [\tau'(\xi)]^2. \tag{2.12}$$

For $\eta > 0$, it then follows from (2.8), (2.10), (2.12) that

$$\begin{aligned} \sigma(\xi,\eta) &\geq \tau(\xi) + \mu(\xi)h(\eta) \\ &\geq \tau(\xi) + [\sqrt{\lambda(\xi)} + o(\sqrt{\lambda(\xi)})]h(\eta) \\ &\geq \tau(\xi) + [\frac{\tau'(\xi)}{N\sqrt{g'(0)}} + o(\frac{\tau'(\xi)}{N\sqrt{g'(0)}})]h(\eta). \end{aligned} \tag{2.13}$$

Now the rightmost term of (2.13) approaches ∞ as $\xi \to 0$, for if it were bounded above, then an easy exercise in differential inequalities would show that τ is bounded below, in violation of (2.6) for $\eta = 0$. □

If $\tau(\xi)$ is asymptotic to $-C\xi^{-\alpha}$ as $\xi \to 0$ where C and α are positive constants, then Proposition 2.9 remains true if we replace (2.10) with

$$\mu(\xi) \geq \lambda(\xi)^{\frac{\alpha+\varepsilon}{2\alpha+2}} + o(\lambda(\xi)^{\frac{\alpha+\varepsilon}{2\alpha+2}}) \text{ as } \lambda(\xi) \to \infty. \tag{2.14}$$

We conjecture that Proposition 2.9 is valid under conditions even more general than (2.10) and (2.14).

We know of no existence theory for (2.1) valid for large time in which σ both depends nonlinearly upon η and satisfies (2.6). Thus the dissipative mechanism for (2.1) is the source of serious analytic difficulties. We remark that Proposition 2.9 does not deprive the pretty work of Dafermos (1969) of its physical significance. His results are perfectly suitable for shearing motions, generalizations of which are described in Section 4.

3. Viscous Shocks

If the function σ introduced in Section 2 is independent of η, then the rod is elastic. If, furthermore, $\sigma_\xi > 0$, then the following degenerate version of (2.1)

$$u_{tt}(x,t) = \frac{\partial}{\partial x}\sigma(u_x(x,t)) \tag{3.1}$$

is hyperbolic. If we set $u_t = v$, $u_x = w$, then (3.1) is equivalent to the system

$$v_t = \sigma(w)_x, \quad w_t = v_x \; . \tag{3.2a,b}$$

If we set $u = (v,w)$, $f(u) = (\sigma(w),v)$, then (3.2) can be written compactly as the hyperbolic conservation law

$$u_t = f(u)_x, \tag{3.3}$$

a form that obscures the basic physics displayed in (3.1).

Let us now regard (3.3) as a general hyperbolic conservation law with u taking values in R^n. Equation (3.3) admits discontinuous solutions, i.e., shocks. To ensure uniqueness of shocks some sort of entropy condition is required (cf. Dafermos (1984)). Gel'fand (1959) proposed that one prove that the entropy condition, having a thermodynamical origin, is equivalent to the requirement that a shock solution of (3.3) of the form

$$u(x,t) = \begin{cases} u_L & \text{for } x < st \\ u_R & \text{for } x > st \end{cases} \tag{3.4}$$

where u_L and u_R are given constant vectors and where s is the shock speed, is a suitable singular limit as $\varepsilon \searrow 0$ of travelling wave solutions of the associated parabolic problem

(3.5) $$u_t = f(u)_x + \varepsilon[D(u)u_x]_x$$

of the form

(3.6) $$u(x,t) = w\left(\frac{x - st}{\varepsilon} \right)$$

with

(3.7) $$\lim_{\xi\to-\infty} w(\xi) = u_L, \quad \lim_{\xi\to\infty} w(\xi) = u_R.$$

In (3.5) $D(u)$ represents a positive-definite matrix. Note that the equation for w is an ordinary differential equation in which the parameter ε is scaled out. Solutions of it that satisfy (3.7) have trajectories in the phase space of w that connect singular points at u_L and u_R. Such solutions are termed <u>viscous shocks</u>.

Conley & Smoller (1973) have studied (3.5) when D is a constant matrix, finding that the qualitative behavior of the phase portrait for w depends crucially on the form of D. Indeed, they exhibited D's for which there are no connecting orbits. Majda & Pego (1985) studied (3.5) when D depends on u. They showed that the D's for which there are no connecting orbits could be excluded by the imposition of a further requirement they call <u>strictly stability</u>. (In their work Conley & Smoller used novel topological arguments to prove the existence of viscous shocks.)

Note that the dissipative term $\varepsilon[D(u)u_x]_x$ of (3.5) has a form reminiscent of that of (1.1). The linearity of $y \to D(u)y$ is responsible for the absence of ε from the travelling wave equations. Note also that if (3.3) is regarded as representing (3.2), then this dissipative term contributes an apparently artifical dissipation to the compatibility equation (3.2b). (Slemrod, however, has observed that this dissipation can be associated with certain strain gradient (or surface tension) effects. cf. Hagen & Slemrod (1983).)

Our goal in the next section is to avoid such artificial models for elastic viscous response as (3.5). Instead, we study certain special "one-dimensional" motions of exact equations of three-dimensional continuum mechanics. We can thus use the highly developed constitutive theory of continuum mechanics to yield the most general physically possible nonlinear models for viscoelasticity.

4. Shearing Motions

Let a viscoelastic medium occupy all of Euclidean three-space. We identify a material point of body by its Cartesian coordinates (X,Y,Z). We restrict our attention to the special shearing motions in which the material point (X,Y,Z) occupies position (x,y,z) at time t, where

$$x = X + u(Z,t),\ y = Y + v(Z,t),\ z = Z. \tag{4.1}$$

We henceforth replace Z with z.

We assume that the medium is an incompressible isotropic homogeneous viscoelastic material of differential type of complexity 1. If the pressure at infinity is independent of X and Y, then a rather lengthy computation shows that the most general equations of motion for this material when (4.1) holds have the form

$$\begin{aligned} u_{tt} &= (\mu u_z + \nu u_{zt})_z, \\ v_{tt} &= (\mu v_z + \nu v_{zt})_z \end{aligned} \tag{4.2}$$

where μ and ν depend on the invariants

$$\eta_0 \equiv u_z^2 + v_z^2\ ,\ \eta_1 \equiv 2(u_z u_{zt} + v_z v_{zt}),\ \eta_2 \equiv u_{zt}^2 + v_{zt}^2. \tag{4.3}$$

Thus the response of this material in the motion (4.1) is characterized by two scalar constitutive functions μ and ν. That their arguments have the form (4.3) is a consequence of the requirements that the material properties are invariant under rigid motions and that the material is isotropic. Note that (4.2), (4.3) admit elegant complex forms in terms of $w = u + iv$.

We assume that μ and ν are twice continuously differentiable. We require that (4.2) be parabolic: The matrix of partial derivatives of $\mu u_z + \nu u_{zt}$ and $\mu v_z + \nu v_{zt}$ with respect to u_{zt} and v_{zt} is positive-definite. We also adopt the coercivity condition that

$$(4.4)\qquad u_{zt}(\mu u_z + \nu u_{zt}) + v_{zt}(\mu v_z + \nu v_{zt}) \to \infty \text{ as } u_{zt}^2 + v_{zt}^2 \to \infty.$$

By assuming suitable two-dimensional generalizations of (2.4), we can obtain global existence of solutions of certain initial-boundary value problems for (4.2) by following Dafermos (1969). (Cf. Antman & Malek-Madani (1986b).) The objections we voiced against (2.4) are irrelevant here because the material is incompressible.

The hyperbolic version of (4.2), obtained by setting $\nu = 0$ and by assuming that μ depends only on η_0, is richer than its scalar analog: It admits a countable infinity of standing waves. (Cf. Antman & Guo (1984).)

In accord with the discussion of Section 3, we seek travelling wave solutions of (4.2) of the form

$$(4.5)\qquad u(z,t) + iv(z,t) = w(z - ct).$$

We wish to determine how these solutions depend on μ and ν. We let $\zeta \equiv z - ct$ and henceforth denote derivatives with respect to ζ by primes. If we now set

$$(4.6)\qquad w' \equiv W \equiv U + iV,$$

then a simple integration shows that travelling wave solutions of (4.2) are determined by

$$(4.7)\qquad \mu W - c\nu W' = c^2 W + a$$

or equivalently by

$$(4.8)\qquad c\nu U' = (\mu - c^2)U - a,\ c\nu V' = (\mu - c^2)V$$

where a is a constant of integration, which without loss of generality we take to be real by virtue of the rotational invariance of the problem. Note that μ and ν depend on U' and V'. But the parabolicity and coercivity conditions show that (4.8) is equivalent to a system in standard form in which U' and V' are expressed as functions of U, V and the parameters a and c.

We now study the qualitative behavior of solutions of (4.7). We begin by examining the singular points. In the case that $a = 0$, the singular points are $(0,0)$ and all the points of circles of radius A centered at the origin where A satisfies

$$\mu(A^2,0,0) = c^2. \tag{4.9}$$

For $a \neq 0$, the singular points are all points of the form $(U_*,0)$ where U_* is a solution of

$$[\mu(U^2,0,0) - c^2]U = a. \tag{4.10}$$

Using the rotational invariance of our problem, we can subsume (4.9) under (4.10) by considering only those singular points of the former that lie on the U-axis. It then follows that given any pair of singular points $(U_L,0)$, $(U_R,0)$, there is a unique pair (a,c^2) of parameters for which U_L and U_R satisfy (4.10). Note that the location of the singular points is determined solely by $\mu(\cdot,0,0)$, i.e., by the elastic response (which is to be expected). If we now linearize (4.8) about a singular point $(U_*,0)$, we obtain

$$\begin{aligned} c(\nu + 2\frac{\partial\mu}{\partial\eta_1}U_*^2)\delta U' &= (\mu - c^2 + 2\frac{\partial\mu}{\partial\eta_0}U_*^2)\delta U, \\ c\nu\delta V' &= (\mu - c^2)\delta V \end{aligned} \tag{4.11}$$

where μ,ν, and their derivatives have arguments $(U_*^2,0,0)$. The parabolicity condition ensures that $\nu > 0$, $\nu + 2\frac{\partial\mu}{\partial\eta_1}U_*^2 > 0$, so that in the generic case that the coefficients of δU and δV in (4.11) do not vanish at the singular point, the singular points are either nodes or saddles with axes parallel to the U and V-axes. Thus the type of singular point is determined by the elastic response.

If we introduce polar coordinates R and Θ by

$$U = R\cos\Theta,\ V = R\sin\Theta,$$

then (4.8) reduces to

$$c\nu R' = (\mu - c^2)R - a\cos\Theta,\quad c\nu R\Theta' = a\sin\Theta.$$

The second of these conditions implies that the phase-plane trajectories of (4.8) move counterclockwise in the upper half plane $V > 0$ when $ac > 0$. This is yet another result independent of viscous response.

Indeed, given the elastic response $\mu(\cdot,0,0)$, the results we have just obtained show that it is a combinatorial problem to determine the qualitatively different phase portraits admitted by (4.8). Now we show how these results can be sharpened when the viscosity is small, i.e., when μ and ν have the form

$$(4.14)\qquad \mu(\eta_0,\eta_1,\eta_2) = \bar{\mu}(\eta_0) + \varepsilon\mu^*(\eta_0,\eta_1,\eta_2),\ \nu(\eta_0,\eta_1,\eta_2) = \varepsilon\nu^*(\eta_0,\eta_1,\eta_2),$$

where ε is a small positive number.

Now let us convert (4.2) into the complex system for $\omega = (u + iv)_z$, $\psi = (u + iv)_t$ and seek travelling wave solutions for it in the form

$$(4.15)\qquad \omega(z,t) = \Omega\left(\frac{z-ct}{\varepsilon^\alpha}\right),\ \psi(z,t) = \Psi\left(\frac{z-ct}{\varepsilon^\alpha}\right)$$

where α will be specified later. The resulting system for Ω and Ψ can be integrated to yield

$$(4.16)\qquad \varepsilon^{1-\alpha}c\nu^*\Omega' = (\bar{\mu} - c^2 + \varepsilon\mu^*)\Omega - a$$

where the arguments of μ^* and ν^* are

$$(4.17)\qquad \eta_0 = |\Omega|^2,\ \eta_1 = -2c\varepsilon^{\alpha}\mathrm{Re}(\Omega\bar{\Omega}'),\ \eta_2 = c^2\varepsilon^{-2\alpha}|\Omega'|^2 .$$

(Had we sought solutions in the form $u(z,t) + iv(z,t) = w((z - ct)/\varepsilon^\alpha)$ and set $w' = \varepsilon^{-\alpha}\Omega$, then we would again recover (4.16) and (4.17).)

Suppose for large values of $|\eta_1|$ and $|\eta_2|$ that μ^* and ν^* have the form

$$(4.18)\qquad \begin{aligned} \mu^* &= M_1(\eta_0)|\eta_1|^k[\mathrm{sign}(\eta_1)]^\beta + M_2(\eta_0)(\eta_2)^{\ell/2} + \ldots \\ \nu^* &= N_1(\eta_0)|\eta_1|^p + N_2(\eta_0)(\eta_2)^{q/2} + \ldots \end{aligned}$$

where $k, \ell, p, q > 0$, where M_1, M_2, N_1, N_2 are prescribed functions, where $\beta = 0$ or 1, and where the ellipses denote lower order terms. (The ensuing argument can be carried out when (4.18) has a far more general representation.) Let us now substitute (4.17) and (4.18) into (4.16) and retain just the leading terms, the choice of which depends on the relative size of the parameters. E.g., if $q > p$, $k - 1$, $\ell - 1$, we find that the leading equation, corresponding to the limit $\varepsilon \to 0$, can be reduced to the explicit form

$$(4.19) \qquad \Omega' = \frac{(\bar{\mu} - c^2)\Omega - a}{N_2 c\,|(\bar{\mu} - c^2)\Omega - a|^{q/(q+1)}} \quad .$$

This complex equation has the property, typical of all such leading equations, that its horizontal and vertical isoclines are determined entirely by the elastic response function $\bar{\mu}$. Therefore, in many cases the viscous response has no effect on the qualitative nature of the phase portrait.

If $\bar{\mu}$ has the form shown in Fig. 4.20, and if $\alpha > 0$ and $c > 0$ are suitably chosen, then the phase portrait of the leading form of (4.16) is given by Fig. 4.21. Note that the consecutive singular points can always be connected by an orbit or by the reversal of an orbit (corresponding to a switch of the sign of c). The same is not true for singular points that are not consecutive. Note also that there are singular points connected by an infinity of orbits.

Let us summarize our results. The qualitative properties of the phase portrait are scarcely affected by the viscous dissipative mechanism. Contrast this result with that of Conley & Smoller (1973). Their phase portraits corresponding to a very limited class of dissipative mechanisms are highly sensitive to changes in the mechanism. The chief distinction between our model and theirs is that our comes from a correctly formulated physical theory and theirs does not. Majda & Pego (1985) showed how the pathological cases of Conley & Smoller could be eliminated by the imposition of their strict stability condition. But our system is not strictly stable. In our problem certain singular points are connected with an infinity of orbits. Admissibility conditions such as those of Liu (1975) illuminate some of the issues but fail to provide a complete resolution of the difficulties. What may be needed is a much deeper study of stability and constitutive restrictions. Note that the hyperbolic version of (4.2) is not genuinely nonlinear as a consequence of the evenness of $u_z \to \bar{\mu}(u_z + v_z)$. The work in this section is based in part on that of Antman & Malek-Madani (1986a).

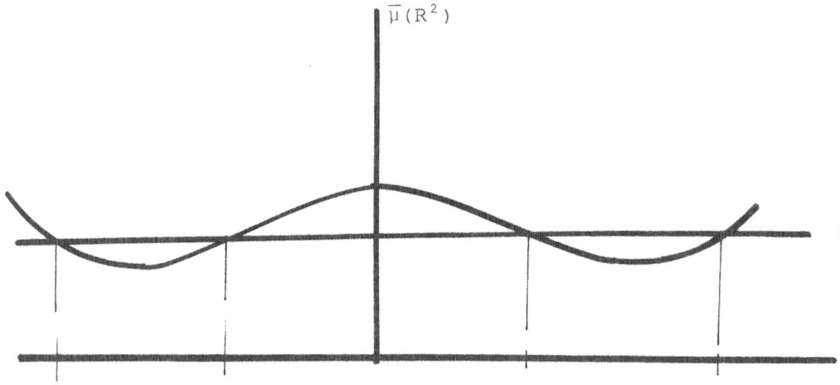

Fig. 4.20

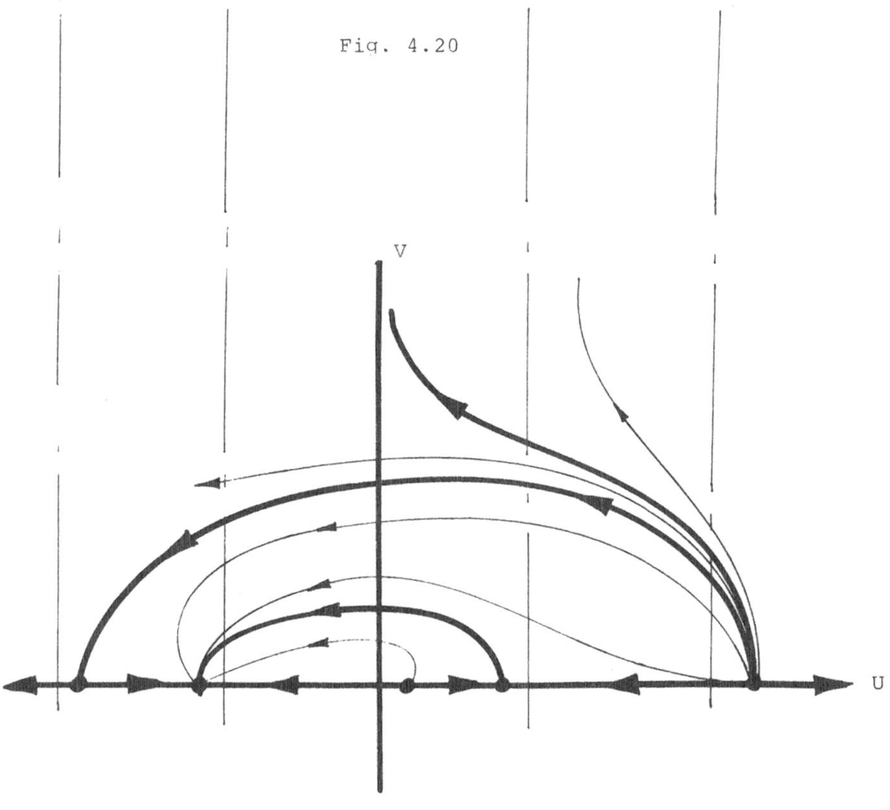

Fig. 4.21

5. Entropy Conditions

In our work above we studied the question of obtaining conditions across shocks by regarding the hyperbolic system admitting shocks as a singular limit of a parabolic system. The dissipative mechanism is embodied in the mathematical requirement that the full system be parabolic. We can regard this parabolicity as a sort of stability condition, obtained by the distillation of a considerable body of work on partial differential equations and on physical systems described by them. In our analysis of Section 4, the role of parabolicity is crucial.

The Clausius-Duhem version of the Second Law of Thermodynamics gives conditions on the constitutive functions weaker than parabolicity. It would merely require that

$$u_{zt}(\mu u_z + \nu u_{zt}) + v_{zt}(\mu v_z + \nu v_{zt}) > 0 \tag{5.1}$$

(cf. (4.4)).

In more general settings, the Clausius-Duhem inequality is not comparable to mathematical requirements like that of parabolicity in Section 4. E.g., in certain problems involving heat conduction, like that of (1.1)-(1.3), the Clausius-Duhem inequality requires that the dot product of the heat flux vector and the temperature gradient be non-negative, a condition analogous to (5.1). But such a condition is insufficient to ensure the parabolicity of the heat equation. On the other hand, the Clausius-Duhem inequality can prevent other dependent constitutive variables, such as the stress, from depending on the temperature gradient. From the mathematical viewpoint such a restriction does little to promote analysis, whereas the absence of parabolicity may greatly impede analysis. (Cf. Antman & Rogers (1986) for a further discussion of this matter.)

A purely mathematical approach to stability embodied in a classification of the spatial part of an operator as elliptic can preclude such interesting phenomena as phase transitions. We observe, however, that the problems for phase transitions for elastic media are mathematically somewhat simpler than the study of travelling waves for hyperbolic equations of motion for elastic media. Moreover, in the study of phase transitions in continuum physics, discontinuous solutions (shocks) can occur. Criteria are needed to identify "admissible" discontinuities, those that are presumably physically reasonable. A technique for constructing such criteria is to embed the equations in a higher order system, in which the role of viscosity is replaced by some other physical process. (Cf. Hagan & Slemrod (1983).) Such an approach is in the spirit of our development of Section 4.

ACKNOWLEDGEMENT. The work of Antman reported here was partially supported by a grant from the National Science Foundation. The work of Malek-Madani was partially supported by a grant from the United States Naval Academy Research Council.

References

G. Andrews, (1980) On the existence of solutions to the equation $u_{tt} = u_{xxt} + \sigma(u_x)_x$, J. Diff. Eq. 35, 200-231.

S.S. Antman & Guo Zhong-heng, (1984) Large shearing oscillations of incompressible nonlinearly elastic bodies, J. Elasticity 14, 249-262.

S.S. Antman & R. Malek-Madani, (1986a) Travelling waves in viscoelastic media and shock structure in elastic media, in preparation.

S.S. Antman & R. Malek-Madani, (1986b) Existence and uniqueness of solutions to an initial-boundary value problem of nonlinear viscoelasticity, in preparation.

S.S. Antman & R.C. Rogers, (1986) Steady-state problems of nonlinear electro-magneto-thermo-elasticity, Arch. Rational Mech. Anal., to appear.

C.C. Conley & J.A. Smoller, (1973) Topological methods in the theory of shock waves, Proc. Symp. Pure Math. 23, Amer. Math. Soc., 293-302.

C.M. Dafermos, (1969) The mixed initial-boundary value problem for the equations of nonlinear one-dimensional viscoelasticity, J. Differential Equations 6, 71-86.

C.M. Dafermos, (1984) Discontinuous thermokinetic processes, in Truesdell (1984), 211-218.

I.M. Gel'fand, (1959) Some problems in the theory of quasilinear equations, Uspekhi Mat. Nauk 14, 87-158; English Transl., Amer. Math. Soc. Transl. (2), 29, 1963.

J.M. Greenberg, (1969) On the existence, uniqueness and stability of solutions fo the equations $\rho_0 X_{tt} = E(X_x)X_{xx} + \lambda X_{xxt}$, J. Math. Anal. Appl. 25, 575-591.

J.M. Greenberg, R.C. MacCamy, & V.J. Mizel, (1968) On the existence, uniqueness and stability of solutions of the equation $\sigma'(u_x)u_{xx} + \lambda u_{xtx} = \rho_0 u_{tt}$, J. Math. Mech. 17, 707-728.

R. Hagan & M. Slemrod (1983), The viscosity-capillarity admissibility criterion for shocks and phase transitions, Arch. Rational Mech. Anal. 83, 333-361.

E. Hopf, (1950) The partial differential equation $u_t + uu_x = \mu u_{xx}$, Comm. Pure Appl. Math. 3, 201-230.

Ya. I. Kanel', (1968) On a model system of equations of one-dimensional gas motion. (in Russian) Diff. Urav. 4, 721-734. Engl. Trans., Diff. Eqs, 4, 374-380.

T.P. Liu, (1975) The Riemann problem for general systems of conservation laws, J. Diff. Eqs. 18, 218-234.

R.C. MacCamy, (1970) Existence, uniqueness and stabilty of $u_{tt} = \frac{\partial}{\partial x}(\sigma(u_x) + \lambda(u_x)u_{xt})$. Indiana Univ. Math. J. 20, 231-238.

A. Majda & R. Pego, (1985) Stable viscosity matrices for systems of conservation laws, J. Diff. Eqs., to appear.

C. Truesdell, (1984), Rational Thermodynamics, 2nd edn., Springer Verlag.

DOES RANK-ONE CONVEXITY IMPLY QUASICONVEXITY ?

J.M. Ball

Department of Mathematics
Heriot-Watt University
Edinburgh, EH14 4AS

1. Background

Let $\Omega \subset R^m$ be a bounded open set. Let $M^{n\times m}$ denote the set of real $n \times m$ matrices and suppose that $W: M^{n\times m} \to \overline{R}$ is Borel measurable and bounded below. (Here $\overline{R}$ denotes the extended real line with its usual topology.) We are interested in the problem of minimizing

$$I(u) = \int_\Omega W(Du(x))dx \tag{1.1}$$

among functions $u \in W^{1,1}(\Omega;R^n)$ satisfying appropriate boundary conditions. An important application is to nonlinear elasticity, when $W = W(A)$ is the stored-energy function of a homogeneous material and $u(x)$ is the deformed position of the particle at $x \in \Omega$ in a reference configuration; in this case we usually take $m = n = 3$, but the cases $1 \leq m \leq n \leq 3$ are also of interest and cover certain string and membrane problems. It is convenient to allow W to take the value $+\infty$ so as to include various constraints. In compressible nonlinear elasticity ($m = n = 3$), for example, we may set $W(A) = +\infty$ for $\det A \leq b$, where $b \geq 0$ is a constant, to reflect the fact that infinite energy is required to make a reflection of the body or to homogeneously compress it to b times its original volume. Similarly, for an incompressible material it is convenient to set $W(A) = +\infty$ if and only if $\det A \neq 1$.

Connected with the existence and properties of minimizers for (1.1) are certain convexity conditions on W. Two of these conditions are rank-one convexity and quasiconvexity, and as we shall see the question raised in the title amounts roughly to asking whether they are the same. This has been an open problem since quasiconvexity was introduced by Morrey [21] over 30 years ago.

Most of the material in the paper is drawn from the existing literature, though the remarks in §6, §8(b), (c) are perhaps new.

2. Definitions

Let $A \in M^{n\times m}$. We say that W is <u>rank-one convex at A</u> if

$$W(A) \leq tW(A_1) + (1 - t)W(A_2) \tag{2.1}$$

whenever $t \in [0,1]$, $A = tA_1 + (1 - t)A_2$ and $A_1, A_2 \in M^{n\times m}$ with $A_2 - A_1 = \lambda \otimes \mu$ for some vectors $\lambda \in R^n$, $\mu \in R^m$. We say that W is <u>rank-one convex</u> if it is rank-one convex at every $A \in M^{n\times m}$; equivalently, W is convex along all line segments in $M^{n\times m}$ whose end-points differ by a matrix of rank one.

Replacing λ by $\frac{-\lambda}{1-t}$ in (2.1) we see that W is rank-one convex at A if and only if

$$W(A) \leq tW(A + \lambda \otimes \mu) + (1 - t)W(A - \frac{t}{1-t}\lambda \otimes \mu) \tag{2.2}$$

for all $t \in (0,1)$, $\lambda \in R^n$, $\mu \in R^m$. If W is finite in a neighbourhood of A and differentiable at A it follows easily that W is rank-one convex at A if and only if

$$W(A + \lambda \otimes \mu) \geq W(A) + DW(A)(\lambda \otimes \mu) \tag{2.3}$$

for all $\lambda \in R^n$, $\mu \in R^m$. If in addition W is twice differentiable at A then (2.3) implies that the <u>Legendre-Hadamard condition</u>

$$D^2W(A)(\lambda \otimes \mu, \lambda \otimes \mu) = \frac{\partial^2 W(A)}{\partial A^i_\alpha \partial A^j_\beta} \lambda^i \mu_\alpha \lambda^j \mu_\beta \geq 0 \quad \text{for all } \lambda \in R^n, \mu \in R^m \tag{2.4}$$

holds. Conversely, suppose that

$$\text{dom } W := \{A \in M^{n\times m} : W(A) < \infty\}$$

is a rank-one convex set (i.e. $tA_1 + (1-t)A_2 \in \text{dom } W$ whenever $A_1, A_2 \in \text{dom } W$, $A_2 - A_1 = \lambda \otimes \mu$, $t \in [0,1]$) and open, that $W \in C^2(\text{dom } W)$ and that (2.4) holds for all $A \in \text{dom } W$. Then by integrating $\frac{d^2}{dt^2} W(A + t\lambda \otimes \mu)$ twice we see that W is rank-one convex.

For $1 \leq p \leq \infty$ and $E \subset R^m$ a bounded open set we denote by $W^{1,p}(E; R^n)$ the Sobolev space of all weakly differentiable mappings $u: E \to R^n$ such that $\|u\|_{L^p(E;R^n)} + \|Du\|_{L^p(E;M^{n\times m})} < \infty$, and by $W_0^{1,p}(E;R^n)$ the subset of $W^{1,p}(E;R^n)$ consisting of those u vanishing in the usual sense (cf. [9, p. 227]) on the boundary ∂E of E.

Let $A \in M^{n\times m}$. We say that W is <u>$W^{1,p}$-quasiconvex</u> at A if

$$\int_E W(A + D\phi(x))dx \geq \int_E W(A)dx = (\text{meas } E)W(A) \tag{2.5}$$

for every bounded open set $E \subset R^m$ with meas $\partial E = 0$ and all $\phi \in W_0^{1,p}(E;R^n)$, and that W is <u>$W^{1,p}$-quasiconvex</u> if it is $W^{1,p}$-quasiconvex at every $A \in M^{n\times m}$. If $p = \infty$ we abbreviate $W^{1,\infty}$-quasiconvex to <u>quasiconvex</u>.

The definition of quasiconvexity was introduced by Morrey [21]; the generalization to $W^{1,p}$-quasiconvexity was made in [9]. The definition is independent of E in the following sense; if (2.5) holds for one nonempty bounded open set $E \subset R^m$, some $A \in M^{n\times m}$ and all $\phi \in W_0^{1,p}(E;R^n)$ then W is $W^{1,p}$-quasiconvex at A (see [20, 9 Prop. 2.3]). Clearly if $1 \leq p \leq q \leq \infty$ and W is $W^{1,p}$-quasiconvex at A then W is $W^{1,q}$-quasiconvex at A.

The open question posed in the title can now be stated precisely: <u>is every rank-one convex function W quasiconvex?</u> Alternatively one can modify the question by adding the hypothesis that W be continuous (or finite and continuous). Whether adding such regularity hypotheses could affect the answer is not obvious.

3. Quasiconvexity Implies Rank-One Convexity

Suppose that in (2.5) we take the function ϕ to be piecewise affine, so that E is the disjoint union of a finite number N of open simplices E_i and a set of measure zero and $D\phi(x) = A_i - A$ for a.e. $x \in E_i$, where $A_i \in M^{n\times m}$

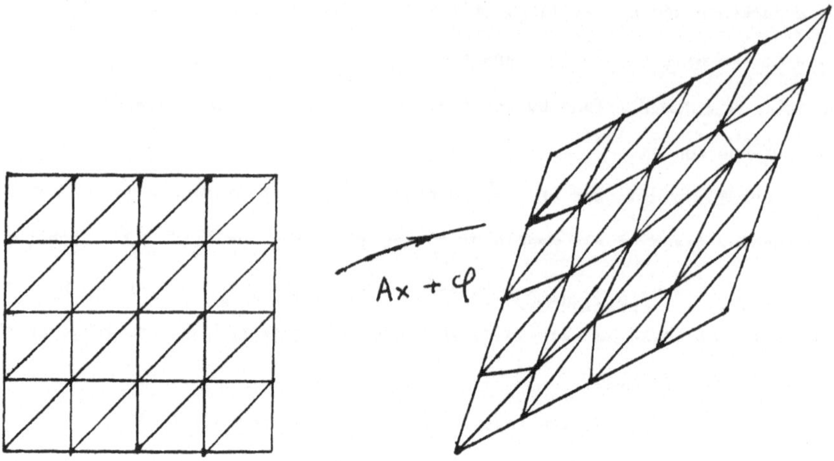

Figure 3.1

is constant. An example with $m = n = 2$ is illustrated in Figure 3.1 above. Let $\lambda_i = \frac{\text{meas } E_i}{\text{meas } E}$, so that $\sum_{i=1}^{N} \lambda_i = 1$. Then

$$A = \frac{1}{\text{meas } E} \int_E D(Ax + \phi(x))dx = \sum_{i=1}^{N} \lambda_i A_i, \tag{3.1}$$

and (2.5) becomes

$$W\left(\sum_{i=1}^{N} \lambda_i A_i\right) \leq \sum_{i=1}^{N} \lambda_i W(A_i). \tag{3.2}$$

Conversely, for a polyhedral domain E and any $\phi \in W_0^{1,\infty}(E;R^n)$ there exists a bounded sequence $\phi^{(j)} \in W_0^{1,\infty}(E;R^n)$ of piecewise affine functions such that $\phi^{(j)} \to \phi$ uniformly and $D\phi^{(j)}(x) \to D\phi(x)$ a.e. in E (cf. [14, Chap. X]). Therefore, for W finite and continuous, quasiconvexity at A is equivalent to the convexity condition (3.2) holding for all piecewise affine functions. If the matrices A_i were independent then (3.2) would be equivalent to convexity of W at A. They are not independent, however, because together they form the gradient of a mapping; to understand the resulting compatibility conditions we recall an observation of Hadamard.

Let S be a smooth $(m - 1)$-dimensional surface with normal μ at the point $x \in S$. Let N be a neighbourhood of x in R^m and suppose that $u:N \to R^n$ is continuous across S and C^1 on either side of S. Let A, B denote the limits at x of Du from either side of S. Equating the tangential derivatives at x we find that

$$B - A = \lambda \otimes \mu \tag{3.3}$$

for some $\lambda \in R^n$. Thus for a piecewise affine function the gradient jumps by a matrix of rank ≤ 1 across the faces of adjoining simplices.

By choosing E to be a rectangular parallelpiped, considering piecewise affine functions with just one interior node, and using the argument of Morrey [21, p. 45], we obtain the following result.

Theorem 3.1

Let $A \in M^{n \times m}$. Suppose that W is quasiconvex at A, that $W(A) < \infty$ and that W is continuous at A. Then W is rank-one convex at A.

As observed in [9, p. 232], it follows that if W is continuous (with values in $\bar{R}$) and quasiconvex then W is rank-one convex. Thus for continuous W quasiconvexity is equivalent to convexity when $m = 1$ or $n = 1$. Without some continuity assumption Theorem 3.1 is false, as shown by the example (see [9, p. 232])

$$W(0) = W(a \otimes b) = 0, \quad W(A) = +\infty \text{ otherwise,}$$

where $a \in R^n$, $b \in R^m$ are given nonzero vectors and $m > 1$. As discussed in [9] the moral of this example is perhaps that for general W taking infinite values some other version of the quasiconvexity condition (for example, one based on weak lower semicontinuity or that in [7]) should be taken as the basic definition. However this issue is not crucial for the main problem discussed in this paper which is unresolved even for smooth integrands.

4. Quasiconvexity as a Necessary Condition Satisfied by a Minimizer

Let $x_0 \in \Omega$. We say that $u \in W^{1,1}(\Omega;R^n)$ is a strong local minimizer of I at x_0 if there are numbers $\rho > 0$, $\varepsilon > 0$ such that $I(v) \geq I(u)$ whenever $v \in W^{1,1}(\Omega;R^n)$ with $v(x) = u(x)$ for a.e. $x \in \Omega$ satisfying $|x - x_0| > \rho$ and $|v(x) - u(x)| \leq \varepsilon$ for a.e. $x \in \Omega$.

A version of the following result was first proved by Meyers [20] (see also

Busemann & Shephard [11]).

Theorem 4.1

Assume W is continuous on dom W. Let $x_0 \in \Omega$ and let u be a strong local minimizer of I at x_0. Suppose further that u is C^1 in a neighbourhood of x_0 with $Du(x_0) = A$ and $W(A) < \infty$. Then

$$\int_E W(A + D\phi(x))dx \geq \int_E W(A)dx \qquad (4.1)$$

for all bounded open subsets $E \subset R^m$ and all $\phi \in W_0^{1,\infty}(E;R^n)$ such that $\operatorname{ess\,sup}_{x \in E} W(A + D\phi(x)) < \infty$.

Idea of proof

We 'blow up' the minimization problem in a neighbourhood of x_0, so that u becomes linear. This is done by defining, for $\varepsilon > 0$ sufficiently small,

$$u_\varepsilon(x) = u(x) + \varepsilon\tilde{\phi}\left(\frac{x - x_0}{\varepsilon}\right),$$

where $\tilde{\phi}$ is ϕ extended by zero outside E, making the change of variables $x - x_0 = \varepsilon y$, and letting $\varepsilon \to 0$ in the inequality $I(u_\varepsilon) \geq I(u)$. □

Refinements of Theorem 4.1, including treatment of the case when $x_0 \in \partial\Omega$, are given in [8]. The condition (4.1) says roughly that W is quasiconvex at A, and this follows if W does not take the value $+\infty$. The proof of Theorem 3.1 still applies and gives the following result.

Corollary 4.2 (Graves [16])

Let the hypotheses of Theorem 4.1 hold. Then W is rank-one convex at A.

In view of the above discussion it would be very interesting (i) to give useful necessary and/or sufficient conditions for W to be quasiconvex at A, and (ii) to identify quasiconvexity at the values of $Du(x)$ as one of a set of sufficient conditions for u to be a local minimizer of I in, say, $W^{1,p}(\Omega;R^n)$.

5. Other Rôles Played by Rank-One Convexity and Quasiconvexity

Quasiconvexity was introduced by Morrey in connection with the direct method of the calculus of variations. In [21] he showed that if W is finite and continuous then quasiconvexity of W is a necessary condition for I to be sequentially weak*lower semicontinuous on $W^{1,\infty}(\Omega;R^n)$. (The same argument shows in general [9, p. 230] that $W^{1,p}$-quasiconvexity of W is necessary for I to be sequentially weakly (weak * if $p = \infty$) lower semicontinuous on $W^{1,p}(\Omega,R^n)$.) He then showed that quasiconvexity is sufficient for I to be sequentially weakly lower semicontinuous on $W^{1,1}(\Omega;R^n)$ provided W also satisfies certain growth conditions. Extensions of this result can be found in [20,1,19], but unfortunately they cannot be used to prove the existence of minimizers for I in nonlinear elasticity since they assume that W is everywhere finite. At present the only existence theorems applying to elasticity [3,7,9] allow W to be singular at the expense of assuming that W is <u>polyconvex</u>, i.e. W can be written as a convex function of minors of A of all orders r, $1 \leq r \leq \min(m,n)$. Polyconvexity implies quasiconvexity, but the converse is false [21,25,23,6].

Quasiconvexity is necessary for the existence of minimizers to certain perturbations of I. In fact the following result is proved in [9, Thm 5.1].

<u>Theorem 5.1</u>

Let $1 \leq p \leq \infty$, $A \in M^{n\times m}$ and $X_A = \{u \in W^{1,p}(\Omega;R^n): u - Ax \in W_0^{1,p}(\Omega;R^n)\}$. Let $\phi:[0,\infty) \to R$ be bounded and continuous with $\phi(0) = 0$, $\phi(t) > 0$ for $t > 0$, and set $\psi(x,u) = \phi(|u - Ax|^2)$. Assume $\text{meas}\ \partial\Omega = 0$. If

$$J(u) := \int_\Omega [W(Du) + \phi(x,u)]dx \qquad (5.1)$$

attains an absolute minimum on X_A then W is $W^{1,p}$-quasiconvex at A.

For a given W and boundary conditions I may or may not attain a minimum. In either case it is of interest to study the behaviour of minimizing sequences of I, and it has been shown by Acerbi & Fusco [1], Dacorogna [12] (see also [2,17,13]) that they possess subsequences converging weakly in $W^{1,1}(\Omega:R^n)$ to minimizers of the relaxed functional

$$T(u) := \int_\Omega QW(Du(x))dx, \tag{5.2}$$

where QW denotes the supremum of all quasiconvex functions less than W. Again these results do not apply to elasticity on account of the strong growth hypotheses made.

The Euler-Lagrange equations for I are given by

$$\frac{\partial}{\partial x^\alpha}\frac{\partial W}{\partial A^i_\alpha}(Du) = 0, \quad i = 1,\ldots,n. \tag{5.3}$$

By definition, these equations are <u>strongly elliptic</u> if (2.4) holds for all A with equality only if $\lambda \otimes \mu = 0$. The slightly weaker condition of <u>strict rank-one convexity</u> (i.e. rank-one convexity with equality in (2.1) only if $\lambda \otimes \mu = 0$ or $t = 0,1$) is necessary, and nearly sufficient, for there to be no piecewise C^1 weak solution of (5.3) whose gradient jumps across a smooth $(m-1)$-dimensional surface (for the details see [4]). Neither strong ellipticity nor strict quasiconvexity are sufficient to prevent weak solutions having other types of singularities, such as that occurring in cavitation [5]. However, recently Evans [15] has proved a partial regularity result for absolute minimizers of I under a strict quasiconvexity hypothesis. Although he assumes W is everywhere finite, his theorem offers the first hope of a regularity theorem applying to nonlinear elasticity.

6. Rank-One Convexity at A Does not Imply Quasiconvexity at A

Let $m > 1$, $n > 1$. Then the closed cone

$$\Lambda := \{\lambda \otimes \mu : \lambda \in R^n,\ \mu \in R^m\}$$

is a proper subset of $M^{n\times m}$. Let $A \in M^{n\times m}$ and let B be an open ball contained in $M^{n\times m} \setminus (A + \Lambda)$. Let $W \in C^\infty(M^{n\times m})$ be negative in B and zero otherwise. Since W is zero on $A + \Lambda$ it follows that W is rank-one convex at A. However, by choosing $\phi \in W_0^{1,\infty}(E;R^n)$ such that $A + D\phi(x) \in B$ on a set of positive measure we can violate (2.5), so that W is not quasiconvex at A. This simple remark shows that in general Theorem 4.1 provides a stronger necessary condition than that of Graves.

The above example is easily adapted so as to apply to isotropic nonlinear elasticity with $m = n > 1$. The isotropy is expressed by the requirement that

$$W(A) = \Phi(v_1,\ldots,v_n),\ A \in M^{n\times m} \tag{6.1}$$

for some symmetric function Φ of the singular values $v_i = v_i(A)$ of A (the eigenvalues of $(A^TA)^{1/2}$). Let $A = 1$, $e = (1,1,\ldots,1) \in R^n$, $\gamma > 1$, $\varepsilon > 0$, and let $\Phi \in C^\infty(R^n)$ be such that $\Phi(v) < 0$ for $|v - \gamma e| < \varepsilon$, $\Phi(v) = 0$ otherwise. We claim that for ε sufficiently small, $W(1 + \lambda \otimes \mu) = 0$ for all $\lambda,\mu \in R^n$. If not, there would exist a sequence $v^{(r)}$ converging to γe in R^n, orthogonal matrices $Q^{(r)}$, $R^{(r)}$, and vectors $\lambda^{(r)},\mu^{(r)} \in R^n$ such that

$$1 + \lambda^{(r)}\otimes \mu^{(r)} = Q^{(r)}\ (\mathrm{diag}\ v^{(r)})R^{(r)}. \tag{6.2}$$

Extracting convergent subsequences and passing to the limit we find that

$$1 + \lambda \otimes \mu = \gamma Q \tag{6.3}$$

for some $\lambda,\mu \in R^n$ and orthogonal matrix Q. This is easily seen to be impossible (for example, by evaluating QQ^T and Q^TQ). We have thus shown that for $\varepsilon > 0$ sufficiently small W is rank-one convex at 1, and the same arguments as before shows that W is not quasiconvex at 1. Note that by adding to Φ a term Φ_0, where $\delta > 0$ is sufficiently small, we can arrange that W satisfies any desired growth conditions as $|A| \to \infty$, $\det A \to 0+$, and that W be strictly rank-one convex, but not quasiconvex, at 1.

7. The Evidence Against

We collect together some remarks which might suggest that rank-one convexity does <u>not</u> imply quasiconvexity.

(a) The inequalities (3.2) arising from writing down the quasiconvexity condition for piecewise affine functions ϕ do not obviously follow from rank-one convexity (for example, in the case discussed in [3, p. 355], where there are 3 interior nodes). A possible riposte to this, suggested by the results of Tartar [24] on separately convex functions, is that to derive (3.2) from rank-one convexity it may be necessary to use values of the deformation gradient other than those taken by $A + D\phi(x)$.

(b) The analogous statement to 'rank-one convexity implies quasiconvexity' for integrands depending on higher derivatives of u is false. It is shown in [7, p. 146] that if $m = 2$, $n = 3$ and

$$W(D^2u) = \varepsilon_{ijk}u^i{}_{,11}u^j{}_{,12}u^k{}_{,22}\,, \tag{7.1}$$

where ε_{ijk} is the usual permutation symbol, then the map

$$t \to W(A + t\lambda \otimes \mu \otimes \mu)$$

is affine for every $A = (A^i_{\alpha\beta})$, $\lambda \in R^3$, $\mu \in R^2$, where $(\lambda \otimes \mu \otimes \mu)^i_{\alpha\beta} := \lambda^i\mu_\alpha\mu_\beta$, but W does not satisfy the quasiconvexity condition

$$\int_E W(A + D^2\phi(x))dx \geq \int_E W(A)dx \tag{7.2}$$

for all $\phi \in W_0^{2,\infty}(E;R^3)$, for any A.

(c) Rank-one convexity does not imply $W^{1,p}$-quasiconvexity in general. For example, if $m = n \geq 3$ and

$$W(A) = \mathrm{tr}(A^TA) + (\det A)^2, \tag{7.3}$$

Then (see [3,9]) W is polyconvex, and thus quasiconvex, but is not $W^{1,p}$-quasiconvex if $1 \leq p < n$.

8. The Evidence in Favour

The following remarks might suggest that rank-one convexity does imply quasiconvexity.

(a) If W is quadratic, that is

$$W(A) = c_{ij}^{\alpha\beta} A_\alpha^i A_\beta^j \tag{8.1}$$

for constants $c_{ij}^{\alpha\beta}$, and rank-one convex then W is quasiconvex. The only proof that seems to be known for this fact (see [26,22]) is to show that

$$I(\phi) = \int_E c_{ij}^{\alpha\beta} \phi^i_{,\alpha} \phi^j_{,\beta} \, dx \geq 0$$

for all $\phi \in W_0^{1,2}(E;R^n)$ by extending ϕ by zero outside E, taking Fourier transforms and using Plancherel's formula. Functions W formed by combining polyconvex and quadratic functions seem to be the only known examples of quasiconvex functions.

(b) For isotropic nonlinear elasticity, rank-one convexity implies that the quasiconvexity inequality holds for radial deformations.

Taking $m = n = 3$, for example, with $B = \{x \in R^3: \ |x| < 1\}$, a radial deformation $u: B \to R^3$ is one having the form

$$u(x) = \frac{r(R)}{R} x, \ R = |x|, \tag{8.2}$$

where $r: [0,1] \to [0,\infty)$. If $u \in W^{1,\infty}(B;R^3)$ with $\det Du(x) > 0$ a.e. $x \in B$ then $r \in W^{1,\infty}(0,1)$ with $r(0) = 0$, $r'(R) > 0$, a.e. $R \in (0,1)$ and $\sup_{R\in(0,1)} \frac{r(R)}{R} < \infty$ (cf. [5, p. 566]). If the stored-energy function $W \in C^1(M_+^{3\times3})$, where $M_+^{3\times3} := \{A \in M^{3\times3}: \det A > 0\}$, then there exists a symmetric function $\Phi = \Phi(v_1,v_2,v_3)$, defined and continuously differentiable for positive arguments, of the singular values v_1,v_2,v_3 of A. For a radial deformation (8.2) these singular values are given by $r'(R)$, $\frac{r(R)}{R}$ and $\frac{r(R)}{R}$. Hence

$$\int_B W(Du(x))dx = 4\pi \int_0^1 R^2\Phi(r', \frac{r}{R}, \frac{r}{R})dR. \tag{8.3}$$

Taking $A = \text{diag}(v_1,v_2,v_3)$, λ, μ parallel to the x^i-axis $(i = 1,2,3)$, it follows from (2.3), as is well known, that W rank-one convex implies that Φ is convex in each v_i separately. Hence

$$\Phi(r', \frac{r}{R}, \frac{r}{R}) \geq \Phi(\frac{r}{R}, \frac{r}{R}, \frac{r}{R}) + (r' - \frac{r}{R})\Phi_1(\frac{r}{R}, \frac{r}{R}, \frac{r}{R}),$$

$$\text{a.e. } R \in (0,1). \tag{8.4}$$

We now note that

$$\frac{d}{dR}\left[\frac{R^3}{z}\Phi(\frac{r}{R}, \frac{r}{R}, \frac{r}{R})\right] = R^2\left[\Phi(\frac{r}{R}, \frac{r}{R}, \frac{r}{R}) + (r' - \frac{r}{R})\Phi_1(\frac{r}{R}, \frac{r}{R}, \frac{r}{R})\right],$$

$$\text{a.e. } R \in (0,1). \tag{8.5}$$

Combining (8.3)-(8.5) we deduce that for a radial deformation $u \in W^{1,\infty}(B;R^3)$ with $\det Du(x) > 0$ a.e. $x \in B$ and satisfying

$$u(x) = \lambda x, \quad x \in \partial B, \tag{8.6}$$

we have

$$\int_B W(Du(x))dx \geq 4\pi \int_0^1 R^2\Phi(\lambda,\lambda,\lambda)dR = \int_B W(\lambda 1)dx, \tag{8.7}$$

which is the required quasiconvexity inequality. For related results see [5, §6.3, 18].

The above argument can be thought of as an application of the field theory of the calculus of variations [10], the extremal $\bar{r}(R) = \lambda R$ being regarded as embedded in the global field of extremals $r(R) = \mu R$, $\mu > 0$. The slope function of this field is given by $p(R,r) = r/R$ and (8.4) expresses the positivity of the corresponding Weierstrass excess function.

It is instructive to note that in (8.4) rank-one convexity is applied at matrices whose choice is not at all evident a priori, and which do not form the gradient of a deformation.

(c) We present a plausibility argument that rank-one convexity implies quasiconvexity in general. For simplicity we suppose that $W \in C^2(M^{n\times m})$. The argument is based on the following interesting result of Knops & Stuart [18].

Theorem 8.1

Let $\Omega \subset R^m$ be a bounded, star-shaped domain with smooth boundary. Let W be rank-one convex, let $A \in M^{n\times m}$ and let $u \in C^2(\bar{\Omega};R^n) \cap C^1(\Omega;R^n)$ be a solution to the Euler-Lagrange
equation

$$\frac{\partial}{\partial x^\alpha}\frac{\partial W}{\partial A^i_\alpha}(Du) = 0, \quad x \in \Omega, \tag{8.8}$$

satisfying

$$u(x) = Ax, \quad x \in \partial\Omega. \tag{8.9}$$

Then

$$I(u) \leq I(Ax). \tag{8.10}$$

Suppose that W is rank-one convex but not quasiconvex. Then there exist $A \in M^{n\times m}$ and $\bar{\phi} \in W_0^{1,\infty}(\Omega;R^n)$ such that

$$I(Ax + \bar{\phi}) < I(Ax). \tag{8.11}$$

Define, for $\varepsilon > 0$,

$$W_\varepsilon(A) = W(A) + \varepsilon \operatorname{tr}(A^TA), \tag{8.12}$$

$$I_\varepsilon(u) = \int_\Omega W_\varepsilon(Du(x))dx. \tag{8.13}$$

Then

$$I_\varepsilon(Ax + \bar{\phi}) < I_\varepsilon(Ax) \tag{8.14}$$

provided ε is sufficiently small. With ε so chosen, and assuming as we may that Ω is star-shaped with smooth boundary, we note that by the argument in (a) the second variation

$$\delta^2 I_\varepsilon(Ax)(\phi,\phi) := \frac{d^2}{dt^2} I_\varepsilon(Ax + t\phi)|_{t=0}$$

$$> 2\varepsilon \int_\Omega |D\phi|^2 dx.$$

Hence [26,8] the linear map Ax minimizes I_ε locally in $W^{1,\infty}(\Omega;R^n)$ subject to the boundary condition (8.9). If we could apply an approximate 'mountain-pass' lemma we could conclude from (8.14) that there exists a critical point u of I_ε with $I_\varepsilon(u) > I_\varepsilon(Ax)$. If we could also assert that u had sufficient regularity for Theorem 8.1 to hold then we would have a contradiction to Theorem 8.1, applied to the rank-one convex function W_ε.

Examination of the proof of Theorem 8.1 shows that, as in (b), the rank-one convexity of W is applied in the above argument at some matrices, A and $Du(x)$, whose choice was not evident a priori.

9. Concluding Quotations

We end by quoting two passages from the work of Morrey concerning the problem discussed in this paper; the terminology has been altered to conform with ours.

(From Morrey [21]) 'It would seem that there is still a wide gap in the general case between the necessary and sufficient conditions for quasi-convexity which the writer has obtained. In fact, after a great deal of experimentation, the writer is inclined to think that there is no condition of the type discussed, which involves W and only a finite number of its derivatives, and which is both necessary and sufficient for quasi-convexity in the general case'.

(From Morrey [22]) '*It is an unsolved problem to prove or disprove the theorem that every rank-one convex fucntion of Du is quasi-convex*'.

ACKNOWLEDGEMENT

This paper was completed following a visit to the Institute for Mathematics and its Applications, University of Minnesota. I would like to thank the members of the Institute, and especially Jerry Ericksen & David Kinderlehrer, for their hospitality and lively interaction. The research was also partially supported by a U.K. Science & Engineering Research Council Senior Fellowship.

References

1. E. Acerbi & N. Fusco, 'Semicontinuity problems in the calculus of variations', Arch. Rat. Mech. Anal., 86 (1984), 125-146.

2. E. Acerbi, G. Buttazzo and N. Fusco, 'Semicontinuity and relaxation for integrals depending on vector valued functions', J. Math. Pure et Appl. 62 (1983), 371-387.

3. J.M. Ball, 'Convexity conditions and existence theorems in nonlinear elasticity', Arch. Rat. Mech. Anal., 65 (1977), 193-201.

4. J.M. Ball, 'Strict convexity, strong ellipticity, and regularity in the calculus of variations', Math. Proc. Camb. Phil. Soc., 87 (1980) 501-513.

5. J.M. Ball, 'Discontinuous equilibrium solutions and cavitation in nonlinear elasticity', Phil. Trans. Royal Soc. Lond., A 306 (1982), 557-611.

6. J.M. Ball, 'Remarks on the paper 'Basic Calculus of Variations', Pacific J. Math., 116 (1985) 7-10.

7. J.M. Ball, J.C. Currie & P.J. Olver, 'Null Lagrangians, weak continuity, and variational problems of arbitrary order', J. Functional Analysis, 41 (1981), 135-174.

8. J.M. Ball & J.E. Marsden, 'Quasiconvexity at the boundary, positivity of the second variation, and elastic stability', Arch. Rat. Mech. Anal., 86 (1984) 251-277.

9. J.M. Ball & F. Murat, '$W^{1,p}$-quasiconvexity and variational problems for multiple integrals', J. Functional Analysis, 58 (1984), 225-253.

10. O. Bolza, "Calculus of variations", reprinted by Chelsea, New York, 1973.

11. H. Busemann & G.C. Shephard, "Convexity on nonconvex sets', Proc. Coll. on Convexity, Copenhagen, Univ. Math. Inst., Copenhagen, (1965), 20-33.

12. B. Dacorogna, 'Quasiconvexity and relaxation of nonconvex problems in the calculus of variations', J. Functional Analysis, 46 (1982), 102-118.

13. B. Dacorogna, 'Remarques sur les notions de polyconvexité, quasi convexité et convexité de rang 1', J. Math. Pure et Appl.,

14. I. Ekeland & R. Témam, "Analyse convexe et problèmes variationnels", Dunod, Gauthier-Villars, Paris, 1974.

15. L.C. Evans, 'Quasiconvexity and partial regularity in the calculus of variations', preprint.

16. L.M. Graves, 'The Weierstrass condition for multiple integral variation problems', Duke Math J., 5 (1939) 656-660.

17. R.V. Kohn & G. Strang, 'Optimal design and relaxation of variational problems, Comm. Pure and Appl. Math 39 (1986) 113-137,

18. R.J. Knops & C.A. Stuart, 'Quasiconvexity and uniqueness of equilibrium solutions in nonlinear elasticity', Arch. Rat. Mech. Anal. 86 (1984) 233-249.

19. P. Marcellini, 'Approximation of quasiconvex functions, and lower semicontinuity of multiple integrals', Manuscripta Math. 51 (1985) 1-28.

20. N.G. Meyers, 'Quasi-convexity and lower semicontinuity of multiple variational integrals of any order', Trans. Amer. Math. Soc., 119 (1965), 125-149.

21. C.B. Morrey, 'Quasi-convexity and the lower semicontinuity of multiple integrals', Pacific J. Math. 2 (1952) 25-53.

22. C.B. Morrey, "Multiple integrals in the calculus of variations", Springer, Berlin, 1966.

23. D. Serre, 'Formes quadratiques et calcul des variations', J. de Math. Pures et Appl., 62 (1983), 177-196.

24. L. Tartar, conference in the workshop, Metastability and incompletely posed problems.

25. F.J. Terpstra, 'Die darstellung biquadratischer formen als summen von quadraten mit anwendung auf die variationsrechnung; Math. Ann., 116 (1938) 166-180.

26. L. van Hove, 'Sur le signe de la variation seconde des intégrales multiples à plusieurs fonctions inconnues', Koninkl. Belg. Acad., Klasse der Wetenschappen, Verhandelingen, 24 (1949).

METASTABLE HARMONIC MAPS

Haim Brezis

Département of Mathématiques
Universite Paris VI
4 Pl. Jussieu
75230 Paris Cedex 05

1. Introduction

The purpose of this lecture is to discuss a simple system with at least two distinct equilibrium states. The first state is an absolute minimum of the energy. The second state is a local minimum which exhibits some of the features of both stable and unstable equilibria.

The system is defined as follows. Let

$$\Omega = \{(x,y) \in R^2 \ ; \ x^2 + y^2 < 1\}$$

and

$$S^2 = \{(x,y,z) \in R^3; \ x^2 + y^2 + z^2 = 1\}.$$

We look for maps $u : \Omega \to R^3$ satisfying

$$(1) \qquad \begin{aligned} -\Delta u &= u|\nabla u|^2 && \text{on} \quad \Omega , \\ u(x,y) &\in S^2 && \text{on} \quad \Omega , \\ u &= \gamma && \text{on} \quad \Omega \end{aligned}$$

where $\gamma: \partial\Omega \to S^2$ is a prescribed boundary condition. Solutions of (1) are called harmonic maps. They arise as critical points of the functional

$$E(u) = \int_\Omega |\nabla u|^2$$

under the constraint

$$u \in \mathcal{E} = \{u \in H^1(\Omega \ ;R^3), \ u \in S^2 \text{ a.e. on } \Omega, \ u = \gamma \text{ on } \partial\Omega\}.$$

[The multiplier rule readily yields the Euler-Lagrange equation for the minimization of E on $\mathcal{E}$. These involve a Lagrange multiplier depending on x,y for the constraint that $|u| = 1$. When this multiplier is evaluated by means of the

Euler-Lagrange equation and the constraint and then substituted into the Euler-Lagrange equation, this equation reduces to the first equation of (1)]. This type of problem also presents also some analogies with questions occurring in the theory of liquid crystals (see [6], [8], [4]).

In what follows we assume that $\gamma \in H^{1/2}(\partial\Omega; S^2)$ so that $\mathcal{E} \neq \emptyset$ (see [9]). It is clear that E achieves its absolute minimum on $\mathcal{E}$, i.e., that there is some $\underline{u}$ (not necessarily unique) such that

$$E(\underline{u}) = \operatorname*{Inf}_{u} E(u) \equiv J_0.$$

Indeed, let (u^n) be a minimizing sequence in $\mathcal{E}$. By passing to a subsequence we may assume that $u^n \rightharpoonup \underline{u}$ weakly in H^1 and $u^n \to \underline{u}$ a.e. on Ω so that $\underline{u} \in \mathcal{E}$ and $E(\underline{u}) \leq J_0$.

The main result concerning the existence of a non-minimal solution is the following.

Theorem 1. (see [3] and also [10]). Assume γ is not a constant. Then, there exists a solution $\overline{u}$ of (1) with $\overline{u} \neq \underline{u}$ which corresponds to a local minimum of E on $\mathcal{E}$.

Remark 1. When $\gamma \equiv C$ is a constant, it is known that $u \equiv C$ is the only (smooth) solution of (1) (see [11]). Here is a simple argument. Take the dot product of (1) with $xu_x + yu_y$ and note that $u \cdot u_x = u \cdot u_y = 0$ (since $|u| = 1$ on Ω). Integrating by parts and using the fact that $\partial\Omega$ is a circle we find that

$$\int_{\partial\Omega} \left| \frac{\partial u}{\partial n} \right|^2 = 0 ,$$

and thus $\frac{\partial u}{\partial n} = 0$ on $\partial\Omega$.

$$\text{Set } \tilde{u} = \begin{cases} u - C & \text{on } \Omega \\ 0 & \text{on } R^2 \setminus \Omega, \end{cases}$$

so that $\tilde{u}$ satisfies

$$-\Delta\tilde{u} = (\tilde{u} + C)|\nabla\tilde{u}|^2 = A \cdot \tilde{u}_x + B \cdot \tilde{u}_y \quad \text{on } R^2$$

with $A,B \in L^\infty$. By using a unique continuation argument (personally communicated by D. Jerison and C. Kenig) we can conclude that $\tilde{u} \equiv 0$. It is an open problem whether weak (H^1) solutions of (1) with $\gamma \equiv C$ are constants.

Theorem 1 also answers a question raised by Giaquinta and Hildebrandt in [7]. They had considered the special case where

$$\gamma(x,y) = (Rx,\ Ry,\ \sqrt{1-R^2}) \text{ with } 0 < R < 1, \tag{2}$$

and they had pointed out that there are two explicit (distinct) solutions of (1), namely

$$\underline{u}(x,y) = \frac{2\lambda}{\lambda^2+r^2}(x,y,\lambda) + (0,\ 0,\ -1) \tag{3}$$

and

$$\overline{u}(x,y) = \frac{2\mu}{\mu^2+r^2}(x,y,-\mu) + (0,0,1) \tag{4}$$

where $r^2 = x^2 + y^2$, $\lambda = \frac{1}{R} + \sqrt{\frac{1}{R^2} - 1}$, $\mu = \frac{1}{R} - \sqrt{\frac{1}{R^2} - 1}$.

Note that $\Gamma = \gamma(\partial\Omega)$ is a (small) circle on S^2, $\underline{u}(\Omega)$ is the small spherical cap determined by Γ while $\overline{u}(\Omega)$ is the large spherical cap determined by Γ.

Incidentally, it is not known whether $\underline{u}$ and $\overline{u}$ are the only solutions of (1) when γ is given by (2). I shall, however, describe some partial results on this problem (see Theorem 3).

In order to prove Theorem 1 one splits $\mathcal{E}$ into (connected) components with the help of degree theory. The main observation is the following.

Set

$$Q(u) = \frac{1}{4\pi}\int_\Omega u \cdot u_x \wedge u_y \qquad \text{for } u \in H^1(\Omega;\ R^3) \cap L^\infty(\Omega;\ R^3). \tag{5}$$

<u>Lemma 2</u> Let $u_1, u_2 \in \mathcal{E}$, then

$$Q(u_1) - Q(u_2) \in \mathbb{Z}.$$

Sketch of the proof. If M denotes a manifold of dimension 2 (without boundary) and $\phi : M \to S^2$ is a smooth map, then its degree is given by (see [12])

$$\deg \phi = \frac{1}{4\pi} \int_M J_\phi$$

where J_ϕ denotes the Jacobian determinant of ϕ, i.e., $\nu \cdot \phi_x \wedge \phi_y$ where (x,y) are orthonormal coordinates in the tangent plane to M and ν is the normal to S^2 at the point ϕ, so that $J_\phi = \phi \cdot \phi_x \wedge \phi_y$.

Set $Q = \Omega \times (0,1)$ and $M = \partial Q$. Consider the map ϕ defined on $\overline{Q}$ - and thus on M - by

$$\phi(x,y,z) = z u_1(x,y) + (1-z)\, u_2(x,y)$$

with $(x,y) \in \overline{\Omega}$ and $z \in [0,1]$. Note that ϕ maps M into S^2 (since $u_1 = u_2 = \gamma$ on $\partial\Omega$) and that $\phi_z = 0$ on $\partial\Omega \times [0,1]$. It follows that $J_\phi = 0$ on $\partial\Omega \times [0,1]$ and therefore

$$\deg \phi = Q(u_1) - Q(u_2) \in \mathbb{Z}.$$

This argument is valid provided u_1 and u_2 are smooth (say C^1). In the general case (when u_1, u_2 are just in H^1) one uses an approximation technique of [13] based upon the fact that smooth functions from M into S^2 are dense in $H^1(M,S^2)$. □

Now, we split $\mathcal{E}$ into (connected) components as follows. For every $k \in Z$, set

$$\mathcal{E}_k = \{u \in \mathcal{E} \,;\, Q(u) - Q(\underline{u}) = k\}.$$

Here $\underline{u}$ plays the role of a reference element and could be replaced by any element in $\mathcal{E}$. Clearly $\mathcal{E} = \bigcup_{k \in Z} \mathcal{E}_k$ and each $\mathcal{E}_k$ is both open and closed. Moreover $\mathcal{E}_k \neq \emptyset$ for all k and $\mathcal{E}_0$ contains $\underline{u}$. In order to find critical points of E on it is tempting to consider

(6) $$J_k = \operatorname*{Inf}_{\mathcal{E}_k} E$$

and to try to prove that J_k is achieved. A major difficulty in trying to carry out this program is the following. Suppose (u^j) is a minimizing sequence for (6). Clearly, we may assume that $u^j \rightharpoonup \bar{u}$ weakly in H^1 and thus (by lower semicontinuity) we have $E(\bar{u}) < J_k$. However $\bar{u}$ need not belong to $\mathcal{E}_k$ since the sets $\mathcal{E}_k$ are not closed under weak convergence (Q is not continuous under weak convergence). It will become transparent, from the analysis in Section 3, that such "accidents" do occur. Next, we examine the special case where γ is given by (2) and we discuss the following

Theorem 3

(a) J_0 is achieved only for $u = \underline{u}$.

(b) $\bar{u} \in \mathcal{E}_{-1}$ and J_{-1} is achieved only for $u = \bar{u}$.

(c) J_k is not achieved for $k \neq 0$ and $k \neq -1$.

Moreover if (u^j) is a minimizing sequence for J_k with $k>0$ (resp. $k<-1$), then $u^j \rightharpoonup \underline{u}$ (resp $\bar{u}$) weakly in H^1.

Remark 2 In principle, there could be other critical points (e.g. saddle points, local minima). This is an open problem.

2. The Existence of a Non-Minimal Solution $\bar{u}$

In [3] we use the same kind of approach as in various other problems where a "lack of compactness" occurs (see [1], [2], [5]). Namely, we divide the argument in two steps:

Step 1 We show that if $J_k < J_0 + 8\pi$ for some k, then J_k is achieved.

Step 2 We show that indeed $J_k < J_0 + 8\pi$ (at least) for $k = +1$ or $k = -1$.

Sketch of the proof of Step 1.

Let (u^j) be a minimizing sequence for J_k so that

(7) $$\int_\Omega |\nabla u^j|^2 = J_k + o\,(1),$$

$$Q(u^j) - Q(\underline{u}) = k . \tag{8}$$

We may assume that $u^j \rightharpoonup \overline{u}$ weakly in H^1 and a.e. We write

$$u^j = \overline{u} + v^j$$

so that $v^j \rightharpoonup 0$ weakly in H_0^1 and by (7) we have

$$\int |\nabla \overline{u}|^2 + \int |\nabla v^j|^2 = J_k + o\ (1). \tag{9}$$

On the other hand we write

$$Q(u^j) = \frac{1}{4\pi} \int u^j . (\overline{u}_x + v_x^j) \wedge (\overline{u}_y + v_y^j)$$

$$= Q\ (\overline{u}) + \frac{1}{4\pi} \int u^j \cdot v_x^j \wedge v_y^j + o(1)$$

since $\int u^j \cdot v_x^j \wedge \overline{u}_y \to 0$ and $\int u^j \cdot \overline{u}_x \wedge v_y^j \to 0$.

(Note that $u^j \wedge \overline{u}_y$ and $u^j \wedge \overline{u}_x$ converge strongly in L^2 by dominated convergence)

It follows that

$$|Q(u^j) - Q(\overline{u})| < \frac{1}{8\pi} \int |\nabla v^j|^2 + o(1) \tag{10}$$

(since $|u^j| = 1$ and $|a \wedge b| < \frac{1}{2} (|a|^2 + |b|^2)$).

Combining (8), (9) and (10) we obtain

$$|Q(\overline{u}) - Q(\underline{u}) - k| < \frac{1}{8\pi} [J_k - \int |\nabla \overline{u}|^2] \tag{11}$$

$$< \frac{1}{8\pi} [J_k - J_0] < 1$$

since $J_0 < \int |\nabla \overline{u}|^2$. On the other hand $Q(u) - Q(\underline{u}) \in \mathbb{Z}$ and therefore $Q(\overline{u}) - Q(\underline{u}) = k$, that is, $\overline{u} \in \mathcal{E}_k$. From (9) we see that $\int |\nabla \overline{u}|^2 < J_k$ and consequently J_k is achieved at $\overline{u}$. $\square$

<u>Sketch of the proof of Step 2</u>

One constructs explicitly an element $v \in \mathcal{E}_1 \cup \mathcal{E}_{-1}$ such that

(12) $$\int |\nabla v|^2 < \int |\nabla \underline{u}|^2 + 8\pi .$$

Fix a point $(x_0,y_0) \in \Omega$ and for small $\varepsilon>0$ set

$$D_\varepsilon = \{(x,y) \; ; \; (x-x_0)^2 + (y-y_0)^2 < \varepsilon^2\}.$$

Set

$$v^\varepsilon(x,y) = \begin{cases} \underline{u}(x,y) & \text{in } \Omega\setminus D_{2\varepsilon} \\ \dfrac{2\lambda}{\lambda^2+r^2}(x-x_0,y-y_0,-\lambda) + (0,0,1) & \text{in } D_\varepsilon \end{cases}$$

with a smooth transition in the annulus $D_{2\varepsilon}\setminus D_\varepsilon$, where $\lambda = c\varepsilon^2$ and $r^2 = (x-x_0)^2 + (y-y_0)^2$. One can prove (see [3] for details) that, as $\varepsilon\to 0$,

$$|Q(v^\varepsilon) - Q(\underline{u})| = 1$$

and

$$\int |\nabla v^\varepsilon|^2 = \int |\nabla \underline{u}|^2 + 8\pi - \alpha\varepsilon^2 + o(\varepsilon^2)$$

where α depends on c and $\nabla\underline{u}(x_0,y_0)$. If $\nabla\underline{u}(x_0,y_0) \neq 0$ (which we can always assume since $\underline{u} \not\equiv$ Const), then one may choose c in such a way that $\alpha>0$. □

<u>Remark 3</u> A similar argument shows that

(13) $$J_{p+q} < J_p + 8\pi\,|q| \qquad \forall\, p,q \in \mathbb{Z}.$$

Indeed, suppose, for example, that $q>0$ and let u be any element in $\mathcal{E}_p$.

Set

$$D_\varepsilon = \{(x,y) \; ; \; x^2 + y^2 < \varepsilon^2\},$$

$$v^\varepsilon(x,y) = \begin{cases} u(x,y) & \text{in } \Omega\setminus\Omega_{2\varepsilon} \\ \dfrac{2\lambda}{\lambda^2+r^{2q}}(r^q\cos m\theta, - r^q\sin m\theta, - \lambda) + (0,0,1) & \text{in } D_\varepsilon \end{cases}$$

with a smooth transition in $D_{2\varepsilon}\setminus D_\varepsilon$, where $\lambda = \varepsilon^{q+1}$ and $x = r\cos\theta$, $y = r\sin\theta$. It is easy to check that, as $\varepsilon\to 0$,

$$Q(v^\varepsilon) - Q(u) = q$$

so that $v^\varepsilon \in \mathcal{E}_{p+q}$, and moreover

$$\int |\nabla v^{\varepsilon}|^2 = \int |\nabla u|^2 + 8\pi q + o(1).$$

Since u is any element in $\mathcal{E}_p$ we conclude that (13) holds. □

3. Detailed Analysis of an Example; Behavior of Minimizing Sequences

For the proof of part (a), (b) and (c) in Theorem 3 we refer to [3]. Consider now any boundary condition γ such that

(14) $\qquad$ only J_0 and J_{-1} are achieved.

We assert that

(15) $$J_k = J_0 + 8\pi k \qquad \forall k > 0$$

(16) $$J_k = J_{-1} + 8\pi|k+1| \qquad \forall k < -1 .$$

Moreover if (u^j) is a minimizing sequence for J_k with $k>0$ (resp $k < -1$), then, for some subsequence, $u^j \rightharpoonup u$ weakly in H^1 where u is a minimizer for J_0 (resp. J_{-1}).

Indeed, we have (7) and (8). As in Step 2 of Section 2 we write

$$u^j = u + v^j$$

so that

(17) $$\int |\nabla u|^2 + \int |\nabla v^j|^2 = J_k + o(1),$$

(18) $$|Q(u^j) - Q(u)| < \frac{1}{8\pi} \int |\nabla v^j|^2 + o(1).$$

Combining (8), (17) and (18) we find

(19) $$|k - (Q(u) - Q(\underline{u}))| < \frac{1}{8\pi} [J_k - \int |\nabla u|^2].$$

Set

$$\ell = Q(u) - Q(\underline{u}) \in \mathbb{Z}$$

so that $u \in \mathcal{E}_\ell$ and consequently

(20) $$\int |\nabla u|^2 \geq J_\ell .$$

From (19) and (20) we have

(21) $$|k-\ell| \leq \frac{1}{8\pi} [J_k - J_\ell].$$

On the other hand, we know, by (13), that

$$J_k - J_\ell \leq 8\pi \, |k-\ell| .$$

Therefore we must have

(22) $$\int |\nabla u|^2 = J_\ell$$

and

(23) $$J_k - J_\ell = 8\pi |k-\ell| .$$

In particular, u is a minimizer for J_ℓ and this can happen only with $\ell = 0$ or $\ell = -1$ (by assumption (14)).

When $k > 0$ (resp. $k < -1$) we must have $\ell = 0$ (resp. $\ell = -1$). Indeed suppose, for contradiction, that $k > 0$ and $\ell = -1$. From (23) we have

$$J_k = J_{-1} + 8\pi \, (k+1).$$

However, by (13) we know that

$$J_k \leq J_0 + 8\pi \, k$$

and thus $J_{-1} + 8\pi \leq J_0$ which is absurd.

Similarly, if we suppose that $k < -1$ and that $\ell = 0$ we find

$$J_k = J_0 - 8\pi k.$$

However, by (13) we know that

$$J_k \leq J_{-1} - 8\pi \, (k+1)$$

and therefore $J_0 \leq J_{-1} - 8\pi < J_0$, which is absurd. ☐

References

[1] H. Brezis, Some variational problems with lack of compactness, Proc. Symp. Nonlinear Functional Analysis, Berkeley, 1984, AMS.

[2] H. Brezis - J.M. Coron, Multiple solutions of H-systems and Rellich's conjecture, Comm. Pure Appl. Math. 37 (1984) p. 149-187.

[3] H. Brezis - J.M. Coron, Large solutions for harmonic maps in two dimensions, Comm. Math. Phys. 92 (1983) p. 203-215.

[4] H. Brezis - J.M. Coron - E. Lieb, in preparation.

[5] H. Brezis - L. Nirenberg, Positive solutions of nonlinear elliptic equations invoving critical Sobolev exponents, Comm. Pure. Appl. Math. 36 (1983) p. 433-477.

[6] J.L. Ericksen - D. Kinderlehrer, eds., Proceedings of the IMA Workshop on the Theory and Applications of Liquid Crystals, Minneapolis, Jan. 1985.

[7] M. Giaquinta - S. Hildebrandt, A priori estimates for harmonic mappings, J. Reine Angew. Math. 336 (1982) p. 124-164.

[8] R. Hardt - D. Kinderlehrer - F.H. Lin, Existence and partial regularity of static liquid crystal configurations, Comm. Math. Phys. (to appear).

[9] R. Hardt - F.H. Lin in preparation.

[10] J. Jost, The Dirichlet problem for harmonic maps from a surface with boundary onto a 2-sphere with nonconstant boundary values, J. Diff. Geom. 19 (1984) p. 393-401.

[11] L. Lemaire, Applications harmoniques de surfaces riemanniennes, J. Diff. Geom. 13 (1978) p. 51-78.

[12] L. Nirenberg, Topics in Nonlinear Functional Analysis, N.Y.U. Lecture Notes 1973-74.

[13] R. Schoen - K. Uhlenbeck, Boundary regularity and miscellaneous results on harmonic maps, J. Diff. Geom. 18 (1983) p. 253-268.

BIFURCATION PROBLEMS IN CONSTRAINED NONLINEAR THERMOELASTICITY

M.C. Calderer

Department of Mathematical Sciences
University of Delaware
Newark, Delaware 19716

1. Introduction

I study the problem of inflation under constant pressure of nonlinear thermoelastic and isotropic spherical shells. The boundary of the shell is kept insulated and the material of the body is supposed to satisfy the internal constraint,

$$\det F(X,t) = f(\tau(X,t)), \tag{1.1}$$

where F denotes the gradient of deformation matrix at the particle X at time $t > 0$ and $\tau > 0$ denotes the absolute temperature. Moreover, $f(\tau)$ in (1.1) is assumed to be smooth and to satisfy the following conditions:

$$f(\tau_0) = 1, \quad f'(\tau) > 0, \tag{1.2a,b}$$

where τ_0 denotes a constant references temperature. According to (1.2a), the constraint expressed by equation (1.1) reduces to incompressibility when thermal effects are ignored and the temperature is taken to be τ_0. When such conditions are met, we will refer to the problem as 'purely mechanical'. Another special situation arises when τ is set to be equal to a constant $\hat{\tau}$ in the governing equations. The corresponding problem will be called 'isothermal'. In this case, the body experiences an increase in volume given by equation (1.1).

It has been experimentally observed (cf [T1]) that certain rubberlike materials, normally regarded as incompressible, increase their volume as the temperature increases. Although this effect is small compared with large deformations consistent with nonlinear elastic constitutive equations, it seems to be appropriately modeled by a constraint of the form (1.1)-(1.2) (cf [A2], [G2], [G3]). It is also well known (cf. [T1]) that at large extensions the stress is an increasing function of the temperature, within a certain temperature range, as predicted by the kinetic theory of elasticity. However at suitably low extensions the stress decreases with τ. This is known as the thermoelastic inversion phenomenon and it is interpreted in terms of the increase in volume of the rubber on heating. A decrease of stress with τ also occurs for values of τ below the glass transition temperature, τ_g. The behavior then corresponds to that of an ordinary hard solid. Such a phenomenon is not taken into account by our model and therefore, we shall take $\tau > \tau_g$ in (1.1). A significant contraction of volume also occurs at large extensions and is due to crystallization. This is also excluded from our model.

The constitutive equations of a constrained material are determined only up to a 'reaction' term depending on the Lagrange multiplier maintaining such a constraint (cf [A1], [G2], [G3]). Here, we will assume that the reaction terms in the constitutive equations give rise to no production of entropy, in any process satisfying (1.1). This approach is consistent with the point of view that a constrained material is, in fact, the limiting case of a family of unconstrained ones (cf. [A]) and is also compatible with the variational formulation of the problem. In this framework, we seek deformations and temperature fields that maximize the entropy (cf. [B], [E], [G1], [K]), an approach that is only meaningful when the entropy is completely determined by the constitutive equations (i.e., if the 'reaction entropy' integrates to zero on the body) (cf. [G2], [C3]).

In this article, I study radial equilibrium solutions to the equations of balance of mass, linear momentum and energy for a spherically symmetry body with the previously described material properties.

The main goal is to discover how the presence of a thermokinetic constraint of the form (1.1) affects the structure of the equilibrium solutions as well as their stability properties, when compared to the purely mechanical problem.

In section 2, the main properties of thermoelastic materials are presented, while in sections 3 and 4 the formulation of the problem is carried out. In section 5, I present a bifurcation analysis of the equilibrium solutions, showing in particular, how the solutions of the thermoelastic problem may bifurcate from those of the corresponding mechanical one. A Liapunov function for the governing equations is constructed and necessary and sufficient conditions for the dynamical stability of the equilibrium solutions are presented. In particular, such conditions reduce to the minimization of the static potential energy associated with the mechanical problem. Examples of materials satisfying such conditions may be found among the class studied by Ogden (cf. [O]) and they may have strong elliptic Helmholtz free energies.

2. Material Properties

Since the material is assumed to be thermoelastic, the internal energy $\varepsilon(X,t)$, the Piola-Kirchhoff stress tensor $T_R(X,t)$, the entropy $\eta(X,t)$ and the Helmholtz free energy $\psi(X,t)$ are functions of $F(X,t)$ and $\tau(X,t)$ as follows,

$$\varepsilon(X,t) = \hat{\varepsilon}(X,F(X,t),\tau(X,t)), \text{ etc.} \tag{2.1}$$

The reference heat-flux vector q_R satisfies

$$q_R(X,t) = \mathbf{q}_R(X,F(X,t),\tau(X,t),\nabla\tau(X,t)). \tag{2.2}$$

Moreover, we assume that the following relations hold

$$\hat{T}_R = \rho_R \frac{\partial\hat{\psi}}{\partial F}, \quad \hat{\eta} = -\frac{\partial\hat{\psi}}{\partial\tau}, \quad \varepsilon = \psi + \eta\tau, \tag{2.3}$$

where ρ_R denotes the density of the material in the reference configuration and it is taken to be a constant.

The second law of thermodynamics in the form of the Fourier inequality, requires that

$$q_R \cdot \nabla\tau < 0. \tag{2.4}$$

Since the material is supposed to satisfy the thermokinetic constraint (1.1), T_R, η and ε are of the form (cf. [G3])

$$T_R = \overline{T}_R + \hat{T}_R, \quad \eta = \overline{\eta} + \hat{\eta}, \quad \psi = \hat{\psi},$$

where $\overline{\eta}$ and $\overline{T}_R$, are the reaction terms described in Section 1 ($\overline{T}_R$ is due to a hydrostatic pressure). The constitutive components $(\hat{T}_R, \hat{\eta}_1 \ldots)$ are defined only for F and τ satisfying (1.1).

The isotropy assumption on the material implies that

$$\hat{\psi}(\tau,F) = \hat{\psi}(\tau,I_B,II_B,III_B), \tag{2.5}$$

where I_B, II_B, III_B denote the principal invariants of the Cauchy-Green deformation tensor $B = F^TF$. Moreover, the Cauchy stress tensor T for an isotropic material satisfying (1.1) has the form

$$T = -pI + \hat{\psi}_1 B + \hat{\psi}_{-1} B^{-1}, \tag{2.6}$$

with

$$\hat{\psi}_1 = 2\rho \frac{\partial\hat{\psi}}{\partial I_B} (\tau,I_B,II_B,f(\tau)^2) \tag{2.7}$$

$$\hat{\psi}_{-1} = -2\rho \frac{\partial\hat{\psi}}{\partial II_B} (\tau,I_B,II_B,f(\tau)^2)f(\tau)^2, \tag{2.8}$$

where p is the hydrostatic pressure due to the presence of the constraint and ρ is the density of the material in the present configuration, satisfying

$$\rho = \rho_R\, f(\tau)^{-1}. \tag{2.9}$$

Equation (2.9) follows from the continuity equation as well as (1.1).

Since at constant temperature (2.5)-(2.8) correspond to a hyperelastic material, we assume that

$$\hat{\psi}_1 > 0 \quad \text{and} \quad \hat{\psi}_{-1} < 0, \tag{2.10}$$

for all $\tau > 0$.

Moreover, experiments on rubberlike materials suggest that $\hat{\psi}_1$ and $-\hat{\psi}_{-1}$ are increasing functions of the temperature, at constant deformation ([M1], [T1]), i.e.,

$$\frac{\partial\hat{\psi}_1}{\partial\tau} > 0, \quad \frac{\partial\hat{\psi}_{-1}}{\partial\tau} < 0. \tag{2.11}$$

Relations opposite to those in (2.11) seem to hold for metals, (cf. [U2]). As is customary, and consistent with experience, we assume that the specific heat

$c(\tau,F)$ must be positive, i.e.

$$c(\tau,F) \equiv \tau \frac{\partial \hat{\eta}}{\partial \tau} > 0. \tag{2.12}$$

Therefore, it follows from (2.3) and (2.12) that we can formally write (2.5) as

$$\hat{\psi}(\tau,F) = -\int_{\tau_0}^{\tau} \mu^{-1}(\tau - \mu)c(\mu,F)d\mu + \tau h(F) + g(F), \tag{2.13}$$

where $c(\tau,F)$ is experimentally obtained (cf. [C1], [M2]) and the dependence on F is through the invariants of B as in (2.5). In particular, if c is independent of F, then (2.13) becomes

$$\hat{\psi}(\tau,F) = M(\tau) + \tau h(F) + g(F). \tag{2.14}$$

For a material of Mooney-Rivlin type, $h(F)$ and $g(F)$ are linear functions of the principal invariants of B. Therefore, it follows from (2.6)-(2.8) and (2.14) that the constitutive equation for T in such case is given by

$$\begin{aligned} T = -pI + f(\tau)^{-1}\left(a_1 + a_2 \frac{\tau - \tau_0}{\tau_0}\right)B - \\ - f(\tau)\left(b_1 + b_2 \frac{\tau - \tau_0}{\tau_0}\right)B^{-1}, \end{aligned} \tag{2.15}$$

with $a_1, a_2, b_1 > 0$ and $b_2 \geq 0$.

The corresponding equation for a neo-hookean material is obtained by letting

$$b_1 = 0 = b_2 \quad \text{in (2.15).}$$

Finally, the well known relation between T and T_R will be needed in later sections,

$$T = \rho F T_R^T . \tag{2.16}$$

3. Boundary Value Problems for Spherically Symmetric Bodies

We consider a cartesian system and a set of spherical coordinates (R,Θ,Φ) in $\mathbb{E}^3$. Let the reference configuration of the shell be the domain $\Omega \subset \mathbb{E}^3$,

(3.1) $$\Omega = \{(R,\Theta,\Phi): 0 < R_1 < R < R_2\},$$

where R_1 and R_2 denote the radii of the inner and the outer surfaces of the shell, respectively. Let the position (r,θ,ϕ) at time $t > 0$ of the particle labelled by (R,Θ,Φ) in the reference configuration be given by

(3.2) $$r = r(R,t), \quad \theta = \Theta, \quad \phi = \Phi .$$

The gradient of deformation matrix in the orthonormal basis associated with the spherical coordinates is given by

(3.3) $$F = \mathrm{diag}\left(\frac{\partial r}{\partial R}, \frac{r}{R}, \frac{r}{R} \right).$$

Since the deformations and temperature fields must satisfy (1.1), it follows that

(3.4) $$r(R,t)^3 = 3 \int_{R_1}^{R} u^2 f(\tau(u,t))du + r_1(t)^3.$$

Let $x = \mathbf{x}(X,t)$ denote the position of the particle X at time t. The balance laws of linear momentum and energy, with zero body force and zero heat supply, are given by

(3.5) $$\mathrm{Div}\, T_R = \rho_R \ddot{x} ,$$

(3.6) $$-\mathrm{Div}\, q_R + \mathrm{tr}(T_R \dot{F}^T) = \rho_R \dot{\varepsilon},$$

where Div represents the material divergence and the superposed dots denote derivatives with respect to t.

Equations (2.9), (3.5) and (3.6) are supplemented with the following boundary conditions on $\partial\Omega$:

<u>Mechanical:</u> The outer surface is free of traction and the inner one is subjected to a constant pressure $P > 0$ measured per unit area in the present configuration. Hence, the radial component of T satisfies

(3.7) $$T_{rr}(r_2,t) - T_{rr}(r_1,t) = P,$$

where

$$r_2 = r(R_2,t) \quad \text{and} \quad r_1 = r(R_1,t), \tag{3.8}$$

denote the outer and inner radii at time t.

Thermal: The boundary is insulated, i.e.,

$$q_R(R_i,t) \cdot N(R_i) = 0, \quad i = 1,2, \tag{3.9}$$

where **N** denotes the unit outward normal to the boundary.

Following steps analogous to those to obtain the equation of the radial motion of an incompressible and hyperelastic spherical shell under pressure (cf. [TN]), we can reduce (3.5) to an integrodifferential equation for the inner radius and th temperature. We first write equation (3.5) in spherical coordinates, using the form for T given by (2.6), together with (3.3) and (3.4). The acceleration $x_{tt} = (r_{tt},0,0)$ in (3.5) can be obtained by differentiating both sides of (3.4) twice with respect to t. Integrating the radial component of (3.5) between R_1 and R_2 and using (2.16) and (3.7), we obtain

$$\begin{aligned}
&\rho_R \int_{R_1}^{R_2} R^2\{2r_1(t)\dot{r}_1(t)^2 + r_1(t)\ddot{r}_1(t) + \\
&\qquad \int_{R_1}^{R} u^2[f''(\tau(R,t))\tau_t(u,t)^2 + + f'(\tau(R,t))\tau_{tt}(u,t)]du - \\
&\qquad 2r(R,t)^{-3}[\int_{R_1}^{R} uf'(\tau(R,t))\tau_t(u,t)du + r_1(t)^2\dot{r}_1(t)]^2\}r(R,t)^{-4}dR = \\
&P + \int_{R_1}^{R_2} 2r(R,t)^{-3}R^2 \cdot f(\tau(R,t))[\hat{\psi}_1(\tau(R,t),I_B(R,t),II_B(R,t),f(\tau(R,t),t)^2) \\
&\qquad \cdot (R^4r(R,t)^{-4}f(\tau(R,t))^2 - r(R,t)^2R^{-2}) + \\
&\qquad + \hat{\psi}_{-1}(\tau(R,t),I_B(R,t),II_B(R,t), f(\tau(R,t))^2) \cdot \\
&\qquad \cdot (R^{-4}r(R,t)^4f(\tau(R,t))^{-2} - R^2r(R,t)^{-2})]dR,
\end{aligned} \tag{3.10}$$

with $r(R,t)$ given by (3.4) and $P > 0$ as in (3.7).

4. Equilibrium Solutions

For a prescribed constitutive function $\phi(F,\tau)$ satisfying (2.5), (2.10), (2.11) and (2.12), we study radially symmetric equilibrium solutions

$$r = r(R), \quad \tau = \tau(R), \quad R \in \Omega$$

of the system (1.1), (2.7)-(2.9), (3.5)-(3.10), with Ω given by (3.1). Moreover, we assume that q_R satisfies Fourier's law

$$q_R = \kappa(R,r,\tau,\nabla\tau)\nabla\tau, \tag{4.1}$$

where $\kappa > 0$ is the conductivity.

It is easy to check that the assumption of radial symmetry of the equilibrium solutions together with equations (3.6), (3.8), and (4.1) imply that

$$\tau = \text{constant in } \Omega. \tag{4.2}$$

Remark: The previous result is still true if one of the surfaces of the shell remains insulated while the other is held at a constant temperature, τ_c. Then, the constant in (4.2) is τ_c. Under the previous conditions, equation (3.4) reduces to

$$r(R)^3 = (R^3 - R_1^3)f(\tau) + r_1^3. \tag{4.3}$$

Therefore, it follows from (3.12) and (4.2) that the problem of finding radially symmetric equilibrium solutions to equations (1.1), (2.7)-(2.9), (3.5)-(3.10), reduces to finding pairs $(r_1,\tau) \in (0,\infty) \times (0,\infty)$ satisfying

$$\begin{aligned} P + \int_{R_1}^{R_2} & 2\, r(R)^{-3}R^2[\hat{\psi}_1(\tau,I_B(R),\, II_B(R),\, f(\tau)^2)\, \cdot \\ & \cdot (r(R)^{-4}R^4 f(\tau)^2 - R^{-2}r(R)^2) + \hat{\psi}_{-1}(\tau,I_B(R),II_B(R),f(\tau)^2)\, \cdot \\ & \cdot (r(R)^4R^{-4}f(\tau)^{-2} - R^2 r(R)^{-2})]f(\tau)dR = 0, \end{aligned} \tag{4.4}$$

with $r(R)$ given by (4.3) and $P > 0$ as in (3.7).

Letting

(4.5) $$\xi = r^3 R^{-3}, \quad y = f(\tau), \quad x = r_1 R_1^{-1}, \quad \delta = (R_2^3 - R_1^3) R_1^{-3},$$

we can write

$$I_B = \xi^{-4/3} + 2\xi^{2/3},$$
$$II_B = \xi^{4/3} + 2y^2\, \xi^{-2/3}.$$

Equation (4.4) becomes

$$y\hat{g}(x,y,\tau) = P, \text{ where}$$

$$\hat{g}(x,y,\tau) = \frac{2}{3} \int_{\frac{x^3+\delta y}{1+\delta}}^{x^3} (\hat{\psi}_1(\tau,y,\xi)\xi^{-7/3} - \hat{\psi}_{-1}(\tau,y,\xi)y^{-2}\xi^{-5/3}) \cdot (y + \xi)d\xi.$$

The parameter $\delta > 0$ is taken to be fixed and is related to the thickness of the shell in the reference configuration. Using relations (2.7)-(2.9), we can write the previous equations

(4.6) $$g(x,y,\tau) = P, \quad \text{with}$$

(4.7) $$g(x,y,\tau) = \rho_R \int_{\frac{x^3+\delta y}{1+\delta}}^{x^3} (\xi - y)^{-1} \frac{\partial\hat{\psi}}{\partial\xi}(\xi,y,\tau)d\xi.$$

5. Bifurcation Analysis of the Solutions

In [C2], I studied the equilibrium solutions of equations (4.6) and (4.7), for $P > 0$ given, under the assumptions of the purely mechanical case. In particular, the bifurcation of solutions with respect to P is studied.

In this article, the temperature enters equations (4.6) and (4.7) through the constitutive equations as well as the special form (4.3) of the radial solutions. The results of this section show how the presence of the temperature in the governing equations affects the structure of the equilibrium solutions, in comparison with the case described in the first paragraph.

For this, let us introduce the following notation

$$(5.1)\qquad \begin{aligned} &G(X,\tau) = g(x,y,\tau) - P, \text{ where} \\ &X \equiv (x,P) \in (0,\infty) \times (0,\infty). \end{aligned}$$

Equation (4.6) then reduces to

$$(5.2)\qquad G(X,\tau) = 0.$$

Before establishing the main results of this section, it is convenient to describe the properties of $g(x,y,\tau)$ implied by the constitutive assumption studied in Section 2. In particular, the smoothness of $\phi(F,\tau)$ together with relations (4.3) and (4.10) imply that

$$g(\cdot,\cdot,\cdot): (0,\infty) \times (0,\infty) \times (0,\infty)$$

is smooth and satisfies the following inequalities:

$$(5.3)\qquad \begin{aligned} &g(x,y,\tau) > 0 \quad \text{for} \quad x > y^{1/3} \quad \text{and} \\ &g(x,y,\tau) < 0 \quad \text{for} \quad x < y^{1/3}, \end{aligned}$$

with $y = f(\tau)$.

As an alternative approach, we could place hypotheses on $g(x,y,\tau)$ rather than on $\phi(F,\tau)$. The relationship between the corresponding sets of assumptions is based on equations (2.14), (3.3), (4.3) and (4.7) (cf. [C2]). In order to establish results on existence and global bifurcation of the equilibrium solutions, we need growth hypotheses on $g(x,y,\tau)$ with respect to x, for x large and small with τ fixed. In particular, we assume that for $g(x,y,\tau)$ prescribed, the following hypotheses are satisfied:

There exist continuous functions $D(\tau) > C(\tau) > 0$, $M(\tau) > N(\tau) > 0$ and constants $a,b > 0$, such that

$$(5.4)\qquad C(\tau)x^{a/3-1} < g(x,f(\tau),\tau) < D(\tau)x^{a/3-1}, \quad \text{for } x \text{ large},$$

$$(5.5)\qquad -M(\tau)x^{-b/3} < g(x,f(\tau),\tau) < -N(\tau)x^{-b/3}, \quad \text{for } x \text{ small}.$$

The previous hypotheses are satisfied by a very large class of physically realistic materials (cf. [O]). In some special examples, the exponents a and b may be related.

According to the previous remarks, one could reformulate the problem in the following way:

For a given function $g(x,y,\tau)$ satisfying inequalities (5.3)-(5.4) and for $P > 0$ fixed, find pairs $(x,\tau) \in (1,\infty) \times (0,\infty)$ such that equations (5.1) and (5.2) holds for (x,P,τ).

The following result is a consequence of the smoothness of $g(x,y,\tau)$ as well as hypotheses (5.3)-(5.4).

<u>Proposition 5.1</u>

Suppose that $g(x,y,\tau)$ satisfies hypotheses (5.3)-(5.5) and let $\tau > 0$ be fixed. Then, one of the following holds,

(i) If $a > 3$ in (5.4), then for every $P > 0$ there exists at least one real number $x > f(\tau)^{1/3}$ such that (x,P,τ) satisfies equation (5.2).

(ii) If $a < 3$ in (5.4), then there exists a critical value $P_c = P_c(\tau)$ such that for every $P \in (0,P_c)$ there are at least two real numbers $x_1,x_2 > f(\tau)^{1/3}$ such that (x_i,P,τ), $i = 1,2$, satisfy equation (5.2). Moreover, if $P = P_c$ there is only one solution and if $P > P_c$, equation (5.2) has no solution. □

Remark: The previous result when $\tau = \tau_0$ reduces to that for an incompressible material with temperature independent constitutive equations as studied in [C2].

As a consequence of the implicit function theorem, we can establish the following result

<u>Proposition 5.2</u>

Let $(\hat{X},\hat{\tau})$ satisfy equation (5.2) and suppose that

$$\frac{d}{d\tau} g(\hat{x},f(\hat{\tau}),\hat{\tau}) \neq 0 \text{ holds.} \tag{5.6}$$

Then there is a neighborhood B of $\hat{X} = (\hat{x},\hat{P})$ in $(0,\infty) \times (0,\infty)$ such that equation (5.2) can be uniquely solved for τ in terms of X, $X \in B$. □

If an equilibrium solution $(\hat{X}, \hat{\tau})$ does not satisfy condition (5.6), then the following alternative result can be established. The proof of such a result can be found in [C3].

<u>Proposition 5.3</u>

Let $g(x, f(\tau), \tau)$ satisfy the hypotheses of Proposition 5.1. Suppose that there exists a $(\hat{X}, \hat{\tau})$ satisfying equation (5.2) and such that

(5.7) $$\frac{d}{d\tau} g(\hat{x}, f(\hat{\tau}), \hat{\tau}) = 0 \text{ holds.}$$

Then $(\hat{X}, \hat{\tau})$ is a bifurcation point, i.e., given any neighborhood B of $\hat{X}$, to each $X \in B$ there corresponds either two solutions (X, τ_i), $i = 1,2$, of (5.2) with $\tau_2 > \tau_1 > \hat{\tau}$, or one solution only with $\tau_1 > \hat{\tau}$, or no solutions at all. □

Remarks:

(i) The results stated in the last two propositions, when $\hat{\tau}$ is taken to be the reference temperature τ_0, show how the structure of the equilibrium solutions of the thermoelastic problem compare locally to that of the purely mechanical problem.

(ii) Relations of the form (5.6) and (5.7) are not genuinely characterized in terms of the constitutive equations. However, examples can be constructed to show that such relations may be satisfied by physically realistic materials having strongly elliptic free energies. Let us consider the two following cases:

1) Set $\psi = \frac{1}{2} (a_1 + a_2 \frac{\tau - \tau_0}{\tau_0}) I_B + h(\tau)$,

with a_1 and a_2 as in (2.15) and $h(\cdot)$ prescribed. This is the free energy function corresponding to a neo-heokean material. Equation (5.2) now becomes

(5.8)
$$\frac{2}{3}\left(a_1 + a_2 \frac{\tau - \tau_0}{\tau_0}\right)\overline{g}(x,y) - P = 0, \text{ where}$$

$$\overline{g}(x,y) = \int_{\frac{x^3 + \delta y}{1+\delta}}^{x^3} \xi^{-7/3}(y + \xi)\,d\xi.$$

The results of this section for equation (5.8) can be summarized as follows.

(a) Let $\tau > 0$ be fixed. Then, there exist positive constants $0 < P_0(\tau) < P_c(\tau)$ such that for each $0 < P < P_c$ one can find $x_i = x_i(P,\tau)$, $i = 1,2$, $x_2 > x_1 > f(\tau)^{1/3}$, with (x_i,P,τ), $i = 1,2$ satisfying equation (5.8).

(b) Let $(\hat{x},\hat{\tau},\hat{P})$, $\hat{P} \neq P_0$ be as in part (a). Then, there is a neighborhood B of $(\hat{x},\hat{P})$ that contains a two-parameter family of solutions $(x,P,\tau(x,P))$, $(x,P) \in B$, with $\tau > \hat{\tau}$.

(c) The one-parameter family of solutions $\{(x_1(P_0,\tau),P_0,\tau),\ P_0 = P_0(\tau),\ \tau > 0\}$ is a curve of bifurcation points, in the sense of Proposition 5.3. Moreover,

(5.9)
$$\frac{\partial\overline{g}}{\partial x}(x_1(P_0,\tau),f(\tau)) > 0.$$

2) The free energy function of a material of the class studied by Ogden (cf [O]) in terms of the principal stretches d_1, d_2, d_3 is of the form

(5.10)
$$\sigma(d_1,d_2,d_3,\tau) = \sum_{i=1}^{N} \alpha_i(\tau)(d_1 + d_2 + d_3 - 3) +$$
$$+ \sum_{j=1}^{M} \kappa_j(\tau)((d_1d_2)^{\nu_j} + (d_1d_3)^{\nu_j} + (d_2d_3)^{\nu_j} - 3) + h(\tau),$$

where $a_i(\cdot)$, $\kappa_j(\cdot)$ and $h(\cdot)$ are prescribed functions of the temperature and μ_i, ν_j are constants. A bifurcation analysis can be carried out in this case as well. However, it is worth emphasizing the existence of bifurcation solutions, as in proposition 5.3, for which

(5.11) $$\frac{\partial g}{\partial x}(x,f(\tau),\tau) < 0.$$

[As I shall point out later, condition (5.11) implies instability of the bifurcation solutions.]

Example: A strongly elliptic Helmholtz free energy, for which there is a bifurcation solution as in Proposition 5.3 satisfying (5.11), can be constructed with the following data:

Set $N = 3$, $M = 0$ in (5.10), with

$$\alpha_1 = 0.157 \qquad \alpha_2 = -0.092 \qquad \alpha_3 = 0.5$$
$$\mu_1 = 2.8 \qquad \mu_2 = 2.5 \qquad \mu_3 = 2.0$$
$$\beta_1 = 0.021 \qquad \beta_2 = 0.007 \qquad \beta_3 = 0.04.$$

6. Stability Analysis

The static stability criterion for the equilibrium solutions of the 'isothermal' problem, obtained by setting $\tau = \hat{\tau}$, $\hat{\tau}$ fixed, in the governing equations, is given in terms of the minimization of a certain potential energy. Such a criterion is also sufficient to ensure dynamical stability of the corresponding equilibrium solutions ([C2], [E], [G1], [K]).

It follows from equation (5.2) that the set of equilibrium solutions $(x,P,\tau) \in (0,\infty)^3$ of the thermoelastic problem determines a surface in R^3, which is in fact the graph of the function G given by (5.1). The intersection of such a graph with the plane $\tau = \hat{\tau}$, determines the curve of equilibrium solutions of the corresponding problem under isothermal conditions, as discussed in the previous paragraph. Therefore, defining

(6.1) $$V(x,\tau,P) = \int_{f(\tau)}^{x^3} g(\lambda, f(\tau), \tau)d\lambda - \frac{P}{3}x^3 ,$$

we find that

1) The elements $(x,P,\tau) \in (0,\infty)^3$ such that

$$(6.2) \qquad \frac{\partial V}{\partial x}(x,\tau,P) = 0$$

are solutions of (5.2).

A necessary condition for $(\overline{x},\overline{P},\overline{\tau})$ satisfying (5.2) to be a (dynamically) stable equilibrium solution is that

$$(6.3) \qquad \frac{\partial g}{\partial x}(\overline{x},f(\overline{\tau}),\ \overline{\tau}) > 0, \text{ with}$$

$$\overline{P} = g(x,f(\overline{\tau}),\ \overline{\tau}).$$

It is easy to check from the definition (6.1) that relations (6.2) and (6.3) imply that $V(\overline{x},\ \overline{\tau},\ \overline{P})$ is a local minimum with respect to x. We are going to show that such relations are also sufficient to ensure dynamical stability of the equilibrium solutions of the governing equations for the thermoelastic problem. For this purpose, we introduce the following notation

$$(6.4) \qquad \mathcal{L}_1(t) = -\int_\Omega \overline{\tau}\eta \, dx, \quad \text{for any } \overline{\tau} > 0 \text{ constant.}$$

$$(6.5) \qquad E(t) = \int_\Omega [\tfrac{1}{2}\rho_R \dot{x}^2 + \varepsilon(F,\tau)]dX - \int_{\partial\Omega} T_R \, \mathbf{N} \cdot \mathbf{x} \, dS$$

(I omit the dependence of F and τ on R and t). Notice that η and ε in (6.4) and (6.5) contain both the constitutive component, depending on F and τ satisfying equation (1.1), together with the one due to the constraint.

The proof of the following lemmas can be found in [C3].

<u>Lemma 6.1</u>

(i) $\mathcal{L}_1$ in (6.4) is a Liapunov function, i.e.,

$$\frac{d}{dt}\mathcal{L}_1(t) \leq 0,$$

on the solutions of the governing equations. (In fact, the equal sign holds due to the boundary conditions (3.9)).

(ii)

$$\int_\Omega \overline{\eta} \, dX = 0. \qquad \square$$

Equation (6.7) is a consequence of the nature of the constraint itself, as mentioned in Section 1. Therefore we can replace η with $\hat{\eta}$ in (6.4).

Lemma 6.2

The total energy E defined by (6.5) is a constant on the solutions of the governing equations, i.e.,

$$\frac{dE}{dt}(t) = 0,$$

and therefore,

(6.6) $$\mathcal{L}(t) \equiv -\int_{\Omega} (\bar{\tau}\hat{\eta}(F,\tau) + E)dX$$

is also a Liapunov function. □

Lemma 6.3 (Ericksen's lemma, cf. [E], [G3])

The Liapunov function $\mathcal{L}$ given by (6.6) can be written as

(6.7) $\mathcal{L}(t) = \int [\hat{\psi}(F,\bar{\tau}) + \kappa(F,\tau)(\tau - \bar{\tau})^2 + \frac{1}{2} p_R \dot{x}^2] - \int T_R N \cdot x \, dS$, with

$$\kappa(F,\tau) = -\frac{\partial^2\hat{\psi}}{\partial\tau^2}(F,\tau) > 0.$$ □

We now establish the main result of this section.

Theorem 6.1. A necessary and sufficient condition for an equilibrium solution $(\bar{x},\bar{P},\bar{\tau}) \in (0,\infty)^3$ of the governing equations to be (neutrally) stable is that relations (6.2) and (6.3) hold.

Proof: necessity has already been established at the beginning of the section. The proof of sufficiency follows from Lemma 6.2 together with the Liapunov theorem (cf. [4]) provided we can show that

(6.8) $$\mathcal{L}(t) > \mathcal{L}_{eq} \equiv V(\bar{x},\bar{\tau},\bar{P}),$$

where $\mathcal{L}_{eq}$ denodes the restriction of $\mathcal{L}$ given by (6.7) to the equilibrium solution $(\bar{x},\bar{P},\bar{\tau})$. Therefore, according to (6.7) the problem reduces to

$$\text{Minimize} \int_\Omega \hat{\psi}(\nabla x, \bar{\tau})dX - \int_\Omega T_R N \cdot x \, dS$$

subject to the constraint $\det \nabla x = f(\bar{\tau})$. Moreover, by equations (2.16), (3.3), (3.7), (3.8), (4.3), (4.7) and (6.1), it reduces to finding the minimum with respect to x of $V(x,\bar{\tau},\bar{P})$. The existence of such a minimum is a consequence of (6.2) and (6.3). Moreover, it takes the value $V(x,\bar{\tau},\bar{P})$ from which (6.8) follows.

□

Remarks:

(20) The proof of global existence in time of solutions in a neighborhood of $(\bar{x},\bar{\tau},\bar{P})$, in some appropriate Banach space, is given in [C4]. I also showed there that Liapunov stability is actually equivalent to stability with respect to the norm of the Banach space used to prove existence of solutions.

(21) From the remarks at the beginning of this section, we can conclude that a solution of (5.2) is unstable if the inequality in (6.3) holds in the opposite sense.

(22) Setting $y = 1$ (and thus $\tau = \tau_0$) in the governing equations, the results of Theorem 6.1 reduce to those mentioned at the beginning of the section.

References

[A1] Antman, S.S., Material Constraints in Continuum Mechanics, Atti Acc. Naz. Lincei, Rend. Cl. Sci. Fis. Mat. Nat., Ser VII 70 (1982), 256-264.

[A2] Alts, T., On the energy-elasticity of rubberlike materials, Progr. Colloid and Polymer Sci. 66, 1979, 367-375.

[B] Ball, J.M., Material instabilities and the calculus of variations, in 'Phase Transformations and Material Instabilities in Solids', ed M. Gurtin, Academic Press, 1984, 1-19.

[C1] Calderer, M.C., Singular Solutions to Boundary-Value Problems of Nonlinear Thermoelasticity, IMA Preprint, 1986.

[C2] Calderer, M.C., Dynamical behavior of nonlinear elastic spherical shells', Journal of Elasticity 13, 1983, 17-47.

[C3] Calderer, M.C., Dynamical stability analysis of bifurcation solutions to the equations of nonlinear thermoelasticity, IMA Preprint, 1986.

[C4] Calderer, M.C., Dynamical problems of constrained nonlinear thermoelasticity, forthcoming.

[E] Ericksen, J.L., Thermoelastic stability, Proc. Fifth U.S. Cong. on Appl. Mech., 1966, 187-193.

[G1] Gurtin, M.E., Thermodynamics and stability, Arch. Rat. Mech. An. 59, 1975, 63-96.

[G2] Green, A.E., Naghdi, P.M., Trapp, J.A., Thermodynamics of a Continuum with Internal Constraints, Int. J. Eng. Sci. 8, 1970, 891-908.

[G3] Gurtin, M.E., Podio Guidugli, P., The Thermodynamics of Constrained Materials, Arch. Rat. Mech. An. 51, 1973, 192-208.

[H] Hale, J.K., Ordinary Differential Equations, Wiley-Interscience, 1969.

[K] Knops, R.J., Wilkes, E., Theory of Elastic Stability, in Handbuch der Physik, VIa/3, C. Truesdell, ed., Berlin: Springer-Verlag, 1973.

[M1] Müller, I., 'Gases and Rubbers', IMA Preprint #167, 1985.

[M2] Müller, I., Personal communication.

[O] Ogden, R.W., 'Large deformation isotropic elasticity', Proc. Roy. Soc. London. A326: 565-584 and A328: 567-583, 1972.

[T1] Treloar, L.R.G., 'The Physics of Rubber Elasticity', Oxford, 1975.

[T2] Truesdell, C., Noll, W., 'The non-linear field theories of mechanics', Handbuch der Physic, Vol. III/3, S. Flügge, ed., Berlin: Springer-Verlag, 1965.

Acknowledgments: This work was carried out during my visit to the Institute for Mathematics and its Applications for the Special year in Continuum Physics and Partial Differential Equations. I wish to thank Professors J.L. Ericksen and D. Kinderlehrer for their support and encouragement. I am very grateful to them as well as to Professors M. Beatty, J.M. Ball, D. Carlson and I. Müller for many useful suggestions. It was partially supported by National Science Foundation, grant number: 8525530017.

THE COMPRESSIBLE REYNOLDS LUBRICATION EQUATION

Michel Chipot

Département de Mathématique
Université de Metz
Ile de Saulcy - 57000 Metz
FRANCE

and

Mitchell Luskin[1]

School of Mathematics
University of Minnesota
Minneapolis, Minnesota 55455
USA

Abstract

We give a general derivation of the Reynolds lubrication equation. We then state and sketch the proofs of some of the authors' recent results concerning the existence, uniqueness, and qualitative behavior of solutions to the compressible Reynolds lubrication equation. Finally, we give an application of our results to a problem in elastohydrodynamics.

1. Introduction

The Reynolds lubrication equation for the pressure, $P = P(x_1,x_2,t)$, that develops in a layer of fluid of thickness, $h = h(x_1,x_2,t)$, which is confined between two solid bodies when the sum of the velocities of the upper and lower bodies is $\underline{V} = (V_1,V_2)$ is [8, p. 60]

1) $$12\mu \frac{\partial}{\partial t}(\rho h) + 6\mu \nabla \cdot (\rho h \underline{V}) = \nabla \cdot (h^3 \rho \nabla P),$$

$$x = (x_1,x_2) \in \Omega,\ t \in R,$$

$$P = P_a\ , \qquad x \in \partial\Omega,\ c\ t \in R,$$

where ρ is density, μ is the dynamic viscosity, $\Omega \subseteq \mathbb{R}^2$ is the region (with boundary, $\partial\Omega$) where the upper and lower boundaries are in proximity, and $P_a > 0$ is the ambient pressure.

---

1 Supported by the NSF, Grants DMS 830-1575 and DMS 835-1080.

If the lubricating film is incompressible (liquid films), then the assumption ρ = constant is valid. In this case, we obtain the classical Reynolds lubrication equation

2) $$12\mu \frac{\partial h}{\partial t} + 6\mu\nabla \cdot (h\underline{V}) = \nabla \cdot (h^3 \nabla P), \quad x \in \Omega,\ t \in \mathbb{R},$$

$$P = P_a, \quad x \in \partial\Omega,\ t \in \mathbb{R},$$

which is a linear elliptic equation for the pressure.

Gas films such as those used to lubricate disk and tape magnetic recording systems [6,7,11] are usually modeled by assuming the gas to be compressible. The temperature of the surfaces of the confining solid bodies (the disk and the head, for example) are assumed to be equal and constant. It is further assumed that the temperature of the gas film remains equal to the temperature of the confining solid bodies (the gas is assumed to be isothermal). The ideal gas relationship

3) $$P/\rho = \text{constant}$$

is then utilized in 1) to derive the compressible Reynolds lubrication equation

4) $$12\mu \frac{\partial}{\partial t} (Ph) + 6\mu \nabla \cdot (Ph\underline{V}) = \nabla \cdot (h^3 P\nabla P),$$

$$x \in \Omega, \quad t \in \mathbb{R},$$

$$P = P_a \ , \quad x \in \partial\Omega,\ t \in \mathbb{R},$$

which is a nonlinear parabolic equation for the pressure.

2. Derivation of the Equation

We will now give a derivation of the Reynolds lubrication equation. We represent our upper surface by

$$\{(x_1, x_2, h_2(x_1,x_2,t)) \mid (x_1,x_2) \in \Omega,\ t \in \mathbb{R}\},$$

and we suppose that it moves with velocity

$$\underline{V}_2 = (V_{2,1}, V_{2,2}, 0).$$

We also represent our lower surface by

$$\{(x_1,x_2,h_1(x_1,x_2,t))|(x_1,x_2) \in \Omega, t \in \mathbb{R}\},$$

and we suppose that it moves with velocity

$$\underline{V}_1 = (V_{1,1}, V_{1,2}, 0).$$

Thus, $h = h_2 - h_1$ and

$$V_i = V_{1,i} + V_{2,i}$$

for $i = 1,2$.

We can integrate the continuity equation to obtain

5) $$\int_{h_1}^{h_2} \frac{\partial\rho}{\partial t} dx_3 + \sum_{i=1}^{3} \int_{h_1}^{h_2} \frac{\partial}{\partial x_i} (\rho u_i)\, dx_3 = 0$$

where $\underline{u}(x_1,x_2,x_3,t) = (u_1,u_2,u_3)$ is the fluid velocity.

Now

6) $$\int_{h_1}^{h_2} \frac{\partial\rho}{\partial t} dx_3 = \frac{\partial}{\partial t} \int_{h_1}^{h_2} \rho\, dx_3 - \sum_{j=1}^{2} (-1)^j \frac{\partial h_j}{\partial t} (x_1,x_2,t)\rho(x_1,x_2,h_j(x_1,x_2,t),t),$$

7) $$\int_{h_1}^{h_2} \frac{\partial}{\partial x_i} (\rho u_i)\, dx_3 = \frac{\partial}{\partial x_i} \int_{h_1}^{h_2} \rho u_i dx_3 - \sum_{j=1}^{2} (-1)^j [\frac{\partial h_j}{\partial x_i} (x_1,x_2,t)][(\rho u_i)(x_1,x_2,h_j(x_1,x_2,t),t)]$$

for $i = 1,2$, and

8) $$\int_{h_1}^{h_2} \frac{\partial}{\partial x_3} (\rho u_3)\, dx_3 = \sum_{j=1}^{2} (-1)^j (\rho u_3)(x_1,x_2,h_j(x_1,x_2,t),t).$$

The kinematic boundary conditions require that

9) $$\frac{\partial h_j}{\partial t} (x_1,x_2,t) =$$

$$- \sum_{i=1}^{2} \frac{\partial h_j}{\partial x_i} (x_1,x_2,t)u_i(x_1,x_2,h_j(x_1,x_2,t),t) + u_3(x_1,x_2,h_j(x_1,x_2,t),t)$$

for $j = 1,2$. Thus, it follows from 5) - 9) that

10) $$\frac{\partial}{\partial t} \int_{h_1}^{h_2} \rho \, dx_3 + \sum_{i=1}^{2} \frac{\partial}{\partial x_i} \int_{h_1}^{h_2} (\rho u_i) \, dx_3 = 0.$$

We now make the assumptions that $P = P(x_1,x_2,t)$, that inertial effects may be neglected, and that viscous effects due to changes in u_1 and u_2 in the x_1 and x_2 directions may be neglected. We then obtain from the Navier-Stokes equations that

11) $$\frac{\partial^2 u_i}{\partial x_3^2} (x_1,x_2,x_3,t) = \frac{1}{\mu} \frac{\partial P}{\partial x_i} (x_1,x_2,t),$$

$$h_1 (x_1,x_2,t) < x_3 < h_2(x_1,x_2,t),$$

for $i = 1,2$. The no-slip boundary conditions give that

12) $$u_i(x_1,x_2,h_j(x_1,x_2,t),t) = V_{j,i}$$

for $i = 1,2$; $j = 1,2$. The solution of 11) and 12) is

13) $$u_i(x_1,x_2,x_3,t) = \left[\frac{1}{2\mu} \frac{\partial P}{\partial x_i} (x_1,x_2,t)(x_3-h_1(x_1,x_2,t)) - \frac{(V_{1,i}-V_{2,i})}{h(x_1,x_2,t)} \right] \cdot [\, x_3 - h_2(x_1,x_2,t) \,] + V_{2,i}$$

for $i = 1,2$.

In either the incompressible case, ρ = constant, or in the compressible case, P/ρ = constant, the density, ρ, is independent of x_3. Thus, it follows from 13) that

14) $$\int_{h_1}^{h_2} \rho u_i dx_3 = \rho \int_{h_1}^{h_2} u_i dx_3 = -\frac{\rho h^3}{12\mu} \frac{\partial P}{\partial x_i} + \frac{\rho h V_i}{2}$$

for $i = 1,2$. Substitution of 14) in 10) gives the Reynolds lubrication equation 1).

When the thickness of a gaseous fluid layer is of the order of the molecular mean-free path of the gas, then the compressible Reynolds equation 4) becomes a poor model for the pressure in the fluid layer. A better model results when we allow the slip flow at the boundary proposed by Maxwell for perfectly diffuse reflection [1,9]

$$15) \qquad (u_i(x_1,x_2,h_j(x_1,x_2,t),t) - V_{j,i}) + (-1)^j \lambda \frac{\partial u_i}{\partial x_3}(x_1,x_2,h_j(x_1,x_2,t),t) = 0 ,$$

for $i = 1,2$; $j = 1,2$, where λ is the molecular mean-free path of the gas. To derive a modified Reynolds equation, we now solve 11) with the boundary conditions 15) and we then compute that

$$16) \qquad \int_{h_1}^{h_2} u_i dx_3 = -\frac{1}{12\mu}\frac{\partial P}{\partial x_i} h^3\left(1 + \frac{6\lambda}{h}\right) + \frac{V_i h}{2}$$

for $i = 1,2$. We substitute 16) into 10) to obtain the equation

$$17) \qquad 12\mu \frac{\partial}{\partial t}(\rho h) + 6\mu \, \nabla\cdot(\rho h \underline{V}) = \nabla \cdot \left(h^3\rho\left(1 + \frac{6\lambda}{h}\right) \nabla P\right).$$

Finally, we recall that

$$\lambda\rho = \text{constant} = \lambda_a \rho_a$$

where ρ_a (resp. λ_a) is the ambient density (resp. ambient molecular mean-free path). Thus, we finally obtain the compressible Reynolds lubrication equation modified for slip flow

$$18) \qquad 12\mu \frac{\partial}{\partial t}(Ph) + 6\mu \, \nabla \cdot (Ph \, \underline{V}) = \nabla \cdot \left(h^3 P\left(1 + \frac{6\lambda_a P_a}{hP}\right)\nabla P\right), \quad x \in \Omega,\ t \in R,$$

$$P = P_a, \qquad x \in \partial\Omega,\ t \in R.$$

In most magnetic recording systems, the medium (disk or tape) is stretched around the head so that the thickness of the gaseous fluid layer is very small at one or more regions. We usually find internal layers (where the pressure has large derivatives) near the points of minimum thickness in the fluid layer [6,7]. The accurate numerical resolution of these internal layers is much improved by appropriate mesh refinements around the internal layers. Now the efficiency and accuracy of mesh refinement algorithms can be greatly increased by *a priori* qualitative information about the solution such as the location of the internal layers and estimates on the magnitude of the layers. Thus, the following results on the qualitative nature of solutions to the compressible Reynolds lubrication equation are the start of a program which we hope will result in improved mesh refinement algorithms for solutions to the compressible Reynolds equation. The first step in this direction is to study the steady state equation, i.e., to look for a solution $P = P(x_1,x_2)$ which does not depend on t and to assume that $h = h(x_1,x_2)$ also does not depend on t. This is the objective of the next section.

3. The Steady State Equation

If we set $\underline{\Lambda} = 6\mu\underline{V}$ and $\lambda = 6\lambda_a P_a$, then the steady state equation corresponding to 18) reads:

$$\nabla\cdot(h^3 P(1 + \frac{\lambda}{hP})\nabla P) = \nabla\cdot(hP\underline{\Lambda}), \qquad x \in \Omega,$$

19)

$$P = P_a, \qquad x \in \partial\Omega,$$

(In the case $\lambda = 0$, 19) is the steady state equation for 1). So we can handle both equations at the same time.). Since P denotes the pressure, we are looking for a positive solution, P, of 19). Introduce the dependent variable

$$u = \frac{1}{2} P^2 + \frac{\lambda P}{h} .$$

If P is a positive solution of 19) - for instance in the classical sense - then it is easy to check that u is a solution of

$$\nabla\cdot(h^3 \nabla u) = \nabla\cdot\underline{\alpha}(x,u), \qquad x \in \Omega,$$

20)

$$u = \phi(x) = \frac{1}{2} P_a{}^2 + \frac{\lambda P_a}{h(x)} , \qquad x \in \partial\Omega,$$

where $\underline{\alpha}(x,u)$ is defined by

21) $$\underline{\alpha}(x,u) = (-\lambda + \sqrt{\lambda^2 + 2h^2(x)u})\ (\underline{\Lambda} - \lambda\nabla h(x)).$$

Conversely, if u is a positive solution of 20), then

$$P = -\frac{\lambda}{h} + \sqrt{\frac{\lambda^2}{H^2} + 2u}$$

is a positive solution of 19). So, instead of solving 19), we can solve 20). The advantage of considering 20) lies in the fact that the degeneracy (when $\lambda = 0$, $P = 0$) of the elliptic part of the operator in 19) has now disappeared. Of course, when $\lambda = 0$ this is at the expense of introducing a singularity in $\underline{\alpha}(x,u)$ (see 21)). Now, since we don't know a priori if, for $\phi > 0$, the solution of 20) is non negative, we change the definition of $\underline{\alpha}(x,u)$ to:

22) $$\underline{\alpha}(x,u) = \begin{cases} (-\lambda + \sqrt{\lambda^2+2h^2u})\ (\underline{\Lambda}-\lambda\nabla h(x)) & \text{when } u>0 \\ 0 & \text{when } u<0. \end{cases}$$

Assuming that Ω is a smooth bounded open set in R^2 we will denote by $L^2(\Omega)$ the space of square integrable functions with the norm

$$|v|_{L^2(\Omega)} = \left(\int_\Omega |v|^2 dx\right)^{1/2}$$

and by $H^1(\Omega)$, $H_0^1(\Omega)$ the usual Sobolev spaces defined by

$$H^1(\Omega) = \{v \in L^2(\Omega) |\ \nabla v \in (L^2(\Omega))^2\},$$

$$H_0^1(\Omega) = \{v \in H^1(\Omega) |\ v = 0 \ \text{ on } \ \partial\Omega\}.$$

We will assume that these two spaces are given the norm:

$$\|v\|^2_{H^1(\Omega)} = |v|^2_{L^2(\Omega)} + ||\nabla v||^2_{L^2(\Omega)}$$

where $|\ |$ denotes the Euclidean norm in $\mathbb{R}^2$.

We will also assume that the function $h = h(x)$ is Lipschitz continuous and satisfies,

23) $$\begin{aligned} &0 < h_1 < h(x) < h_2, && \text{a.e. } x \in \Omega, \\ &|\nabla h(x)| < h_3, && \text{a.e. } x \in \Omega, \end{aligned}$$

where h_1, h_2, h_3 are strictly positive constants.

For $\phi \in H^1(\Omega)$, we will say that u is a weak solution to 20) if

$$u - \phi \in H_0^1(\Omega) ,$$

24)

$$\int_\Omega h^3 \nabla u \cdot \nabla \xi dx = \int_\Omega \underline{\alpha}(x,u) \cdot \nabla \xi dx, \quad \forall \xi \in H_0^1(\Omega).$$

Note that the above integrals make sense for $u, \xi \in H^1(\Omega)$. For the first one this is by 23), for the second one this is due to the estimate:

25)
$$|\underline{\alpha}(x,u)| < C(-\lambda + \sqrt{\lambda^2 + 2h^2|u|}) < C\sqrt{2h^2|u|} < C\sqrt{|u|}$$

where C denotes some constants (see 23)). Indeed, if $u \in L^2(\Omega) \subset L^1(\Omega)$, then the above estimate implies that $|\alpha(x,u)| \in L^2(\Omega)$ and thus $|\underline{\alpha}(x,u)| \cdot |\nabla \xi| \in L^1(\Omega)$. Under the above assumptions we first can prove:

<u>Theorem</u> 1 [3] If $\phi \in H^1(\Omega)$, then there exists a weak solution to 24).

<u>Sketch of proof</u>. For $v \in L^2(\Omega)$ we consider $u = T(v)$ the solution of

$$u - \phi \in H_0^1(\Omega),$$

$$\int_\Omega h^3 \nabla u \cdot \nabla \xi dx = \int_\Omega \underline{\alpha}(x,v) \cdot \nabla \xi dx, \quad \xi \in H_0^1(\Omega).$$

Due to the growth of $\underline{\alpha}(x,u)$ at infinity in u (see 25)) we can show that for R large enough T maps compactly the ball

$$B_R = \{v \in L^2(\Omega) | \ |v|_{L^2(\Omega)} < R\}$$

into itself. It results then, by the Schauder fixed point theorem, that T has a fixed point in B_R. But such a fixed point $u = T(u)$ is a solution to 24). This concludes the proof. We refer the reader to [3] for the details.

It is physically reasonable that P (and thus u) is increasing when its boundary values increase, and thus it is not surprising that we have:

Theorem 2 [3] Assume that u_1 is a weak solution to 24) corresponding to boundary data ϕ_1 and that u_2 is a weak solution to 24) corresponding to boundary data ϕ_2. If $\phi_1 \geq \phi_2$ a.e. on $\partial\Omega_1$, then $u_1 \geq u_2$ a.e. on $\partial\Omega$.

Proof. We refer to [3].

As a consequence we have:

Theorem 3 [3] For $\phi \in H^1(\Omega)$ there exists a unique solution to 24). Moreover, if $\phi \geq 0$ a.e. on $\partial\Omega$, then $u \geq 0$ a.e. on Ω.

Proof Let u,v be two solutions of 24). Applying Theorem 2 with $\phi_1 = \phi_2 = \phi$ we get $u \geq v$ and $v \geq u$ so that $u = v$ in Ω and thus the problem 24) has a unique solution. For $\phi \equiv 0$ it is clear (since $\underline{\alpha}(x,0) = 0$) that $u \equiv 0$ is the only solution to 24). So if $\phi \geq 0$ and if u is the solution to 24), then we can apply Theorem 2 with $u_1 = u$, $u_2 = 0$ to get $u \geq 0$ a.e. in Ω. This completes the proof of the theorem.

Remark: If $\phi \geq 0$ (which is the case when $\phi = \frac{1}{2} Pa^2 + \frac{\lambda p_a}{h(x)}$), then we can define

26) $$p = -\frac{\lambda}{h} + \sqrt{\frac{\lambda^2}{h^2} + 2u}$$

and clearly p is a weak solution to 19). So we have proved that 19) has a weak solution.

Next, we would like to know when u and p defined by 26) are classical solutions to 20) and 19) respectively. First, in the general case ($\lambda > 0$) we have:

Theorem 4 [3] Let u be the solution to 24). If Φ is a constant such that

$$\phi \geq \Phi > 0 \text{ a.e. in } \Omega, \quad \nabla \cdot \underline{\alpha}(x,\Phi) \leq 0 \quad \text{in } \Omega$$

(this last inequality is meant for instance in the distributional sense), then we have

$$u \geq \Phi \quad \text{a.e. in } \Omega.$$

Moreover, if Ω is smooth, ϕ, $h \in C^\infty(\overline{\Omega})$, then u (and p defined by 26)) are in $C^\infty(\overline{\Omega})$. ($C^\infty(\overline{\Omega})$ denotes the space of functions of class C^∞ whose derivatives are extendable in a continuous way up to the boundary).

Proof: (See [3]). The idea is to show that thanks to the inequality $\nabla\cdot\underline{\alpha}(x,\Phi) < 0$, the function $v = \Phi$ is a "subsolution" to 24). Arguing then as in Theorem 2 we can show that this implies that $u \geq \Phi$ a.e. in Ω. This proves the first part of the theorem. To prove the second, we remark that the function

$$u \to \underline{\alpha}(x,u)$$

is C^∞ in u on $[\Phi,+\infty)$. Thus for $u \in H^1(\Omega)$, $\underline{\alpha}(x,u) \in (H^1(\Omega))^2$ (note that $\underline{\alpha}(x,u)$ is smooth in x since we assume that h is smooth). By usual elliptic regularity theory, this implies $u \in H^2(\Omega)$ and thus $\underline{\alpha}(x,u) \in (H^2(\Omega))^2$. So, $u \in H^3(\Omega)$ and we can keep going, proving that $u \in H^k(\Omega)$ for all k (see for instance [5] for a definition of these spaces). This completes the proof of the theorem.

Remark: In the physical case that we are considering we have (see 20))

$$\phi(x) = \frac{1}{2} P_a^2 + \frac{\lambda P_a}{h(x)} \geq \frac{1}{2} P_a^2 > 0 .$$

In the case where $\lambda = 0$, that is to say, in the case of the classical Reynolds equation, we can get rid of the hypothesis on $\underline{\alpha}(x,\Phi)$. More precisely we have:

Theorem 5 [2] Let u be the solution to 24) and assume that for some constant Φ we have

$$\phi \geq \Phi > 0 \quad \text{a.e. on} \quad \partial\Omega .$$

If $h \in W^{2,\infty}(\Omega)$, then there exists a constant C, depending on the data, such that

$$u \geq C > 0 \quad \text{a.e. in} \quad \Omega.$$

Moreover, if Ω is smooth, ϕ, $h \in C^\infty(\overline{\Omega})$, then u (and P defined by 26)) are in $C^\infty(\overline{\Omega})$.

Proof: We refer to [2]. The idea is again to find a positive subsolution to 24). However, in this case, the function $v = \Phi$ is not a subsolution in general and some more work is needed. The fact that $u \in C^{\infty}(\overline{\Omega})$ results from the same arguments that were given in Theorem 5.

As we mentioned in the preceeding section, it would be helpful for the numerical analysis of the problem to know the region of Ω where p has large derivatives. In other words, a certain knowledge of the shape of p or u is needed. We will assume in the following of this section that $\lambda = 0$, i.e., we will deal with the usual Reynolds equation, and $P = P_a > 0$ on $\partial\Omega$ where P_a is a positive constant.

When the scale of Ω in the x_1 direction is small compared to its scale in x_2, when h doesn't depend on x_2 and $\underline{V} = (V_1,0)$, we can reduce 19) to an ordinary differential equation and we are then able to deduce some information about the shape of P. We refer the interested reader to [2] for details. The problem is much more difficult for the general two dimensional problem. For the sake of simplicity, we are going to assume that $\underline{V} = (V_1,0)$ and that h and Ω are smooth so that P will be the smooth solution to

27)
$$\begin{aligned} \nabla\cdot(h^3 P \nabla P) &= \Lambda \frac{\partial}{\partial x_1}(hP) , \qquad & x = (x_1,x_2) \in \Omega , \\ P &= P_1 & x = (x_1,x_2) \in \partial\Omega , \end{aligned}$$

where $\Lambda = 6\mu V_1$ (see 19)).

Let us denote by $n = (n_1,n_2)$ the outward normal to $\partial\Omega$ and by $d\sigma$ the surface measure on $\partial\Omega$. Then we have:

Theorem 6 [2] Assume that

28)
$$\int_{\Omega} hn_1 \, d\sigma = 0,$$

29)
$$\exists x_0 \in \partial\Omega \text{ such that } \frac{\partial h}{\partial x_1}(x_0) > 0,$$

and

30)
$$\exists x_0' \in \partial\Omega \text{ such that } \frac{\partial h}{\partial x_1}(x_0') < 0,$$

then P, the solution of 27), achieves a maximum strictly greater than P_a and a minimum strictly less than P_a in Ω.

Proof: Let us prove for instance that P achieves a maximum strictly greater than P_a. If not, then we have $P \leq P_a$ in Ω. This implies $\frac{\partial P}{\partial n} \geq 0$ on $\partial\Omega$. Now around x_0 we have by 27), 29):

31) $$\nabla\cdot(h^3P\nabla P) - \alpha h \frac{\partial P}{\partial x_1} = \alpha \frac{\partial h}{\partial x_1} P > 0.$$

Let us denote by N a neighborhood of x_0 in Ω where 31) holds. If $P < P_a$ in N, then by the strong maximum principle we have

32) $$\frac{\partial P}{\partial n}(x_0) > 0.$$

Now integrating 27) over Ω and applying the divergence theorem we get

$$0 < \int_{\partial\Omega} h^3P_a \frac{\partial P}{\partial n}\, dx = \int_{\partial\Omega} \alpha h P_a n_1 d\sigma = \alpha P_a \int_{\partial\Omega} h n_1 d\sigma = 0\,.$$

This is impossible and thus one cannot have $P < P_a$ in N. Since $P \leq P_a$ in Ω, there is a point in N where P achieves its maximum P_a. Thanks to 31) and the strong maximum principle this would imply that $P \equiv P_a$ in N. Hence, 31) would read

$$0 = \alpha \frac{\partial h}{\partial x_1} P_a > 0$$

in N which is impossible. So, we cannot have $P \leq P_a$ in Ω and P achieves a maximum strictly greater than P_a in Ω. The proof that P achieves a minimum strictly less than P_a uses the same arguments around x_0'. This completes the proof of the theorem.

Remark: We have that 28) holds, for instance, when Ω and h are symmetric with respect to the x_2-axis. Using the maximum principle and 31) we can easily see that P can only achieve its maximum (resp., minimum) in the region

$$[\frac{\partial h}{\partial x_1} < 0] = \{(x_1,x_2) \in \Omega \mid \frac{\partial h}{\partial x_1}(x_1,x_2) < 0\}$$

$$(\text{resp., } [\frac{\partial h}{\partial x_1} > 0] = \{(x_1,x_2) \in \Omega \mid \frac{\partial h}{\partial x_1}(x_1,x_2) > 0\}) \ .$$

For this and complementary results we refer the reader to [2].

4. Application to a system governing the deflection of a floppy disk

In magnetic recording systems such as "floppy" disk systems and "Winchester" disk systems the following model equations for the deflection of a rotating disk by the pressure load developed between the recording head and the disk have been successfully utilized [10]:

33) $$(E\Delta^2 - T\Delta + \rho\omega^2 \frac{\partial^2}{\partial\theta^2} + \alpha\omega \frac{\partial}{\partial\theta})v = P - P_a, \qquad x \in \Omega \ ,$$

$$v = \frac{\partial v}{\partial n} = 0, \qquad x \in \partial\Omega,$$

34) $$\nabla \cdot (h^3(v)P\nabla P) = 6\mu \underline{V} \cdot \nabla(h(v)P), \qquad x \in \Gamma,$$

$$P = P_a, \qquad x \in \partial\Gamma.$$

Here $\Omega \subset \mathbb{R}^2$ is an annulus, $\Gamma \subset \Omega$ is an open set in $\mathbb{R}^2$ which represents the region where the head and the disk are in close proximity (see the figure below),

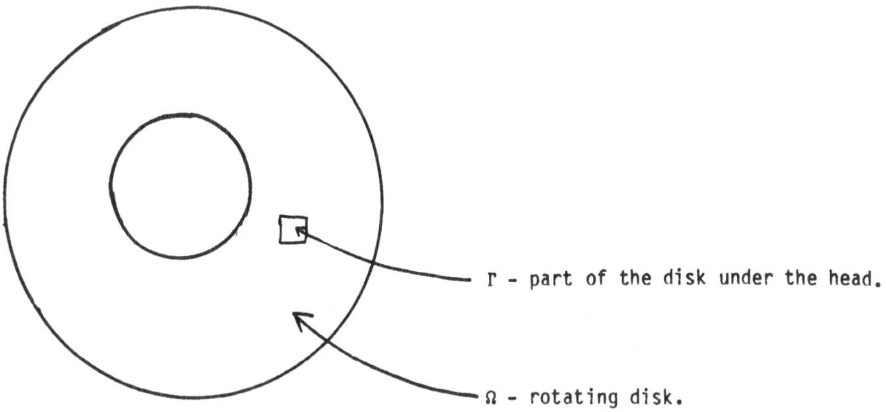

v is the vertical displacement of the disk, and P is the pressure which develops between the head and the disk. So 33), 34) couples the equation of the deflection of the disk with the Reynolds equation under the head. We wish to solve 33), 34) for v and P when we are given the positive constants $E, T, \rho, \omega, \alpha, P_a$ and μ. Here E is the stiffness coefficient, T is the tension, ρ is the density of the disk, ω is the angular speed of rotation of the disk, α is the damping coefficient, P_a is the ambient pressure, and μ is the dynamic viscosity of the air layer between the head and the disk. Further, $\underline{V} = \underline{V}(x)$ denotes the velocity of the disk at the point $x \in \Omega$, so we have:

$$\underline{V}(x) = \omega(-x_2, x_1), \quad x = (x_1, x_2) \in \Omega.$$

Also, $h(v) = h(v)(x)$ is the thickness of the fluid layer between the head and the disk and we have

$$h(v) = \psi - v$$

where ψ is the vertical position of the head. Finally, note that 34) gives the pressure P in Γ. We assume that P is extended by P_a outside of $\underline{\Gamma}$ in such a way that 33) makes sense. We assume that ψ is a Lipschitz continuous function such that

$$0 < m < \psi(x) < M, \quad x \in \Omega.$$

We then have:

<u>Theorem 7</u> [4] If ω or $|\Gamma|$ - the measure of Γ - is small enough, then there exists a weak solution to 33), 34).

<u>Proof:</u> We refer the interested reader to [4]. Using the existence results of the preceding section the method consists in applying the Schauder fixed point theorem on a suitable convex subset of $L^2(\Gamma)$.

Bibliography

1. A. Burgdorfer, "The influence of the molecular mean-free path on the performance of hydrodynamic gas lubricated bearings," Trans. ASME, Part D (J. Basic Engineering), vol. 81, 94-100 (1959).

2. M. Chipot, "On the Reynolds lubrication equation", to appear.

3. M. Chipot and M. Luskin, "Existence and uniqueness of solutions to the compressible Reynolds lubrication equation", to appear in SIAM J. of Math. Anal.

4. M. Chipot and M. Luskin, in preparation.

5. G. Folland, Introduction to Partial Differential Equations, Princeton University Press, Princeton, N.J., (1976).

6. H.J. Greenberg, "Flexible disk-read/write head interface", IEEE Transactions on Magnetics, Vol. 14, No. 5, 336-338 (1978).

7. H.J. Greenberg, "Study of head-tape interaction in high speed rotating head recording, IBM J. Res. Develop., Vol. 23, No. 2, 197-205 (1979).

8. W. Gross, L. Matsch, V. Castelli, A. Eshel, J. Vohr, M. Wildmann, Fluid Film Lubrication, Wiley & Sons, New York, (1980).

9. E.H. Kennard, Kinetic Theory of Gases, McGraw-Hill Book Company, Inc., New York and London, (1938).

10. M. Luskin, M. Chipot, S. Kistler, D. Perry, "Elastohydrodynamical problems in magnetic recording", Proceedings of the 18th French Numerical Analysis Congress, (1985).

11. B. Wolf, N. Deshpande, and V. Castelli, "The flight of a flexible tape over a cylinder with a protruding bump", AMSE Journal of Lubrication Technology, vol. 105, 138-142 (1983).

TWINNING OF CRYSTALS (I)

J.L. Ericksen

Departments of Aerospace Engineering and Mechanics
and School of Mathematics
University of Minnesota

Abstract: Twinning is a kind of defect, commonly observed in crystalline solids. We explore some implications of this, with respect to the theory of crystal elasticity.

1. Introduction

When the crystallographer uses mathematics, it is more likely to be elementary algebra, geometry or group theory than sophisticated analysis. However, some of the things he or the metallurgist observes and describes can be viewed as realizations of solutions of partial differential equations of a rather unusual kind, involving deep analytical difficulties. Twinning and like phenomena are included among the defects commonly seen in crystalline solids. Nonlinear elasticity theory is being used, with some success, to analyze such phenomena. In a very formal sense, the general equations are old and well-known, Euler-Lagrange equations of a somewhat special kind, there being a great number of studies of special cases. Yet the simplest considerations of these phenomena have made clear that we need to revise our thinking about the subject. My purpose is to elaborate these vague remarks.

2. Elastic Crystals

Here, we ponder what might be considered to be the simplest possible kind of problem in elasticity theory. Consider a homogeneous elastic body, with a shape as simple as you like, and seek to determine how it can be in stable equilibrium, when it is subject to no forces or kinematic constraints.

Recall that such theories involve a fixed configuration of a body as a reference, all others being related to this by deformations (mappings), usually fairly smooth and one-to-one. If F denotes the gradient of the map from the reference to another configuration, we generally have

$$\det F > 0, \tag{2.1}$$

implying that

$$C = C^T = F^T F \quad \text{is positive-definite} \tag{2.2}$$

If W denotes the strain energy per unit reference volume, we have a constitutive equation of the form

$$W = \hat{W}(C) = \hat{\hat{W}}(F) \, . \tag{2.3}$$

That W depends on F, in the combination C, reflects the fact that we don't expect the energy to change, if we merely subject the material to a static translation and rotation, motions which don't affect C. Tacitly, it is understood that W will also depend on temperature, commonly thought of as a control parameter. I will keep the discussion fairly informal. Consider $\hat{W}$ to be as smooth as you wish, defined on the entire domain indicated by (2.2). Generally, we know that that function $\hat{W}$ depends on the material and the choice of reference configuration.

By conventional reasoning, in the absence of any constraints on the deformation, and with no forces applied, the stable equilibria should be minimizers of the total energy

$$E = \int_B W \, , \tag{2.4}$$

B denoting the region in E_3 occupied by the body in the reference configuration. As the usual formal conditions for extremals, we have the Euler-Lagrange equations and natural boundary conditions

$$\nabla \cdot \frac{\partial \hat{\hat{W}}}{\partial F} = 0 \quad , \quad \frac{\partial \hat{\hat{W}}}{\partial F} N = 0 \, , \tag{2.5}$$

where N is normal to ∂B. For other static problems, experience indicates that we are likely to be dealing with (2.5), with different boundary or other side conditions, or related minimization problems.

Actually, it is easier to note that E will be minimized if W is, so the common practice is to look for a symmetric positive definite tensor C_1 such that

$$\hat{W}(C) \geq \hat{W}(C_1). \tag{2.6}$$

Given such a constant C_1, we can find a constant $F = F_1$ satisfying

$$F_1^T F_1 = C_1 \text{ , } \det F_1 > 0, \tag{2.7}$$

and integrate this gradient to get a mapping giving a minimizer. Given C_1, the mapping is determined to within a translation and rotation, so we have a nice smooth minimizer, a trival solution of (2.15). One might question the tacit assumption that $\hat{W}$ attains its minimum but, really, I don't find fault with the reasoning, as far as it goes. Rather, it is the possibility that $\hat{W}$ might have more than one such minimizer, and questions related to this, which we will consider.

So far, we have said nothing about material symmetry, which certainly has some relevance to the crystals, in particular. General theory for this introduces an invariance condition of the type

$$\hat{W}(H^T C H) = \hat{W}(C) \text{ , } H \in G \text{ ,} \tag{2.8}$$

the consensus of opinion being that G should contain only H such that $\det H = \pm 1$. Also, since $H = -1$ satisfies (2.8) for any $\hat{W}$, there is no real loss of generality in assuming that

$$\det H = 1. \tag{2.9}$$

Then, if C_1 is a minimizer, so is $H^T C_1 H$, for any $H \in G$. In almost all expositions of elasticity theory, it is assumed, often tacitly, that, for solids,

$$H^T C_1 H = C_1 \qquad \forall \quad H \in G \tag{2.10}$$

and, clearly, this must hold, if C_1 is to be the only minimizer of W. Automatically, this forces G to be a compact group, conjugate to some subgroup of the orthogonal group. Then, the minimizer becomes the obvious choice for the reference, and using it gives $C_1 = 1$, (2.9) and (2.10) then implying that G is a subgroup of the orthogonal group. All this is compatible with much successful experience with linear elasticity theory. Essentially, Noll [1] relied on such experience, in defining solids by the condition that the maximal G leaving W invariant is such a group. Certainly, I accepted the basic line of though for a time and, no doubt, some still do. If one accepts it, it remains possible that W has other minimizers, not so related by symmetry. As a physicist might argue this, this could well occur at a particular temperature. Change the temperature a little and, because there is no reason for the energies to stay equal, it is almost certain that they won't, so one will take over as the minimizer. So, for most values of the temperature, all minimizers are related by symmetry. The other configuration might continue to have some status as a relative minimizer of $\hat{W}$, perhaps be stable enough to be observed. One might even see both, coexisting in the same body as a metastable configuration. Such things do happen frequently. Austenite and martensite are names long used to describe more and less symmetric phases occurring together in common steels, with the martensite twinned, both phases containing other defects. One needs a microscope to discern this structure. In some other alloys, the metallurgist sees simpler morphologies, cleaner arrangements of austenitic and twinned martensitic phases. So, very commonly in crystalline solids, one sees configurations which must be somewhat stable, since we see them persist for long periods of time, but can hardly be more than metastable. Usually, they are not so smooth. Clearly, it is not so easy to give a precise definition of relative minima of the energy which covers these. Some thoughts concerning such questions can be found in the work of James [2,3].

Some of these ideas about elasticity theory need to be revised to accomodate phenomena like twinning. To abstract some important features of such phenomena, we need to have at least two symmetry-related minimizers, so we must give up (2.10), and replace it by the statement that there is at least one $H \epsilon G$ such that

$$C_2 = H^T C_1 H \neq C_1 . \qquad (2.11)$$

It is then elementary that, if F_1 and F_2 are any two values of F corresponding to C_1 and C_2, as indicated by (2.1) and (2.2), there must be some rotation R,

$$R^{-1} = R^T , \; \det.R = 1 , \qquad (2.12)$$

such that

$$F_2 = RF_1H . \qquad (2.13)$$

Actually, we want more, that two such configurations be able to coexist in the same body, coherently. That is, we want there to be two adjacent regions, with F taking the value F_1 in one, F_2 in the other, with F the gradient of a continuous deformation. If you like, when those interested in crystals speak of twins, they mean such Siamese twins. At the surface where the parts join, the assumption of continuity implies the well-known conditions of compatibility,

$$F_2 = (1 + a \otimes n)F_1 , \qquad (2.14)$$

where n is the unit normal to the interface and a is the so-called amplitude vector, a measure of the size of the jump in F. Eliminating F_2 between (2.13) and (2.14), we get

$$RF_1H = (1 + a \otimes n)F_1 , \qquad (2.15)$$

With F_1 and F_2 constants, a and n must be constant, so the interface is a plane, or part thereof, and the interfaces seen in many crystals do conform pretty well to this. When these seem to be distorted a bit, I would interpret this as evidence of residual stresses, other configurations which might well be interpretable as extremals of E, stable enough to be observed.

Now, let us try to adhere to convention, by selecting the first configuration as a reference, taking

$$F_1 = 1 \Rightarrow RH = 1 + a \otimes n \; . \tag{2.16}$$

Then, the conventional choice of G for a crystal would take this to be the crystallographic point group associated with this configuration, which makes H orthogonal. Elementary analysis then shows that (2.16) can hold only if

$$R = H^T \; , \; a = 0 \; , \tag{2.17}$$

and this won't do. So, either we give up the attempt to use elasticity theory, or we revise some of our ideas concerning G.

What has happened, as I see it, is that the indicated conventional estimate of G is faulty; it is good enough for classical linear theory, but such theory can't cope with the phenomena being considered.

3. Molecular Considerations

To clear the air a bit, we first consider a molecular picture of the simplest kinds of crystal configurations, identical atoms arranged on what is sometimes called a simple or a Bravais lattice, filling all of space. Take three constant linearly independent vectors $a_K (K = 1,2,3)$. Put identical atoms at all of the points with position vectors of the form

$$\sum_{K=1}^{3} n^K a_K \; , \tag{3.1}$$

where the n^K are integers, positive or negative. In common jargon, the a_K are lattice vectors for the configuration, characterizing the periodic structure demanded by crystallographer's definition of crystals. As a rather simple special case, suppose that

$$a_K \cdot a_L = 0 \quad , \quad K \neq L, \tag{3.2}$$

and

$$\| a_1 \| \neq \| a_2 \| \; . \tag{3.3}$$

Consider the possibility that such a configuration corresponds to our first minimizer, for some choice of the function $\hat{W}$, and that we take it as the reference. If you like, use the rectangular Cartesian coordinates, with axes parallel to the a_K, to represent F etc. as matrices. Obviously, a 180° rotation with a_1 as axis takes the infinite set of physically indistinguishable points to itself. Orthogonal transformations of this kind form the point group associated with the configuration, the conventional estimate of G. Of course, this reflects the notion that, had someone so rearranged the identical atoms, unbeknownst to us, we could design no experiment to detect this. We now construct a different kind of rearrangement which, by the same kind of reasoning, should belong to G.

Consider a simple shearing deformation, a homogeneous deformation with gradient

$$F = S = 1 + \bar{a} \otimes \bar{n} \quad , \bar{a} \cdot \bar{n} = 0 , \tag{3.4}$$

wherein

$$\bar{a} = \alpha(a_1 + a_2) \tag{3.5}$$

$$\bar{n} = a_3 \wedge (a_1 + a_2)/\| a_3 \wedge (a_1 + a_2) \| , \tag{3.6}$$

α being a scalar. Let

$$\bar{a}_K = F\, a_K \tag{3.7}$$

be the lattice vectors describing positions of atoms in the new configuration. By sketching a picture of this, you can easily see that if you choose α correctly, the vectors $\bar{a}_K$ will also be orthogonal, with

$$\|\bar{a}_1\| = \|a_2\| \; , \; \|\bar{a}_2\| = \|a_1\| \; , \; \|\bar{a}_3\| = \|a_3\| , \tag{3.8}$$

so use this value of α. Further, the sketch then indicates that if you subject this deformed configuration to a 180° rotation of the form

$$R_1 = -1 + 2\bar{n} \otimes \bar{n} \quad , R_1^2 = 1 , \tag{3.9}$$

you can make the set of atomic positions come back to what it was originally. Individual atoms do not return to their starting positions, but we have accepted the view that this doesn't matter. So we should have that

$$H = R_1 S \in G \, , \tag{3.10}$$

and it is easy to verify that

$$H^2 = 1 \, . \tag{3.11}$$

Next consider (2.16), which now reads

$$\begin{aligned} RH &= RR_1 S = RR_1(1 + \bar{a} \otimes \bar{n}) \\ &= 1 + a \otimes n \, . \end{aligned} \tag{3.12}$$

Clearly, we can satisfying this by taking

$$R = R_1^T = R_1 \, , \; \bar{a} = a \, , \; \bar{n} = n \, , \tag{3.13}$$

and use this to construct an example of a pair of coexisting configurations of the kind discussed earlier. It is an example of twinning, as the crystallographer uses the word. When he uses it, he has in mind some such array of atoms, fitting his concept of a crystal, as a periodic structure. Some reinterpretation must be made, to use any continuum theory to analyze such things. To relate macroscopic motion to microscopic motion, one must introduce some hypothesis. Our (3.7) states the most common hypothesis, relating lattice vectors to the macroscopic deformation gradient, what is commonly called the Born or Cauchy-Born hypothesis. It seems so natural that it is easy to overlook the fact that it is an assumption, but it has its limits, failing to apply to motions encountered in some phase transitions, for example. For a given configuration, infinitely many sets of lattice vectors can serve as lattice vectors. Resulting ambiguities in the Cauchy-Born hypothesis are discussed in detail by Ericksen [4]. They do cause real difficulties for crystallographers and metallurgists seeking to relate observations of lattice vectors to deformation. Variants of the hypothesis are used in a rather informal way, to handle some particular kinds of exceptions, but it is hard to find one umbrella to cover all. My experience is that elasticity theory does poorly in describing what occurs, in such exceptional cases.

Having found reason to think that the classical estimates of G are in need of revision, we can try to make a different kind of estimate, by appealing to molecular theories of elasticity. For the simplest, most classical molecular theories, it is rather easy to do this, as is discussed by Ericksen [5]. This involves picking a reference configuration, with reference lattice vectors A_K, describing the periodicity of the crystal, considered as filling all of space. As is well known from crystallography, and mentioned above, there are infinitely many ways of choosing these. To be more specific, another set of vectors $\bar{A}_K$ is eligible if and only it is related to the possible set A_K by relations of the form

$$\bar{A}_K = \sum_{L=1}^{3} m_K^L \; A_L \; , \tag{3.14}$$

where

$$m = \| m_K^L \| \tag{3.15}$$

is a matrix of integers, with

$$\det m = \pm 1 \; . \tag{3.16}$$

Clearly, these form a group G'. Given m, we can define a linear transformation $H(m)$ by

$$\bar{A}_K = H(m) A_K = m_K^L \; A_L \; , \tag{3.17}$$

and such H are the elements of a conjugate group $\hat{G}$. The aforementioned molecular theory leads to the conclusion that

$$\hat{G} = G \; , \tag{3.18}$$

i.e., that this is the group leaving $\hat{W}$ invariant. Molecular theory has its difficulties, in describing real crystals, but I believe that this prediction is quite good, as long as elasticity theory can reasonably be applied. Some kinds of twins involve what the experts call shuffling, rearrangements of atoms in a unit cell, not involving macroscopic deformation. Elasticity theory is too coarse to deal properly with such phenomena, for example. Pitteri [6,7] discusses more general ideas of kinematics and symmetry which apply to such crystals. James' [8] study of quartz serves to illustrate these ideas.

Use (3.18) and you automatically include the element indicated by (3.10), for the example considered. In translating ideas used by crystallographers and metallurgists into the language of elasticity theory, one encounters various other kinds of elements, enough so that I feel quite comfortable with this choice. With this view, the invariance of the energy is essentially the same for all crysals, although minimizing configurations can display different symmetries.

In brief, nonlinear elasticity theories involving strain energies invariant under G, as just estimated, seem to be capable of describing some of the transition phenomena observed in crystals. As is discussed by Parry [9] and Pitteri [10], select subgroups can suffice, if one is only concerned with suitably small deformations of a particular reference. Make a more conventional choice of G, and, from kinematic considerations, you can make an estimate of the limits of validity of the assumption. From this perspective, the ideas of invariance used in linear theory are seen as sound enough, but one can go wrong, in extrapolating them to nonlinear theory.

Here, I have sketched some of the reasoning which has led us to take seriously a general kind of mathematical theory, deserving deeper study. As mentioned earlier, we know that it can't cope with all of the phenomena observed in crystals, but it covers some which are well outside the range of linear theory. Strategies for constructing strain energies invariant under G are discussed by Parry [9], Pitteri [10] and Fonseca [11], the latter borrowing more ideas from molecular theory.

4. Unloaded Bodies.

It is well known and easy to show that the group G' is infinite, discrete and not compact. Since G is conjugate to G', it shares these properties. One consequence is that no positive-definite symmetric tensor is invariant under $\hat{G}$. Said differently, if W is invariant under $\hat{G}$, it is impossible to satisfy (2.10).

Given the nature of this group, we can't expect $\hat{\hat{W}}$ to grow large as F gets large, in general, although it might happen for particular sequences. Roughly, there is some limit to the size of shear stresses which such a body can tolerate. Said differently, global existence theory for traction boundary value problems, problems of the Neumann type, is not likely to be had. Studies of local existence could help us better understand what happens, as we approach data for which existence fails but, for this discussion, we ignore the more complex problems hinted at here. Assuming $\hat{W}$ has a minimizer C_1, there are infinitely many ways of choosing H to satisfy (2.11). Of particular interest is the subset for which (2.15) or the equivalent (2.16) can be satisfied, enabling configurations to coexist in unloaded bodies.

Included among the possiblities for satisfying (2.16) are the so-called lattice invariant shears, discussed in some detail by Ericksen [4], cases of the type

$$R = 1, \quad H = 1 + a \otimes n \ , \quad a \cdot n = 0 \ , \tag{4.1}$$

G admitting an infinite number of elements of this kind. From this, it follows that the Euler-Lagrange operators indicated by (2.5) cannot always satisfy the conditions of strong ellipticity, as noted in Ericksen [5]. Thus, nature provides us with realizations of solutions of equations of this kind. That the equations degenerate on a rather complicated set can be seen as follows. Pick any case where (2.16) is satisfied and set

$$\phi(x) = \hat{\hat{W}}(1 + x\, a \otimes n). \tag{4.2}$$

From our assumptions, ϕ has minima at $x = 0$ and $x = 1$, so

$$\phi'(0) = \phi'(1) = 0 \ , \ \phi''(0) > 0, \ \phi''(1) > 0 \ . \tag{4.3}$$

The strong ellipticity condition implies that

$$\phi''(x) > 0 . \tag{4.4}$$

Clearly, (4.3) can't hold, if (4.4) holds for $x \in [0,1]$. Physically, we expect the inequalities in (4.3) to be strict, at least in general, at the minimizers, $x = 0$ and $x = 1$. Then the inequality in (4.4) must in fact be reversed for some $x \in (0,1)$, so it doesn't help to just allow equality in (4.4); the Legendre-Hadamard inequality also fails. By similarly using the example involving (3.10), with lattice vectors nearly equal in length, one can get such degeneracies to occur at deformations which are very small, albeit finite. The observations indicate that, depending on the temperature and type of crystal, such deformations can be very small or large. These and other quirks of $\hat{W}$, induced by invariance, are discussed in some detail by Fonseca [11].

The more common kinds of twinning are covered by solutions of (2.15) such that

$$R^2 = H^2 = 1, \tag{4.5}$$

including as a special case the example discussed in §3. As is discussed in detail by Pitteri [12], $H^2 = 1 \Rightarrow R^2 = 1$. He gives an example to show that $R^2 = 1 \not\Leftrightarrow H^2 = 1$, but notes that, if $R^2 = 1$, then either $H^2 = 1$ or H^n is unbounded, as $n \to \infty$. Earlier, Gurtin [13] showed that, if H is orthogonal and $R^2 = 1$, then $H^2 = 1$. Given F_1 and H satisfying (4.5), one solves (2.15) for possible values of R, a and n, details being discussed by these writers. So, one can calculate F_1 and F_2, arrange any number of parallel strips of arbitrary width, with F alternating between one value and the other, minimizers with any number of planes of discontinuity. One sees a great variety of such arrays in real crystals. In specimens which might seem otherwise identical, the number and width of these layers can be quite different, so such nonuniqueness of minimizers is not unrealistic.

If F_1 and H be given, not restricted by (4.5), the analysis of Ericksen [14] shows that either (2.15) has no solutions for R etc. or it has two, depending on whether F_1 and H satisfy a certain equation. Cases where

H satisfies (4.5) are exceptional, in that there is no restriction on F_1, except what is implied by (2.11). This might be viewed as a mathematical explanation of why such twins are so commonly observed in crystals. Actually, the model is simplistic. Physically, one expects there to be some surface energy associated with the discontinuity, making the total energy higher than it would be if the discontinuity were absent. Commonly, a crystallographer or metallurgist will acknowledge this, with a comment to the effect that such surface energies tend to be unusually small. It does reinforce the notion that, very often, we observe solids in configurations which are only metastable, and there is a paucity of good theory for dealing with such things. If H belongs to a finite subgroup of G, $H^N = 1$, where $N = 2,3,4$ or 6. Then, when $N>2$, some rather mild restrictions on F_1 are required, for (2.15) or (2.16) to be soluble for R etc. These conclusions come from analysis of Ericksen [14]. Configurations with $N > 2$ occur in nature but do seem to be less common. Some workers regard some of these as twins, others don't. It isn't hard to construct examples of possibilities not of this form and not included in (4.1), but we don't yet have a good way of characterizing all of the possibilities.

With the numerous possibilities for satisfying (2.15), it might be expected that different planes of discontinuity could intersect, for example to give regions fitting together like the sections of an orange, and such things are observed. James [2,15,16] has developed algorithms useful for analyzing situations of this general kind, not restricted to cases where the deformations involved are symmetry-related, including analyses of some particular cases encountered in practice. Ericksen [17] presents some fairly general schemes for constructing symmetry-related configurations of this kind. It will take more effort to characterize all minimizers, but it seems hopeful that this can be done. It seems rather evident that it doesn't much matter what we assume for the size and shape of the body, since we satisfy (2.5) by having

$$\frac{\partial \hat{W}}{\partial F} = 0 \tag{4.6}$$

as a consequence of minimizing $\hat{W}$. For a start, we can think in terms of what might occur in a crystal filling all of space, taking restrictions to find some possibilities for a crystal of finite size. In doing this, one might overlook possibilities which would be acceptable for bodies with special shapes or topologies.

On occasion, one does see configurations which agree well with our picture of minimizers. Parallel twin planes pretty well cover the complications seen in common observations of high temperature superconductors, for example. Generally, the discontinuities don't move freely about, although one might get them to shift, by applying and removing forces, if the force is large enough. In this sense, a particular minimizer seems to be stable with respect to small enough disturbances. If it were otherwise, linear elasticity theory might not do as well it does, in describing small motions. Clearly, one should bear this in mind, in attempting to analyze changes of minimizers. Pego [18] has developed dynamical stability theory, using a one-dimensional viscoelastic model, predicting that analogous discontinuous static solutions can be stable, when the disturbances are small enough, an interesting beginning step toward constructing relevant dynamical stability theory.

The metallurgist might take one of our minimizers and treat it to produce a more complicated configuration, something he often does to make a specimen harder, for example. Practically, such specimens can be quite stable; we don't expect our chisels to lose their hardness if we treat them well, although we can make them lose their temper, by overheating them while grinding them, for example. Simple possibilities, of some interest, involve cases where $\hat{W}$ also has another relative minimum, of different symmetry. In various crystals, one has more symmetric and less symmetric phases of this kind, often called austenite and martensite, respectively and, commonly, the martensite contains twins. One might see some such combination occurring in the same specimen. Loading and unloading can change the relative proportions. In the so-called memory alloys, this pretty well describes the configurations commonly observed. Even during the loading, the parts are much like this, not far from homogeneous. If, when unloaded, the parts seem to be homogeneous, with planar interfaces, it is at least tempting to think that (4.6) applies. James [19] considers some simpler problems of this kind, involving loading, more detailed analysis of what occurs in particular crystals. As he mentions, he assumes invariance only under a finite subgroup of our group. As is discussed by Fonseca [11], there are theoretical indications that configurations supporting shear stresses cannot be very stable, according to the type of theory being discussed here. As James mentions, some possible arrangements of discontinuity planes mitigate against the applicability of (4.6), so it might be necessary to then consider more complicated solutions of (2.5). In any event, it seems reasonable to try to develop fairly detailed theory of the parts, if the number is not large, to improve understanding of their interactions. Specimens of interest can contain a great number of parts, so one would also like a coarser and simpler theory for these, perhaps a statistical theory somewhat like that considered by Müller and Wilmanski [20].

As is covered in the exposition of Nishiyama [21, Ch. 6], metallurgists and crystallographers have long used a kind of geometric or kinematic theory to analyze configurations which seem to be piecewise homogeneous. This involves jump conditions like (2.14) and ideas of crystal symmetry. Now and then, strain energy might be mentioned, in a casual way. However, one doesn't see any formula for it, and it is not involved in such analyses. The ideas used there fit quite well with the ideas of elasticity theory, including our estimate of the invariance of W. As I interpret, there is a tacit assumption that (4.6) applies. That is, the manner in which rotations are treated then makes sense and, in general, it would not otherwise. On occasion, they do have difficulty in rationalizing an observation, as is discussed by Nishiyama. It would be nice to know whether some more complicated solutions of (2.5) would better fit such observations, but this question remains to be settled. It would seem that one would need to have more specific information about the functions W which might apply, to do much with such problems and, currently, we know very little about this.

In the various steels and other alloys, one encounters a great variety of morphologies, as can be seen by glancing through the pictures presented by

Nishiyama [21], for example. Various other kinds of defects occur and are of concern, for example point or line defects. There is a bulky literature involving use of linear elasticity theory, attempts to understand parts of what is seen. There is no good way to summarize such work in a few words. Of course, it lends support to the notion that complicated singular solutions of (2.5) can be used to model at least some of what is seen. Also, it becomes evident that our "simplest problem" can be exceedingly complicated. The practical metallurgist has learned to control coarser features of such things, to relate these in a rather empirical way to properties which are desirable for particular practical uses of his metals. This does suggest that some kind of control theory is to be had. It does seem plausible enough that a simplistic form could be developed, and it should be instructive to do this. The equations used by Pego [18] seem to be unusually tractable, and one could pose non-trivial control problems for them. Ignore all but our minimizers, and seek a strategy to induce a given one to shift to another such target, compatible with the notion that it is the same lump of matter. Certainly, we will not get the same set of minimizers, if we impose some kinematical constraints, and we might tailor these to fit the initial and target configurations. It seems clear enough that we can use the idea to at least partially control the outcome and that we would do best by controlling the displacement of every point on the boundary, perhaps with some judicious choice of control path. It is a bit risky to guess about the nature of hard theorems which need to be proved, to explore such lines of thought, so it doesn't seem worthwhile to speculate more.

Acknowledgement: This material is based on work supported by the National Science Foundation under Grant No. MEA-8304750.

References

1. Noll, W., A mathematical theory of the mechanical behavior of continuous media, Arch. Rat'l. Mech. Anal. 2, 197-226 (1958).

2. James, R.D., Finite deformation by mechanical twinning, Arch. Rat'l. Mech. Anal. 77, 143-176 (1981).

3. James, R.D., A relation between the jump in temperature across a propagating phase boundary and the stability of solid phases, J. Elasticity **13**, 357-378 (1983).

4. Ericksen, J.L., The Cauchy and Born hypotheses for crystals, in Phase Transformation and Material Instabilities in Solids (ed. M.E. Gurtin), Academic Press, New York (1984).

5. Ericksen, J.L., Special topics in elastostatics, Adv. Appl. Mech. (ed. C.-S. Yih) **17**, 189-244 (1977).

6. Pitteri., On $\nu+1$ lattices, J. Elasticity **15**, 3-25 (1985).

7. Pitteri, M., On crystallographic space groups and generalized lattice groups, pending publication.

8. James, R.D., The stability and metastability of quartz, in these proceedings.

9. Parry, G.P., On the elasticity of monatomic crystals, Math. Proc. Camb. Phil. Soc., **80**, 189-211 (1976).

10. Pitteri, M., Reconciliation of local and global symmetries of crystals, J. Elasticity **14**, 175-190 (1984).

11. Fonseca, I.M.Q.C., Variational methods for elastic crystals, Ph.D. Thesis, Univ. Minnesota (1985), (to appear in Arch. Rat. Mech. Anal.)

12. Pitteri, M. On type II twins, to appear in Int. J. Plasticity.

13. Gurtin, M.E, Two-phase deformations of elastic solids, Arch. Rat'l. Mech. Anal. **84**, 1-29 (1983).

14. Ericksen, J.L., Some surface defects in unstressed thermoelastic solids, Arch. Rat'l. Mech. Anal. **88**, 337-345 (1985).

15. James, R.D., Displacive phase transitions in solids, to appear J. of Mech. and Physics of Solids.

16. James, R.D., Stress-free joints and polycrystals, Arch. Rat'l. Mech. Anal. **86**, 13-37 (1984).

17. Ericksen, J.L., Stable equilibrium configurations of elastic crystals, Arch. Rat'l. Mech. Anal. 94, 1-14 (1986).

18. Pego, R.L., Phase transitions: stability and admissibility in one dimensional nonlinear viscoelasticity, pending publication. c.f. also, Dynamical problems in continuum physics, I.M.A. Volumes, (ed. J. Bona, J.L. Ericksen, C. Dafermos, and D. Kinderlehrer) Springer (1986).

19. James, R.D., The arrangement of coherent phases in a loaded body, in Phase Transformations and Material Instabilities in Solids (ed., M.E. Gurtin), Academic Press, New York (1984).

20. Nishiyama, Z., Martensitic Transformations, Academic Press, New York - San Francisco - London (1978).

QUASICONVEXITY AND PARTIAL REGULARITY IN THE CALCULUS OF VARIATIONS

by

Lawrence C. Evans

Department of Mathematics
University of Maryland
College Park, MD 20742

1. Quasiconvexity and Regularity

This note is a brief and mostly heuristic explanation of some of the main ideas in my recent paper [6], concerning the partial regularity of minimizers for certain natural problems in the calculus of variations.

Let n, N denote positive integers, assume $\Omega \subset R^n$ is open, bounded, and consider then the functional

(1.1) $$I[u] \equiv \int_\Omega F(Du)dx,$$

where

$$Du \equiv \left(\left(\frac{\partial u^i}{\partial x_\alpha} \right)\right) \qquad (1 \leq i \leq N \,,\ 1 \leq \alpha \leq n)$$

is the gradient of $u : \Omega \to R^N$, and

$$F: M^{n \times N} \to \mathbb{R}$$

is given, $M^{n\times N}$ denoting the space of real $n \times N$ matrices. We are interested in minimizing $I[\cdot]$ among all suitable functions u satisfying certain given, but here unspecified, boundary conditions. In particular we ask: (1) do minimizers of $I[\cdot]$ exist? and (2), if so, how smooth are these minimizers?

For the case $N>1$ the major breakthrough regarding the existence question (1) was Morrey's paper [12]. Morrey defined a function $F : M^{n\times N} \to R$ to be <u>quasi-convex</u> provided

(1.2) $$|0|\ F(A) \leq \int_0 F(A + D\phi)dy$$

for all open $0 \subset \mathbb{R}^n$, $A \in M^{n\times N}$, $\phi \in C_0^1(0;\ R^N)$, and worked out then some connections between this notion and the lower continuity of $I[\cdot]$. Acerbi and Fusco [1] have recently refined Morrey's methods and proved:

Theorem 0. Assume $1 \leq q < \infty$ and F is continuous, $0 \leq F(P) \leq C(1+|P|^q)$. Then $I[\cdot]$ is weakly sequentially lower semicontinuous on the Sobolev space $W^{1,q}(\Omega; R^N)$ if and only if F is quasiconvex.

With this theorem in hand it is easy to generate various existence theorems (see for example, [1], [4], etc.), and thereby answer question (1) fairly satisfactorily.

Progress concerning the regularity problem (2) has been less striking. One approach is to note that a minimizer is a weak solution of the Euler-Lagrange equations

(1.3) $$\frac{\partial}{\partial x_\alpha}\left(\frac{\partial F}{\partial p^i_\alpha}(Du)\right) = 0 \qquad (i = 1, N).$$

If F is uniformly strictly convex, that is, if

(1.4) $$\frac{\partial^2 F(P)}{\partial p^i_\alpha \partial p^j_\beta}\, q^i_\alpha q^j_\beta \geq \gamma |Q|^2$$

for all $Q, P \in M^{n\times N}$ and some $\gamma > 0$, then (1.3) falls within the known partial regularity theory for elliptic systems; see for example [9], [11], [14]. But convexity is in some sense an unnaturally strong hypothesis, particularly in light of Theorem 0's identification of quasiconvexity as the crucial assumption on the nonlinearity for the existence theory. See also Ball's arguments [4] against convexity as a natural hypothesis in nonlinear elasticity theory.

These considerations motivate our defining F to be uniformly strictly quasiconvex if there exists $\gamma > 0$ such that

(1.5) $$\int_O F(A) + \gamma|D\phi|^2 dy \leq \int_O F(A+D\phi)dy$$

for all O, A, ϕ as in (1.2). The idea is that uniform strict quasiconvexity is to quasiconvexity as uniform strict convexity is to convexity. And indeed, by analogy with (1.4), it turns out that

$$\frac{\partial^2 F(P)}{\partial p^i_\alpha \partial p^j_\beta} \xi^i \xi^j \eta_\alpha \eta_\beta \geq \gamma |\xi|^2 |\eta|^2 \tag{1.6}$$

for all $P \in M^{n\times N}$, $\xi \in \mathbb{R}^N$, $\eta \in \mathbb{R}^n$, should F satisfy (1.5). But now standard regularity theory for elliptic system fails: there are no known smoothness results for a solution of (1.3) should the Legendre-Hadamard condition (1.6), rather than (1.4)), hold. Such a regularity theory is extremely unlikely in view of some examples constructed by V. Scheffer in [15].

The conclusion so far is that if we assume condition (1.5), we will not be able to use the system (1.3) alone to prove smoothness of u. We presumably need to exploit in addition the fact that u is a minimizer of $I[\cdot]$ (and not just a critical point satisfying (1.3)).

I shall present in § 2 below the outline of just such a regularity assertion. Before doing so, however, let me explain some very rough heuristic considerations which suggest that such a regularity theory is possible. In other words, let me try to answer philosophically at least the question: why should the quasiconvexity (or, for that matter, convexity) of F imply partial smoothness for minimizers?

I suggest the following vague answer. Suppose F satisfies (1.5) and, say, $u \in H^1(\Omega; R^N)$ is a minimizer of $I[\cdot]$. Look at any ball $B(x,r) \subset \Omega$. We would like to show u is really C^1 near x and so want to study the behavior of Du on $B(x,r)$. Now suppose that it happened by chance that u agreed with the plane $\pi(y) \equiv Ay + a$ on $\partial B(x,r)$ (for some $A \in M^{n\times N}$, $a \in R^N$). Then we could set $\phi \equiv u - \pi$ in (1.5) to find

$$|B(x,r)|\ F(A) < \int_{B(x,r)} F(Du)dy,$$

unless $u \equiv \pi$ in $B(x,r)$. But this says that unless $u \equiv \pi$ we could redefine u to equal π in $B(x,r)$, and thereby strictly lower the energy. Since u is a minimizer, this is impossible; so that in fact u equals π in $B(x,r)$. Continuing, we may therefore speculate that if u were "close to" some plane π on $\partial B(x,r)$ or on the annulus $B(x,r)-B(x,r/2)$, then u will be close to π in all of $B(x,r)$. In other words, it seems reasonable to hope that the deviation of u

from a plane on $B(x,r)-B(x,r/2)$ may somehow control the deviation of u from the same plane on all of $B(x,r)$. And this, if true, is interesting, since it suggests that the standard technical device called "hole filling" (see the proof of Lemma 2 below) might be employed to show that the distance of u from its "average tangent plane" lessens as we pass from $B(x,r)$ to $B(x,r/2)$. This fact, suitably interpreted, would imply u to be C^1 near x. Once we know u to be C^1 in some region, standard bootstrap procedures based on (1.3) force u to be C^∞ (assuming F is).

2. Partial Regularity for Minimizers

I outline in this section some of the methods developed in [6] to convert the incoherent ideas above into a proof.

Assume $F : M^{n\times M} \to \Omega$ is at least C^2, F is uniformly strictly quasiconvex, and

$$(2.1) \qquad |D^2F(P)| \leq C \qquad (P \in M^{n\times N}).$$

Theorem 1 Suppose $u \in H^1(\Omega; R^N)$ is a minimizer of $I[\cdot]$. Then there exists an open subset $\Omega_0 \subset \Omega$ such that

(i) $\quad |\Omega - \Omega_0| = 0,$

(ii) $\quad Du \in C^\alpha(\Omega_0; M^{n\times N}) \qquad (0 < \alpha < 1).$

Remark Hypotheses (2.1) excludes many interesting examples: see [6] for partial smoothness under less restrictive hypotheses.

We sketch the proof as follows.

For $B(x,r) \subset \Omega$, let us write

$$(Du)_{x,r} \equiv \fint_{B(x,r)} Du\, dy ,$$

$$U(x,r) \equiv \fint_{B(x,r)} |Du-(Du)_{x,r}|^2 dy,$$

the slash through integrals denoting the average.

Lemma 1. There exists $0 < \tau < 1$ and, for each $L > 0$, $\varepsilon = \varepsilon(L) > 0$ such that

$$|(Du)_{x,r}| \leq L, \ U(x,r) \leq \varepsilon$$

imply

$$U(x,\tau r) \leq \frac{1}{2} \ U(x,r).$$

Outline of proof Were Lemma 1 false there would exist balls $B(x_m, r_m) \subset \Omega$ such that

$$|(Du)_{x_m, r_m}| \leq L, \quad U(x_m, r_m) \to 0,$$

but

$$U(x_m, \tau r_m) > \frac{1}{2} \quad U(x_m, r_m) \qquad (m = 1,2,\ldots), \tag{2.2}$$

where some small $0 < \tau < 1$ has been fixed. We "blow up" from u on $B(x_m, r_m)$ to appropriately rescaled finding v^m on the unit ball $B = B(0,1)$. Now it turns out that the v^m (or perhaps some subsequence) converge weakly in $H^1(B;R^N)$ to a smooth function v, which satisfies "good" estimates. On the other hand the v^m each satisfy a rescaled version of the "bad" estimate (2.2). All this leads to the desired contradiction, assuming that we can somehow get slightly better control on the mode of convergence of the v^m to v.

The next lemma provides a kind of local energy or so-called "Caccioppoli" inequality, which is just enough to make the argument above work. I provide details of the derivation, since it's here that the heuristics from § 1 are transformed into a rigorous proof.

Lemma 2. There exists a constant C_1 such that

$$\int_{B(x,r/2)} |Du - A|^2 \, dy \leq \frac{C_1}{r^2} \int_{B(x,r)} |u-A(y-x)-a|^2 dy \tag{2.3}$$

for all $B(x,r) \subset \Omega$, $A \in M^{n \times N}$, $a \in R^N$.

Proof. We may assume $x = 0$. Let $\frac{r}{2} \leq t < s \leq r$ and choose $\zeta \in C_0^\infty(\Omega; R)$ to satisfy

$$\begin{cases} \zeta \equiv 1 \text{ on } B(t), & \zeta \equiv 0 \text{ on } \Omega - B(s) \\ 0 \leq \zeta \leq 1, & |D\zeta| \leq \frac{C}{s-t} . \end{cases}$$

Define

$$\phi \equiv \zeta(u-a-Ay), \quad \psi \equiv (1-\zeta)(u-a-Ay);$$

then

(2.4) $$D\phi + D\psi = Du - A.$$

As $\zeta = 0$ on $\partial B(s)$, (1.5) and (2.1) imply

(2.5)
$$\begin{aligned} \int_{B(s)} F(A) + \gamma|D\phi|^2 dy &\leq \int_{B(s)} F(A+D\phi)dy \\ &= \int_{B(s)} F(Du-D\psi)dy \quad \text{by (2.4)} \\ &\leq \int_{B(s)} F(Du)-DF(Du)D\psi + C|D\psi|^2 dy. \end{aligned}$$

Since u is a minimizer, we have

$$\begin{aligned} \int_{B(s)} F(Du)dy &\leq \int_{B(s)} F(Du-D\phi)dy \\ &= \int_{B(s)} F(A+D\psi)dy \quad \text{by (2.4)} \\ &\leq \int_{B(s)} F(A) + DF(A)D\psi + C|D\psi|^2 dy. \end{aligned}$$

This inequality combined with (2.5) gives

$$\gamma \int_{B(s)} |D\phi|^2 dy \leq \int_{B(s)} [DF(A)-DF(Du)]D\psi + C|D\psi|^2 dy.$$

Then (2.1) and the definition of ϕ imply

(2.6) $$\int_{B(t)} |Du-A|^2 dy \leq C \int_{B(s)} |Du-A|\ |D\psi| + |D\psi|^2 dy.$$

Now $\psi \equiv 0$ on $B(t)$ and

$$|D\psi| = |(1-\zeta)(Du-A)-D\zeta\otimes(u-a-Ay)| \leq C|Du-A| + \frac{C}{s-t}|u-a-Ay|$$

on $B(s) - B(t)$; whence (2.6) yields

$$\int_{B(t)} |Du-A|^2 dy \leq C \int_{B(s)-B(t)} |Du-A|^2 dy + \frac{C}{(s-t)^2} \int_{B(r)} |u-a-Ay|^2 dy.$$

We "fill the hole" by adding $C \int_{B(t)} |Du|^2 dy$ to both sides to obtain

$$\int_{B(t)} |Du-A|^2 dy \le \theta \int_{B(s)} |Du-A|^2 dy + \frac{C}{(s-t)^2} \int_{B(r)} |u-a-Ay|^2 dy$$

for

$$\theta \equiv \frac{C}{C+1} < 1.$$

This inequality is valid for all $\frac{r}{2} \le t < s \le r$. We may therefore apply Lemma V.3.1 from Giaquinta [p. p. 161] to find

$$\int_{B(r/2)} |Du-A|^2 dy \le \frac{C}{r^2} \int_{B(r)} |u-a-Ay|^2 dy,$$

as required. □

Notice, by the way, that our derivation of (2.3) uses neither the maximum principle nor integration by parts, these sometimes said to be the only tools available for making nonlinear elliptic estimates.

Completion of proof of Theorem 1.

As noted before, it is easy to use Lemma 2 to finish the proof by contradiction of Lemma 1. Now let

$$\Omega_0 \equiv \{x \in \Omega |\ \lim_{r\to 0} (Du)_{x,r} = Du(x),\ \lim_{r\to 0} U(x,r) = 0\}.$$

Then $|\Omega - \Omega_0| = 0$, and Lemma 1 implies Ω_0 is open. Furthermore, our iterating the inqualities from Lemma 1 shows that each point $x \in \Omega_0$ belongs to some neighborhood $0 \subset \Omega_0$ such that

$$U(y,r) \le Cr^{2\alpha} \qquad (y \in 0,\ 0 < r < r_0)$$

for certain constants C, r_0 and $0 < \alpha < 1$. This in turn implies, by a result of Morrey, that $Du \in C^{\alpha}(0,M^{n\times N})$: see [9,p.70]. A refinement of this argument shows that we can take α to be any number < 1.

□

Remarks Fusco-Hutchinson [8] and Giaquinta-Modica [10] have extended the results to more general integrands, and R.F. Gariepy and I [7] have recently found a rather different proof, which does not make use of the Caccioppoli inequality (2.3).

The partial regularity theorem described here bears strong resemblance to some of Almgren's results ([2], [3]) concerning the smoothness of minimizing currents for elliptic integrands in geometric measure theory. See also Bombieri [5].

References

1. E. Acerbi and N. Fusco, Semicontinuity problems in the calculus of variations, to appear in Arch. Rational Mech. Anal.

2. W.K. Allard and F.J. Almgren, Jr., An introduction to regularity theory for parametric elliptic variational problems, in Proc. Symposia Pure Math. XXIII (1973), American Math. Soc., 231-260.

3. F.J. Almgren, Jr., Existence and regularity almost everwhere of solutions to elliptic variational problems among surfaces of varying topological type and singularity structure, Ann. Math. 87 (1968), 321-391.

4. J.M. Ball, Convexity conditions and existence theorems in nonlinear elasticity, Arch. Rational Mech. Anal. 63 (1977), 337-403.

5. E. Bombieri, Regularity theory for almost minimal currents, Arch. Rational Mech. Anal. 78 (1982), 99-130.

6. L.C. Evans, Quasiconvexity and partial regularity in the calculus of variations, to appear in Arch. Rational Mech. Anal.

7. L.C. Evans and R.F. Gariepy, Blow-up, compactness and partial regularity in the calculus of variations, to appear.

8. N. Fusco and J. Hutchinson, Partial regularity of functions minimizing quasiconvex integrals, to appear.

9. M. Giaquinta, Multiple Integrals in the Calculus of Variations and Nonlinear Elliptic Systems, Princeton U. Press, Princton, 1983.

10. M. Giaquinta and G. Modica, Partial regularity of minimizers of quasiconvex integrals, to appear.

11. E. Giusti and M. Miranda, Sulla regolarità delle soluzioni deboli di una classe di sistemi ellittici quasi-lineari, Arch. Rational Mech. Anal. 31 (1968), 173-184.

12. C.B. Morrey, Jr., Quasiconvexity and the lower semicontinuity of multiple integrals, Pac. J. Math. 2 (1952), 25-53.

13. C.B. Morrey, Jr., Multiple Integrals in the Calculus of Variations, Springer, New York, 1966..

14. C.B. Morrey, Jr. Partial regularity results for nonlinear elliptic systems, J. Math. and Mech. 17 (1968), 649-670.

15. V. Scheffer, Ph.D. Thesis, Princeton, 1974.

INTRODUCTION TO PATTERN SELECTION IN DENDRITIC SOLIDIFICATION

Nigel Goldenfeld

Institute for Theoretical Physics
University of California
Santa Barbara, CA 93105, USA

Abstract

After reviewing the basic physics of dendritic solidification, I present a simplified model of this process, the boundary-layer model (BLM). The BLM assumes that the latent heat of solidification diffuses only in a thin boundary layer at the interface between the liquid and the solid, and in addition incorporates the effects of surface tension, molecular attachment kinetics and crystalline anisotropy. The latter is crucial for dendritic growth to occur; with insufficient anisotropy, the interface evolves via tip-splitting and the dendritic morphology is no longer produced. This prediction has been verified in experiments on viscous fingers, which are an analogue for solidification. It is also found that the allowed values of the tip velocity and radius of dendrites in the BLM form a discrete set, satisfying a non-linear eigenvalue problem. The largest tip velocity is dynamically selected. I conclude with speculations on the origin of sidebranching and by presenting some incompletely posed problems.

1. Introduction

In this talk, I will discuss dynamical systems far from thermal equilibrium which seem to posses not just one stable steady state, but an entire family or continuum of such states. What is remarkable about these systems is that when an experiment or careful computer simulation is performed, a unique state is reproducibly selected regardless of the initial conditions. The central problem of interest is to identify the selected state, and to discover what is special about it. Typically, we would be talking about an initial value problem for a partial differential equation, which tends towards a particular solutions asymptotically as time increases; what one wants is to be able to solve the corresponding steady state boundary value problem (usually much simpler), and then find the correct final steady state using a so-called selection principle. A second problem is to provide a physical interpretation to the selection principle. From the physicist's point of view, one is seeking a selection principle which can be stated as a variational principle. It is the hope that if this exists and is sufficiently general, then one will learn something about the laws of physics as they apply to non-equilibrium systems.

The systems I am interested in are those which exhibit propagating pattern selection; that is, a stable state propagates into an unstable state and leaves behind a pattern with a well-defined wavelength. Typically one is concerned with calculating the velocity at which the stable state propagates, and the wavelength of the resulting pattern. This situation is not uncommon, and arises in various diverse branches of science: some examples are the propagation of flames down a tube[1], the propagation of Taylor vortices in a Taylor-Couette experiment[2], the spread of advantageous genes[3] and the propagation of signals in nerves[4]. All of these phenomena can be modelled by partial differential equations in one space and one time dimension.

Here I shall mainly discuss recent advances in our understanding of dendritic growth, the process by which most metals and alloys solidfy[5]. A familiar example is the growth of snowflakes, where the solid grows into an undercooled liquid or vapour by developing long fingers decorated by sidebranches. I will also have occasion to mention the related topic of viscous fingering at the interface between two immiscible fluids in a two-dimensional geometry[6]. These phenomena are realistically described by partial differential equations in one time and two or three spatial dimensions. Nevertheless, it has traditionally been assumed that the one-dimensional nature of the propagation enables analogies to be drawn with the cases mentioned in the previous paragraph[5,7,8]. I will discuss at some length the extent to which the analogy is correct later in this talk.

In the following section, I shall describe the physical phenomenon of dendritic growth, and formulate it as a so-called Free Boundary Problem (FBP). In Section III I will review the theoretical picture of the selection of the tip characteristics as it existed until recently. In particular, I will mention the marginal stability hypothesis, and note in passing its application to partial differential equations in 1+1 dimensions. Section IV will introduce the boundary-layer model (BLM), a phenomenological model of solidification which purports to apply to solidification at large undercooling. The BLM requires the presence of anisotropy (introduced through the lattice of the crystalline solid phase) in order for dendritic growth to occur. Previously, it was thought that this effect was negligible. In Section V I will describe the phenomenon of viscous fingering, show its relation to the solidification problem, and report the result of an experiment which has verified in this system the importance of anisotropy for dendritic growth. Section VI is devoted to a discussion of selection in the BLM, and to speculations regarding the dynamical origin of sidebranching. I conclude by mentioning some important mathematical problems which remain unresolved.

II. Physics of Dendritic Growth

Dendritic growth occurs when a solidification front advances into a supercooled liquid, so that the solid is at a higher temperature than the liquid. The motion of the interface is controlled by the parameter

$$\Delta = \frac{-(T_\infty - T_M)}{L/C} \tag{II-1}$$

where T_∞ is the temperature in the liquid far from the solidification front, T_M is the melting temperature of a planar interface, L is the latent heat associated with the first order solidification transition and C is the specific heat of the liquid.

Dendrites are single crystals of the solid phase and consist of a smooth, nearly parabolic tip which propagates without measurable change of shape at constant velocity, followed by a periodic (in time) emission of small bumps which enlarge to form side brances and eventually dendrites (see Fig. 1). The fact that they are single crystals is responsible for the degree of symmetry seen in real snowflakes:- the growth laws are deterministic, and if the environment is homogeneous then each arm of a snowflake will grow in the same way. The six-fold nature of snowflakes is a manifestation of the unit cell of the solid phase, and can be understood by considerations of quantum chemistry which do not concern us here.

I shall only be interested in understanding the behaviour near the tip of a dendrite, such as the tip velocity and shape, the factors governing the initial sidebranch spacing, the sensitivity to variations in the growth conditions, and the approximate scaling laws which are observed. The tip velocity v_0 and tip radius ρ_0 approximately satisfy[9]

$$\begin{aligned} v_0 &\approx \Delta^{2.6} , \\ \rho_0 &\approx \Delta^{-1} \end{aligned} \tag{II-2}$$

for $\Delta < 0.2$.

Roughly speaking[9], three different growth regimes can be distinguished as Δ is varied. For $0 < \Delta < 0.05$, the dendrites are sufficiently large that fluid convection effects are important in the growth, whilst for $\Delta > 0.3$ nucleation and molecular attachment kinetics dominate the growth. In the intermediate regime $0.05 < \Delta < 0.3$, the motion of the interface is determined by how rapidly the latent heat diffuses away from it.

This situation is described by the free boundary problem (FBP) for the dimensionless temperature field

$$u(x,t) = \frac{T(x,t) - T_\infty}{L/C} \tag{II-3}$$

where x is a point in the liquid. For convenience, we will restrict the following discussion here and below to two spatial dimensions. It is also convenient to ignore thermal diffusion in the solid; in reality this is not correct, but no new points of principle emerge with the inclusion of this effect. For the situation where a binary alloy is solidified isothermally, $u(x,t)$ may be taken to be the dimensionless concentration of the solute, and the assumption of no diffusion in the solid is now realistic. In the liquid $u(x,t)$ satisfies the diffusion equation

$$\frac{\partial u}{\partial t} = D\nabla^2 u \tag{II-4}$$

where D is the thermal diffusion coefficient. There are two boundary conditions in addition to the boundary condition at infinity. The first states that the rate of generation of latent heat at the interface is the rate at which heat flows into the liquid:

$$v_n = -D\nabla u \cdot \hat{n} \ . \qquad \text{(II-5)}$$

Here v_n is the speed of each point on the interface as it moves along its outward normal $\hat{n}$. The second boundary condition relates the temperature field at the interface, u_s, to the curvature κ of the interface:

$$u_s = \Delta - d_0\kappa. \qquad \text{(II-6)}$$

Here, d_0 is the capillary length, typically about 10-20Å, and is proportional to the surface tension. κ is defined to be positive for an interface which bulges into the liquid: i.e. $\kappa = -\hat{n}\,\dfrac{\partial^2 r}{\partial s^2}$, where $r(s)$ is the position vector of the interface as a function of the arclength s along the interface, measured from the tip of the dendrite. In the absence of surface tension, or for a planar interface, the surface temperature is just Δ. The latter boundary condition is a statement of equilibrium thermodynamics[10], and follows from assuming that the interface is locally in equilibrium. In other words, the time scale for equilibration of molecular motion of diffusion is much shorter than the time scale over which the temperature changes appreciably. For a moving interface, this assumption is not strictly justified, and later I will mention a modification of this boundary condition which takes into account the departure from local equilibrium.

The FBP possesses solutions which correspond to the motion of a planar interface with a velocity $v(t) \sim t^{-1/2}$ as $t \to \infty$. At $\Delta = 1$, however, steady state motion is possible at an arbitrary velocity, and the temperature profile ahead of the interface is a decaying exponential, with a decay length

$$\ell = \frac{D}{v} \ . \qquad \text{(II-7)}$$

ℓ is known as the diffusion length. For the case of dendrites grown in controlled studies the diffusion length is typically 10 to 100 times as great as the tip radius.

The planar solutions are unstable, as first pointed out by Mullins and Sekerka[11]. For a perturbation proportional to $\exp(ikx + \Omega t)$ the dispersion relation is sketched in Fig. 2. For wavelengths less than the stability length $2\pi(d_0\ell)^{1/2}$ the interface is linearly stable, whilst for wavelengths larger than this amount the interface is linearly unstable: This stability length sets the scale for the resulting patterns. The physical content of the stability analysis is as follows. If we neglect surface tension for the time being, and consider a small bulge in an otherwise flat interface, we see that the thermal gradient steepens in the vicinity of the tip of the bulge. This causes the bulge to grow faster than the surrounding interface, and thus it extends further and further into the supercooled liquid. Thus diffusion tends to maximize the surface area of the interface. Now let us include surface tension: as the tip of the bulge is more curved than any other point on the interface, it is also the coldest point by virtue of the boundary condition eq. (II-6). Thus heat diffuses towards the tip, raising the temperature and reducing the curvature. Surface tension acts as a regulator at short distances and tends to minimize the surface area of the interface. The non-linear competition between these two effects is responsible for the patterns observed in nature.

III. Selection of the Tip and the Marginal Stability Hypothesis

An important ingredient of theories of dendritic growth is the needle-crystal solution, defined to be a uniformly translating steady state solution of the equations of motion for the interface. The first such solution is due to Ivantsov[12] for the case when $d_0 = 0$. In this case, the FBP is separable when written in parabolic coordinates, and the interface is an isotherm belonging to a continuous family of parabolas whose tip velocity and radius are undetermined except for the condition:

$$\rho_0 v_0 = 2D\Lambda(\Delta) \qquad \text{(III-1)}$$

where $\Lambda^{-1}(x) = (\pi x)^{1/2} e^x \operatorname{erfc}(x)$ and $\operatorname{erfc}(x)$ is the complementary error function.

Since surface tension has been omitted this family is dynamically unstable and has no direct physical significance. Various attempts were made to introduce the capillary term by regular perturbation theory[5]; as we shall see below, this is not possible to do in a consistent manner so the various attempts were not systematic. Inclusion of the capillary term has two consequences[13]. The first is a correction of the computed interface profile of the family of needle-crystals, and is rather unimportant. The second effect is to linearly stabilize those members of the family with tip radius below a critical value ρ^*. Here and below, the term stability is defined with respect to the comoving frame of the tip of the dendrite. It turns out that in the limit $\ell \gg \rho_0$ the family is parametrised by the dimensionless combination

$$\sigma = \ell d_0/\rho_0^2 \qquad \text{(III-2)}$$

and all properties of the needle-crystals depend only on σ. In this picture, the sidebranches are instabilities in the laboratory frame which grow away from the tip in such a way as to maintain the stability of the needle crystal in the comoving frame of the tip. At this stage of the argument, a conundrum arises: Which of the family of stable steady states corresponds to the tip of the observed dendrite?

The answer which has achieved the best agreement with experiment to date is that the selected state is the one which is marginally stable[13,14]. That is, one calculates the eigenvalues spectrum about each needle-crystal. The eigenvalues are functions of σ. The eigenvalue with the largest real part is identified, and σ adjusted until at $\sigma = \sigma^*$ the real part of this eigenvalue is zero. This determines the operating point of the dendrite. There are technical difficulties with this procedure which I will merely mention in passing. Firstly, it is not clear exactly how to write down a well-posed eigenvalue problem for the stability of a needle-crystal, since the domain of the problem is infinite and the operator is not Hermitian or even self-adjoint. Secondly, it has been too difficult to implement the marginal-stability calculation in a realistic situation because of the necessity to include the symmetry of the crystal lattice. For the purposes of comparison with experiment, what is usually done is to supplement the steady state relation eq. (III-1) with the marginal stability condition obtained from the spectrum of excitations with a particular symmetry about a growing circle[9,14]. This approximate procedure has conceptual difficulties too. Firstly, it is not a priori obvious what is the desired symmetry of the perturbation, and it is best to regard this as an adjustable parameter. Secondly, linear stability analysis about a growing circle is perhaps reasonable if one is seeking exponentially growing modes, since they will evolve much faster than the $t^{1/2}$ growth of the circle. However, the marginally stable modes will not grow exponentially, and so it is not clear how to interpret the results of such a calculation.

Nevertheless, the marginal stability hypothesis has been used successfully to predict growth properties of dendrites to within a few percent (circular approximations) or to within 30 percent (paraboloid of revolution approximation, no symmetry breaking). The interested reader is referred to the papers of Huang and Glicksman, and the review article by Langer for a more complete discussion of this topic.

The marginal stability hypothesis is purely ad hoc. Originally it was thought that thermal fluctuations drive the systems to this state[13], but they are too small. In addition, the operating point of the dendrite is observed to adjust extremely rapidly when the growth conditions are changed, suggesting that the dynamics of the selection mechanism are deterministic, not diffusive. Indeed, such a mechanism is known to occur in certain partial differential equations in 1+1 dimensions, such as a form of Fisher's equation[3]:

$$\frac{\partial u}{\partial t} = \frac{\partial^2 u}{\partial x^2} + u - u^3 , \qquad u(-\infty,t) = 1 , \; u(\infty,t) = 0 \tag{III-3}$$

where the initial conditions decay sufficiently fast at infinity, and the Hohenberg-Swift equation[15]:

$$\frac{\partial u}{\partial t} = (\varepsilon - (\partial_x^2 + 1)^2)u - u^3 , \qquad \partial_x u = \partial_x^3 u = 0 \text{ at } x = 0, L \text{ as } L \to \infty . \tag{III-4}$$

In Fisher's equation, travelling waves exist for speeds $c > 2$, yet for physical initial conditions (i.e. decaying faster than exponential at infinity) the form of the solution as $t \to \infty$ is the wave moving at $c = 2$. In the Hohenberg-Swift equation, a localized initial disturbance generates a stationary periodic structure contained within an envelope whose front moves at a uniform velocity. Here, there is a band of stationary stable periodic solutions; the marginal stability hypothesis correctly identifies the wavelength of the selected pattern and predicts the speed of propagation of the front[7]. In these and similar examples, the ordered or broken symmetry phase propagates into an unstable state, whereas in solidification, the supercooled liquid is really metastable. This suggests that the application of the marginal stability hypothesis to solidification might not be straightforward.

IV. The Boundary - Layer Model (BLM)

Despite the apparent success of the marginal stability hypothesis in predicting the tip characteristics, a number of important questions remained unanswered. Perhaps the most surprising omission from the theory is the effect of crystalline anisotropy. Clearly the symmetry of the underlying crystalline solid has a marked effect on the growth. How does the magnitude of the anisotropy, as measured by the surface free energy for example, influence the growth dynamics? In addition the boundary condition eq. (II-6) assumes that the interface is locally in equilibrium and is stationary: How do we correct for the fact that it is in fact in motion, and does this make a significant difference to the growth?

The formulation of the free boundary problem had traditionally been assumed to embody the essential physics, but this assumption had neven been tested. That is, no numerical or analytical studies of the solutions to the FBP existed to demonstrate that dendrites are solutions of the FBP equations[16]. What is required is a model simple enough to be at least numerically tractable, yet complex enough to contain the elements of the physics of dendritic growth.

The boundary-layer model (BLM) is an attempt in this direction[17]. The aim is to replace the FBP in two spatial dimensions by a diffusion problem in one dimension. The basic idea is the observation that if the diffusion length is much smaller than the local radius of curvature of the interface in some region, then diffusion is effectively confined to a narrow region around the interface, which we will call the boundary-layer. We assume that the diffusion current is parallel to the interface, and account for diffusion in the perpendicular direction by allowing the thickness of the boundary-layer to vary with time and arclength position along the interface, s. Thus the model will exhibit a restricted form of non-locality in its dynamics, and will preserve in a crude way the competition between diffusion and surface tension which we argued was a necessary condition for dendritic growth in section II.

Right at the outset, two comments are in order. Firstly, the experiments referred to earlier are performed in the regime where this boundary-layer approximation is completely invalid. However, there is no reason in principle why experiments cannot be performed at higher undercoolings or on different systems so that the approximation is a description of reality. Furthermore, we expect that there will be no new qualitative behaviour in the boundary-layer limit, so that we we should be able to learn qualitative information about the structure of the FBP. Secondly, the boundary-layer approximation ignores interactions between points on the interface which are close in real space, but distant along the arclength of the interface. Thus we do not expect to be able to say anything about the behaviour well away from the tip of the dendrite, where sidebranches compete for the diffusion field in a very non-local two- (really three-) dimensional manner. We shall not even be concerned if the interface intersects itself in these regions.

The formulation of the BLM consists of two parts, kinematics and dynamics. We choose to represent the interface not by its Cartesian coordinates, but implicitly through the curvature as a function of the arclength and time. If θ is the angle between the normal to a point on the interface and a fixed direction in space (which we shall take to be the direction of propagation of the dendrite), then

$$\kappa = \frac{d\theta}{ds} \quad . \tag{IV-1}$$

The interface may be chosen to evolve by letting each point on the interface move outwards along its normal with speed ν_n. If we sit on each point and measure how the curvature and the arclength position from some arbitrary origin evolve in time, we find:

$$\frac{\partial\kappa}{\partial t}\Big|_n = -\left(\kappa^2 + \frac{\partial^2}{\partial s^2}\right)\nu_n ,$$

$$\frac{ds}{dt}\Big|_n = \int_0^s \kappa\nu_n ds' \qquad \text{(IV-2)}$$

where the subscript n implies that the derivative follows each point along its outward normal. These equations are exact. In order to know how to compute the evolution of the interface, we need a prescription for calculating ν_n.

One way of proceeding is to simply write down an ansatz for ν_n as a functional of the curvature of the interface[18]. In the Geometrical Model (GM),

$$\nu_n = \kappa + A\kappa^2 - B\kappa^3 + \sigma\frac{\partial^2\kappa}{\partial s^2} . \qquad \text{(IV-3)}$$

The polynomial terms mimic the Mullins-Sekerka instability, whilst the derivative term mimics the surface tension. The GM is much simpler than the BLM, but lacks the physical interpretation of the BLM. In particular, it does not exhibit the competition between diffusion and surface tension, as can be seen from the fact that a flat interface does not move in the GM.

In the BLM, ν_n is obtained from the dynamics of the boundary-layer thickness. It is convenient to define the heat content per unit length of the interface

$$h(s,t) = u_s \ell \qquad \text{(IV-4)}$$

and to write an equation of motion for h by considering the heat balance in the boundary-layer. We find

$$\frac{\partial h}{\partial t}\Big|_n = \nu_n(1 - u_s) - \nu_n\kappa h + D\frac{\partial}{\partial s}\left(\ell\frac{\partial u_s}{\partial s}\right) . \qquad \text{(IV-5)}$$

This equation is to be regarded as phenomenological; to the best of my knowledge it is not exact, although the first two terms can be rigorously derived from the FBP for the steady state motion of a convex interface[19]. The first term is the fraction of the generated latent heat which enters the boundary-layer - the remainder heats up the liquid to the temperature of the interface. The second term accounts for the redistribution of heat as the interface expands. The third term is the divergence of a thermal current along the arclength of the interface. Its form has been carefully chosen so as to exhibit the correct driving force for curvilinear diffusion, namely the temperature of the interface. The boundary condition eq. (II-5) is expressed by approximating the normal derivative by the forward difference to lowest order i.e.

$$\nu_n = -\nabla u \cdot n \simeq \frac{u_s}{\ell} = \frac{u_s^2}{h} \; . \qquad \text{(IV-6)}$$

For the moment, we will take over boundary condition eq. (II-6) unmodified.

Note that when the interface is an isotherm, the diffusion term vanishes; solving for the steady state of the resulting equation yields a parabolic interface - the BLM analogue of the Ivantsov solution. The BLM equations can be solved for simple geometries such as planes and circles[20], and yield results which are asymptotically close to the corresponding solutions of the FBP as $\Delta \to 1$. In addition, the stability properties of the BLM resemble those of the FBP: The Ivantsov solution is completely unstable, and the planar interface dispersion relation of the BLM is similar to that of the FBP with the minor addition of a stable branch corresponding to out-of-phase motions of the interface and the boundary-layer.

At this point, I wish to discuss the boundary condition eq. (II-6). The first point is that an interface only moves in response to an externally imposed chemical potential difference across it. For small departures from equilibrium, the usual argument of linear response theory can be applied. In condensed matter physics this is the statement that dynamical quantities can be expressed in terms of other quantities calculated <u>in equilibrium</u>. Thus we expect that the interface velocity will be proportional to the chemical potential difference across the interface, with the constant of proportionality, and <u>kinetic coefficient</u> being a function of thermodynamic quantities. The result of the detailed calculation is

$$\nu \approx \frac{L^2}{CT_M \chi} ([\Delta - d_0\kappa] - u_s) \qquad \text{(IV-7)}$$

where χ is the ratio of the surface free energy to the interface thickness[10,21]. The calculation relies on the fact that the interface is molecularly rough; faceted crystal growth does not obey linear response theory because there is a nucleation barrier for the attachment of new molecules to the facet. Secondly, the crystalline anisotropy, which is an automatic consequence of the crystal lattice, enters into the problem through the boundary condition eq. (IV-7). In general, both the surface tension and the kinetic coefficient will depend on the orientation of the surface. Thus we end up with the boundary condition

$$u_s = \Delta - d_0(\theta)\kappa - \beta(\theta)\nu_n \ . \qquad \text{(IV-8)}$$

The precise form of the angular dependence is probably not crucial, and we shall simply assume a sinusoidal variation superimposed on the uniform value of the appropriate coefficient.

Let me now discuss some numerical results from the BLM. The BLM equations are solved using an implicit Crank-Nicholson scheme at each time step. Since the equations follow each point on the interface, the grid points become distributed along the interface in a non-uniform manner after each time step, which can lead to errors. To circumvent this problem, we perform a cubic spline fit to the dependent variables, and redistribute the grid points uniformly after each time step. Since the interface also expands, we continuously add new grid points during the redistribution in order that the mesh size remains constant during the integration. We perform the integration using zero-derivative boundary conditions, and the time for integration scales linearly with the system size.

Fig. 3 shows the evolution of an interface in the limit of very small Δ where three-fold symmetry was externally imposed. The kinetic coefficient $\beta = 0$ and there is no angular dependence of d_0. The initial seed has not continued to grow outward along the tips, but instead the tips have broadened and flattened, and in the final stage shown, have bifurcated. This behaviour will be referred to as tip-splitting. In Fig. 4 the effects of anisotropy have been included in the kinetic coefficient, and a six-fold anisotropy has been imposed, although only a hectant is displayed. The initial seed grows a well defined tip which propagates without change of shape at a uniform velocity, with a trail of sidebranches in the later stages. The question of sidebranches, and whether or not they are transients cannot be decided from these limited simulations. This will be discussed further in section VI. In the later stages shown, only a portion of the interface was evolved; the regions furthest from the tip developed a very fine structure on small scales which makes it almost impossible to properly follow the evolution. This is an artifact of the BLM and arises because a grooved portion of the interface between two sidebranches is driven by the undercooling at infinity, whereas in reality the driving force in the groove is much reduced because of the screening effect of the neighbouring sidebranches. Nevertheless, the computed solution shown is probably accurate to within 0.1%, this estimate arising from a comparison between the measured tip velocity and that predicted by the selection mechanism to be discussed below.

The very clear conclusion from this is that anisotropy is crucial in order for dendritic growth to occur. In retrospect, it is not hard to see why this should be true. The instability of the interface lacks directional coherence in the absence of anisotropy; with anisotropy it is continually focussed in one direction, allowing a steady state tip to grow. I emphasize that this is a dynamical effect, and involves a competition between the diffusive instability of the interface and the anisotropy. For example, the Ivantsov solution has a well defined tip, but is dynamically unstable. We have verified that the importance of anisotropy is not a special result of the way we have included it in the equations. We have included it in the surface tension d_0 and also as a multiplicative prefactor in the velocity and find the same qualitative conclusions.

The GM also exhibits a similar scenario[22]. If anisotropy is introduced by multiplying the ansatz for v_n by $1 + \varepsilon \cos(m\theta)$, then it is found that for $\varepsilon < \varepsilon_c$ tip-splitting occurs, whereas for $\varepsilon > \varepsilon_c$ a tip develops, although this is not necessarily a dendrite. This is discussed in more detail in section VI.

V. Viscous Fingers - an Analogue of Solidification

The models discussed in the previous section are known as local models. The FBP can be posed by writing down an effective equation for the interfacial dynamics in terms of the Green function for the diffusion field[5]. The equation obtained demands that the interface responds to the thermal field from all points in space at all earlier times, weighted by an appropriate factor. Hence the statement that the FBP is nonlocal and has memory. The question then arises: to what extent are the conclusions of the previous section dependent on the locality of the GM and the primitive non-locality of the BLM? Secondly, is it possible to find an experimental system where the role of anisotropy can be tested in a controlled manner for long times, so that questions about transience of sidebranches can be resolved beyond reasonable doubt?

In the absence of any systematic, controlled way of varying anisotropy in solidification, we have turned instead to the phenomenon of viscous fingering[6]. We have found a rather dramatic demonstration of the effects of anisotropy, which I will describe below. At first sight these may not seem related, but I shall show below that there is a rather close similarity between the two problems.

Consider two immiscible fluids of different viscosity confined between two horizontal parallel closely separated plates[23], a system know as a Hele-Shaw cell. When pressure $P_{applied}$ is applied to the less viscous fluid, fingers of the less viscous fluid protrude into the more viscous fluid. This is well known to occur during oil recovery from a porous medium, where D'Arcy's law flow is equivalent to the frictional forces exerted by the plates in a Hele-Shaw cell, and the two fluids are oil and water. In the classic investigations of Saffman and Taylor[6], the initial interface is perpendicular to the walls of a long channel, and the finger which develops propagates at constant velocity and without change of shape, with a well-defined, reproducible finger width to channel width ration, λ, which depends on the viscosity ratio of the two fluids, and the surface tension γ between the fluids. For small surface tension, and when the viscosity of the driven fluid is negligible, λ is found experimentally to be $1/2$. For zero surface tension there is a family of steady state fingers, parametrised by λ. When surface tension is included[24] the resultant perturbation theory does not seem to select a value for λ; however, numerical solutions of the exact steady-state equations do indeed select the correct λ. This situation is, of course, rather reminiscent of the conundrum of pattern selection in dendrites. The stability of the Saffman-Taylor fingers is currently a subject of research, so I shall merely remark that it seems probable[25] that McLean and Saffman's stability analysis, which claimed that the fingers are unstable even with surface tension, is in error.

Why is there a similarity between dendritic solidification and viscous fingering in two dimensions? The two-dimensional velocity field v in the viscous fluid and at the boundary, averaged across the gap, satisfies

$$v = -\frac{b^2}{12\eta} \nabla P \tag{V-1}$$

where b is the plate thickness, P is the pressure in the viscous fluid and η its viscosity. Here and below we shall neglect the viscosity of the driven fluid. The assumed incompressiblity of the viscous fluid implies

$$\nabla^2 P = 0 . \tag{V-2}$$

The boundary condition on the interface between the fluids is that the pressure at the boundary of the fliuds P_s is given by

$$P_s = P_{applied} - \frac{\gamma b^2}{12\eta} \kappa \quad . \tag{V-3}$$

Thus we see that the viscous fingering problem is described by a free boundary problem akin to that of solidification with pressure playing the role of temperature; the difference is that in the Hele-Shaw cell the pressure adjusts instantaneously everywhere to the boundary conditions. Thus there is no memory effect in the Hele-Shaw problem, which makes it much more tractable.

In the experiments performed at the University of Michigan[26], we have demonstrated the effects of anisotropy by engraving a grid on one of the confining plates of the Hele-Shaw cell. This imposes anisotropy through the plate spacing b. We performed experiments in two different geometries, each time comparing the results with and without the imposed anisotropy.

In the first set of experiments, air is pumped through a small orifice in the centre of one of the plates; the plates are filled with glycerine. Without anisotropy, a growing circle of air develops initially, but becomes unstable and develops fingers which grow by tip splitting. In the anisotropic case, the initial circle becomes unstable in the crystallographically preferred directions of the grid and grows a dendritic pattern. In the latter case, the driving pressure has to be greater than a threshold, otherwise tip splitting still occurs.

The second set of experiments are a modification of the Hele-Shaw cell. Instead of applying pressure to the air-glycerine interface, we slowly lift one edge of the upper confining plate at a uniform rate[26,27]. Thus the interface is drawn at a uniform rate through the system. Without anisotropy, the interface develops an array of Saffman-Taylor fingers which are smooth. With anisotropy, an array of dendrites is formed. This situation can be shown to be an analogue for directional solidification[5], the process in which a binary alloy is drawn at constant speed through a strong external temperature gradient such that the alloy is molten on one side of the system, and solid on the other side.

The conclusion which emerges from these experiments is that anisotropy is indeed crucial in real systems for dendritic pattern formation.

VI. Tip Selection and Stability of Needle-Crystals

Having digressed somewhat from dendritic solidification, I will return in this section to the subject of section III: what is the speed and shape of a dendrite? To answer this question[28,29], we have to reexamine the needle-crystal solutions which we mentioned in section III. Recall that a needle crystal is a smooth uniformly translating solution of the steady state equations, such as a parabola in the case of no surface tension. In the BLM and GM, the construction of the needle-crystals is rather easy. The kinematic equations simply give the geometrically obvious fact that

$$\nu_n = v_0 \cos\theta \tag{VI-1}$$

where ν_0 is the tip velocity. The dynamical equations for the h variable have zero time-derivative in the moving frame of the tip. The transformation of the moving frame (i.e. derivatives evaluated at fixed arclength s) is accomplished using the identity

$$\frac{\partial}{\partial t}\Big|_n = \frac{\partial}{\partial t}\Big|_s + \nu_0 \sin\theta \frac{\partial}{\partial s} . \tag{VI-2}$$

The steady state equation for h can then be written in the form of three first order autonomous differential equations:

$$\frac{d\theta}{ds} = (1 - w - \beta\nu_0\cos\theta)/\Delta^2 ,$$

$$\frac{dw}{ds} = \lambda , \tag{VI-3}$$

$$\frac{d\lambda}{ds} = \mathrm{mess}(w,\lambda,\theta,\Delta,\varepsilon;\nu_0)$$

where $w = u_s/\Delta$, $\beta = \varepsilon\Delta^4(1 - \cos(m\theta))$ is the anisotropic kinetic coefficient and we will for definiteness take $m = 4$ corresponding to a four-fold symmetric crystal, such as succinonitrile. The explicit form[28] of the function 'mess' does not matter for present purposes. Note that once the material parameters have been specified, the only variable parameter in these equations is the tip velocity ν_0. These equations have relevant fixed points at $\theta = \pm\pi/2$, $\lambda = 0$, $w = 1$, corresponding to the tail of the needle crystal. It is important to realize that the tip of the needle crystal is not a fixed point. Nevertheless, we will still wish to impose the constraint that the solution of the equation is reflection symmetrical about the tip, in the absence of any symmetry breaking terms in the equation or boundary conditions or initial conditions. In Fig. 5 is sketched the phase portrait of the steady state equations. The trajectory which corresponds to a needle crystal connects both fixed points and passes through the $\lambda - \theta$ plane at $\lambda = 0$. It can be shown that there is only a single trajectory which enters or leaves each fixed point, as the flow field in the phase space is highly singular near the fixed points[20,28]. In general, the trajectory exiting from one of the fixed points will fail to intersect the $\lambda - \theta$ plane and enter the other fixed point. Only for special values ν_0^* can all the boundary conditions be satisfied. Thus the velocity of needle crystals in the BLM satisfies a nonlinear eigenvalue problem, and has a discrete spectrum, not a continuous spectrum as assumed in section III for the FBP. In Fig. 6 is plotted $\lambda|_{\theta=0}$ vs. ν_0 for a particular set of parameters in the BLM. The zeros of this function occur at the eigenvalues ν_0^*. At least two are clearly identified, and there are indications of a possibly infinite set becoming more closely as $\nu_0 \to 0$.

The surprising and most important feature of Fig. 6 is the existence of a maximum eigen-velocity whose value coincides exactly with the dynamically selected tip velocity seen in the fully time-dependent simulations of the BLM. The tip curvature and shape near the tip of the needle crystal also match the tip of the dynamically selected dendrite as shown in Fig. 7. It thus appears that the selection in dendrites is independent of stability; all one needs to predict the tip are the steady state solutions and the maximum-velocity criterion. Note, however, that having found the steady state solution, we do still have to investigate its stability in order to determine to which growth mode it corresponds. The relation between stability and whether or not dendrites or tip-splitting occurs is discussed below. The same scenario also occurs in the GM[29]. Thus, at least in the GM and GLM, the marginal stability hypothesis is not the correct selection criterion.

The origin of the discreteness in the velocity spectrum is that inclusion of surface tension in the BLM equations is a singular perturbation. Setting $\beta = 0$ in eq. (IV-8) and substituting into the equation for h, eq. (IV-5), we see that the diffusion term is

$$-Dd_0 \frac{\partial}{\partial s} \left(\ell \frac{\partial \kappa}{\partial s} \right)$$

so that the surface tension is the coefficient of the highest derivative in the equation. Thus it is not correct to treat it as a regular perturbation as in section III for the FBP. In the FBP, surface tension is also a singular perturbation, although this is not a priori obvious in this formulation. Preliminary results of a careful study of the steady state equations for the FBP seem to indicate a discrete velocity spectrum in this case too[30]. Finally, in the Saffman-Taylor problem, it is known that there is only a discrete velocity spectrum for the steady states, with the largest velocity corresponding to the selected finger[24,31].

Having constructed steady state solutions of the BLM we must now discuss their stability. We have been able to perform a limited form of stability analysis by studying the stability eigenfunctions near the tail of the needle crystal[28]. We write

$$\begin{aligned} k(s,t) &= \kappa_0(s)\,\hat{k}(s,t) \,, \\ h(s,t) &= h_0(s) + \hat{h}(s,t) \end{aligned} \qquad \text{(VI-4)}$$

where the subscript zero denotes the steady-state solution and the caret denotes the perturbation about the steady state. We linearize in the perturbations and assume that as $s \to \infty$

$$\begin{aligned} \hat{k} &\sim s^{\alpha} e^{iqs+\omega t} \,, \\ \hat{h} &\sim s^{\beta} e^{iqs+\omega t} \end{aligned} \qquad \text{(VI-5)}$$

where α and β are to be determined. Denoting the column matrix $(\hat{k},\hat{h})$ by X we find that the linear stability operator is of the form:

$$\omega X = \left[A \frac{\partial^2}{\partial s^2} + B \frac{\partial}{\partial s} + C\right] X \quad . \qquad \text{(VI-6)}$$

Evaluating A, B and C in the limit $s \to \infty$ and demanding that they remain bounded implies

$$\beta - \alpha = 1. \tag{VI-7}$$

Finally we solve for ω to find

$$\omega = -iqv_0. \tag{VI-8}$$

Thus it appears that the continuous spectrum of the needle crystals has zero real part for all needle crystals. The remainder of the spectrum is presumably a set of discrete modes localized near the tip. We have not been able to calculate these analytically or in a reliable fashion numerically. Below I shall offer some speculations which I will support with the results of the stability calculation for the GM.

A preliminary observation is that the dendrite can be considered to be a limit cycle if we neglect the late-stage coarsening effects away from the tip. That is to say, if two photographs of the region near tip are taken an appropriate time interval apart, then they will appear identical. This seems to be approximately true in experiments. How can we understand the fact that the tip of the dendrite is the tip of the fastest needle crystal? At first blush it would seem that the dendrite, with its sidebranches along the body, has completely different properties for large s to the needle crystal.

Let us denote the curvature of the needle crystal with largest velocity by $\kappa^*(s)$, and that of the dynamically selected dendrite by $\tilde{\kappa}(s,t)$. We can consider these functions as points in a function space defined rather loosely as follows. Take a non-denumerable infinity of orthogonal unit vectors which span a space, and label each vector by a real number. Construct the vector whose components along these axes are the value of the curvature at the arclength corresponding to the label of the axis. The function $\kappa^*(s)$ is a point in this space, and the function $\tilde{\kappa}(s,t)$ is a closed curve in this space, but lies close to κ^* in an appropriate sense. The closeness reflects the fact that the tips of the dendrite and the needle crystal are very similar for a non-zero region of arclength. The simplest hypothesis for the dynamical origin of this situation is that the limit cycle was produced via a Hopf bifurcation of the needle crystal. That is to say, as some parameter varies (which we anticipate to be the anisotropy strength), the needle crystal undergoes a transition in which only a discrete mode of the spectrum becomes unstable. If this mode is actually a complex conjugate pair of modes, then a limit cycle will be generated. This discrete complex mode will presumably control the sidebranching period.

Let Ω be the real part of the complex conjugate pair of modes which are localized near the tip. If Ω is negative, then the needle crystal is stable _in the moving frame_. This means that side branches may still be formed during the time-dependent evolution of the needle crystal, but the position along the dendrite where the sidebranches begin to have appreciable amplitude will not remain at a constant distance from the tip, but will move further and further back down the dendrite. This is the sense in which a needle crystal is said to be stable. If Ω is positive, the needle crystal is unstable towards a solution which is periodic in time and everywhere different from the needle crystal. However, the periodic solution may execute oscillations about the needle crystal. This situation corresponds to tip-splitting. If $\Omega = 0$, the needle crystal is not actually unstable towards the periodic solution but coexists with it. Formally, trajectories in the function space will diverge away from the needle crystal algebraically rather than exponentially. In this case, we would not expect tip splitting, but would expect a periodic oscillation superimposed on the needle crystal. This would correspond to dendritic growth. In this picture, dendritic growth can only occur when the discrete mode responsible for the sidebranching is marginally stable.

Note that the use of this term here is very different to the use in Section III. There marginal stability selected a particular needle crystal velocity. The present usage is more tautological, and is not capable of predicting the growth rate and shape of a dendrite. If this picture is correct, then we need to understand how the dendrite can exist over a non-zero range of the parameter controlling the bifurcation.

A concrete but partial realization of this scenario is provided by the GM[29]. There, the needle crystal with the largest velocity does indeed exhibit a discrete pair of complex conjugate modes localized near the tip. As the anisotropy strength is increased from zero, the real part of this mode, Ω, decreases, becomes zero at a particular anisotropy strength, and is negative for greater anisotropy strength. This corresponds to the needle crystal making a transition from tip-splitting behaviour to stable needle crystal behaviour via a dendrite. This only exists for one value of the anisotropy strength, when the real part of the discrete mode is zero. Presumably, in the FBP or at least real life, as the control parameter (anisotropy strength?) is varied, Ω remains zero over a non-zero range of the control parameter. It seems that the GM is not rich enough to exhibit dendritic behaviour over a non-zero range of the anisotropy strength. This may also turn out to be the case in the BLM, in which case the observed sidebranches must be transients. A firm resolution of this question will be forthcoming in the near future when the stability analysis is complete.

To finish this section, I would like to remark that I do not see a conflict between the new selection criterion and experiment. In my opinion, the presently available experiments are not capable of testing the selection hypothesis because they are performed at too low an undercooling. I refer the interested reader to Fig. 11 on page 14 of the review article by Langer, for example. The best experiments seem to have been performed in a regime where the tip of the dendrite matches very well with the Ivantsov needle crystal. In this regime, because the diffusion length is so much larger than the radius of curvature of the tip, there is only one dimensionless group of parameters which can control the operating point, namely σ. A good but not rigorous estimate of σ, such as is obtained from the circular approximation, is thus almost certain to enable good predictions to be made, given the proximity of the operating point to the Ivantsov needle crystal. Secondly, the marginal stability calculations which have been compared to experiment did not take into account crystalline anisotropy, so their success can only be taken as fortuitous. Indeed, the same calculation which has been used successfully by metallurgists to explain experimental data on dendrites gives only a ballpark figure when applied to the time-dependent simulations of the BLM.

When the results of the steady state-analysis for the FBP are available, and experiments with well characterized anisotropy are performed, I will be quite surprised if the new selection criterion does not work well!

VII. Outstanding Problems

Let me finish by posing a few outstanding problems which may be of interest to mathematicians. I will not set out the problems in any concrete or well-posed way, mainly because I am unable to do so, and also because I am not encouraged to do so by the title of this conference. Nevertheless, I hope that this talk will stimulate some people to think about these questions, chase the references and solve some of the following problems.

(1) Prediction of the largest eigenvalue of velocity given Δ, m, ε. How does it scale with Δ? How does the scaling change if the problem is formulated in three dimensions rather than two?

(2) How to provide a systematic framework for non-Hermitian stability eigenvalue problems. What do you do about boundary conditions?

(3) Hopf bifurcation in PDE's. Are the ideas in section VI halfway sensible? If so, what is the mathematical mechanism for Ω not to change monotonically as the control parameter is varied? If not, are real dendrites long-lived transients?

(4) Formal derivation of BLM equations. Can one consistently make a boundary layer approximation, or does the system alway drive itself into a regime where BLM equations are invalid? I.e., do all BLM's develop grooves in which the boundary layers on opposite sides of the groove intersect?

(5) Minimum anisotropy. In GM is there a minimum ε below which needle crystals do not exist? What about the BLM? Is the critical anisotropy which controls the sidebranching mode different from the critical anisotropy for the existence of needle crystals? What is the behaviour in the intermediate regime?

Acknowledgements

It is a pleasure to acknowledge my collaborators on most of the work I have presented: E. Ben-Jacob, G. Kotliar, J. Langer and G. Schon. The work on viscous fingering was in collaboration with E. Ben-Jacob, R. Godbey, J. Koplik, H. Levine, T. Mueller and L. Sander. I would like to thank the organizers of the conference "Metastability and Incompletely Posed Problems" and the Institute for Mathematics and its Applications at the University of Minnesota for inviting me to present this paper. I acknowledge the financial support of the Department of Energy grant DE-FG0384ER45108, National Science Foundation Grant No. PHY 82-17853, supplemented by funds from the National Aeronautics and Space Administration.

References

1. J. Buckmaster and G.S.S. Ludford, Theory of Laminar Flames (Cambridge University Press, New York 1982).
2. G. Ahlers and D. Cannell, Phys. Rev. Lett. **50**, 1583 (1983).
3. R.A. Fisher, Ann. Eugenics **7**, 355 (1937); A. Kolmogorov, I. Petrovsky, N. Piscunov, Bulletin Mathematique de Universite d'Etat a Moscou, Section A: Mathematiques et mechanique **1**, 1 (1937).
4. See, for example, J.P. Pauwelussen, Physica (Utrecht) **4D**, 67 (1981) and references therein.
5. J.S. Langer, Rev. Mod. Phys. **52**, 1 (1980).
6. P.G. Saffman and G.I. Taylor, Proc. Roy. Soc. A **245**, 312 (1958).
7. G. Dee and J.S. Langer, Phys. Rev. Lett. **50**, 383 (1983). E. Ben-Jacob, H.R. Brand and L. Kramer, unpublished.
8. J.S. Langer and H. Muller-Krumbhaar, Phys. Rev. A **27**, 499 (1983).
9. S.C. Huang and M.E. Glicksman, Acta Metall. **29**, 701 (1981); **29**, 717 (1981).
10. B. Chalmers, Principles of Solidification, (Wiley, New York, 1964); D.P. Woodruff, The Solid-Liquid Interface (Cambridge University Press, Cambridge, England 1973).
11. W.W. Mullins and R.F. Sekerka, J. Appl. Phys. **34**, 323 (1963).
12. G.P. Ivantsov, Dokl. Akad. Nauk SSSR **58**, 567 (1947).

13. J.S. Langer and H. Muller-Krumbhaar, Acta Metall. **26**, 1681; 1689; 1697; ibid. **29**, 145 (1981).

14. W. Oldfield, Mater. Sci. Eng. **11**, 211 (1973).

15. J. Swift and P.C. Hohenberg, Phys. Rev. A **15**, 319 (1977).

16. J.B. Smith, J. Comp. Phys. **39**, 112 (1981). See also Ref. 14 for an early attempt to solve numerically the FBP. Judging from the graphs, it seems likely that anisotropy is being introduced through the grid in the simulations. Oldfield has produced a film of the simulation also.

17. E. Ben-Jacob, N.D. Goldenfeld, J.S. Langer, G. Schon Phys. Rev. Lett. **51**, 1930 (1983).

18. R. Brower, D. Kessler, J. Koplik, H. Levine, Phys. Rev. Lett. **51**, 1111 (1983).

19. N.D. Goldenfeld, unpublished.

20. E. Ben-Jacob, N.D. Goldenfeld, J.S. Langer, G. Schon, Phys. Rev. **29**, 330 (1984).

21. J.B. Collins and H. Levine, Phys. Rev. B **31**, 6119 (1985); J.M. Deutsch, unpublished.

22. R.C. Brower, D. Kessler, J. Koplik, H. Levine, Phys. Rev. A **29**, 1335 (1984); D. Kessler J. Koplik, H. Levine, ibid. **30**, 3161 (1984).

23. H.S.S. Hele-Shaw, Nature **58**, 34 (1898).

24. J.W. McLean and P.G. Saffman, J. Fluid Mech. **102**, 455 (1981).

25. D. Kessler and H. Levine, submitted to Phys. Rev. lett.

26. E. Ben-Jacob, R. Godbey, N.D. Goldenfeld, J. Koplik, H. Levine, T. Mueller, L.M. Sander, Phys. Rev. Lett. 55, 1315 (1985).

27. An equivalent experimental arrangement is described in R.J. Fields and M.F. Ashby, Phil. Mag. 33, 33 (1976).

28. E. Ben-Jacob, N.D. Goldenfeld, B.G. Kotliar, J.S. Langer, Phys. Rev. Lett. 53, 2110 (1984).

29. D. Kessler, J. Koplik, H. Levine, Phys. Rev. A **31**, 1712 (1985).

30. D. Meiron, private communication.

31. J. Vanden-Broeck, Phys. Fluids **26**, 2033 (1983).

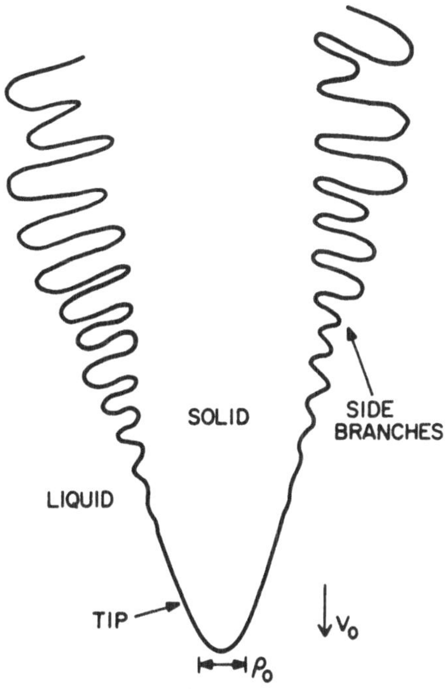

Figure 1: Artist's impression of a dendrite

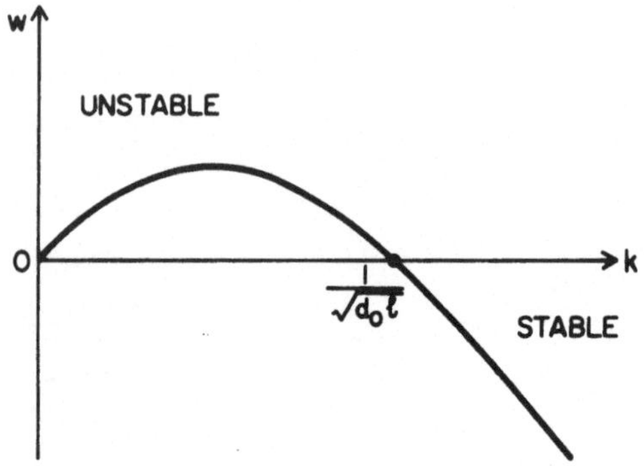

Figure 2: Form of dispersion relation for a planar interface at unit undercooling.

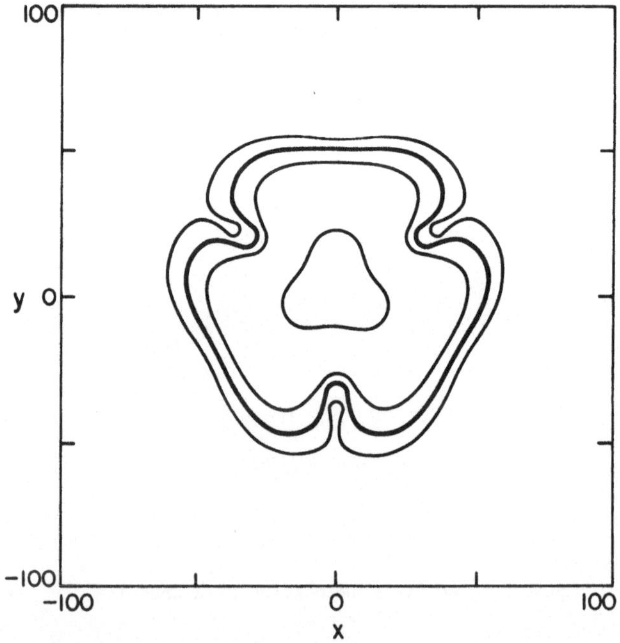

Figure 3: Time-ordered sequence of the evolution of an interface in the limit of very small Δ, with three-fold symmetry. Anisotropy is not included.

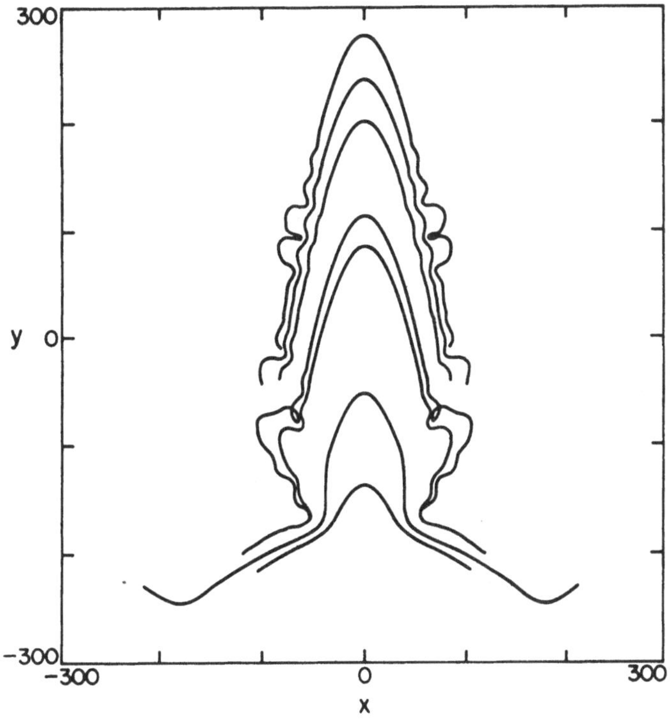

Figure 4: Time-ordered sequence of the evolution of an interface at $\Delta = 0.8$, $\varepsilon = 0.1$, $m = 6$. The interface was initially a circle and is shown at times 3500, 3750, 4100, 4175, 4400, 4500, 4600. Time is in units of $d_0^2/D\Delta^8$ and space is in units of d_0/Δ^3.

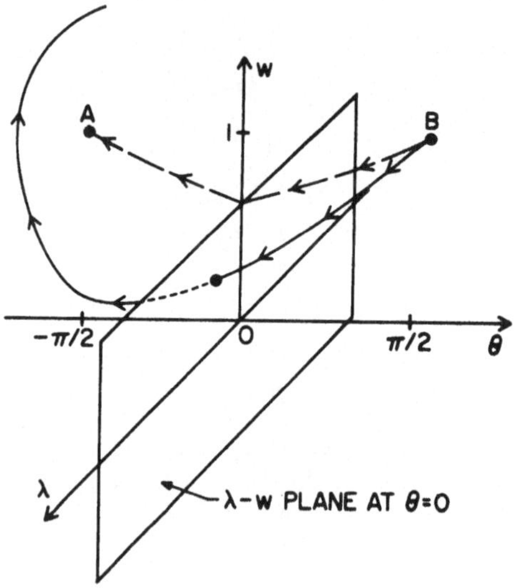

Figure 5: Qualitative phase portrait of the steady state BLM equations. The dashed curve corresponds to a need crystal and connects the fixed points A and B. The full curve is the trajectory for a generic tip velocity.

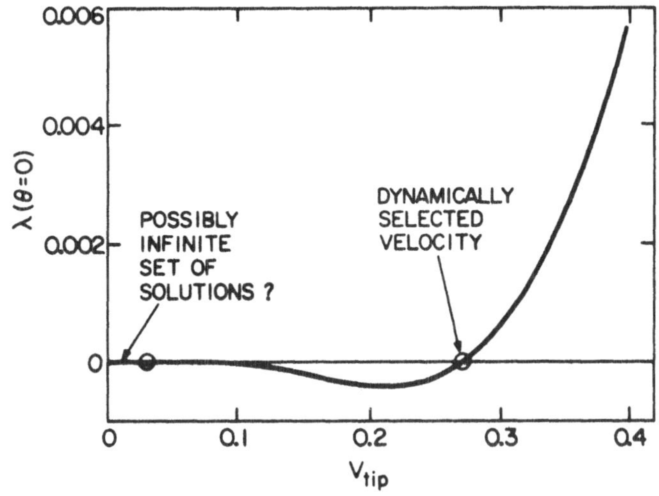

Figure 6: Graph of λ vs. ν_0 for $\Delta = 0.75$, $\varepsilon = 0.1$, $m = 4$.

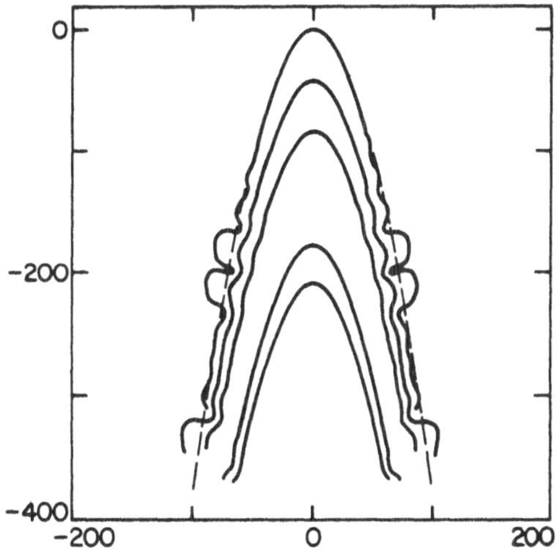

Figure 7: The dynamical solution of Fig. 4 (full curve) compared with the maximal velocity needle crystal (dashed curved).

SOME RESULTS AND CONJECTURES IN THE GRADIENT THEORY OF PHASE TRANSITIONS

Morton E. Gurtin

Department of Mathematics
Carnegie-Mellon University
Pittsburgh, PA 15213

Abstract

In the van der Waals-Cahn-Hilliard theory of phase transitions the energy depends not only on the density, but also on the <u>density gradient</u>, a dependence introduced to account for the interface between phases. Within this theory the stable density-distributions $u(x)$ for a fluid confined to a container Ω are characterized by the variational problem: (P_h) minimize

$$E_h(u) = \int_\Omega \{ W(u(x)) + h^2|\mathrm{grad}u(x)|^2\}dx$$

subject to the constraint

$$\int_\Omega u(x)dx = m.$$

Here $W(u)$ is the coarse-grain energy, assumed nonconvex, h is a constant which characterizes capillary effects, and m is the total fluid mass.

In this paper we discuss recent results and conjectures for problem P_h; in particular, those relating P_h to the classical problem with $h = 0$. We also discuss a generalization of P_h which includes contact energy between the fluid and the container walls.

1. Classical Theory

Consider a fluid confined to a container which occupies a bounded, open region Ω in R^n. Assume that the energy $W(u)$, per unit volume, is a smooth function of the density u, and that the total mass of fluid in Ω is m, so that the admissible density-distributions satisfy the **constraint**

$$\int_\Omega u(x)dx = m. \tag{1.1}$$

Then, if there are no other contributions to the energy, the total energy $E_0(u)$

in any distribution $u(x)$ is given by the functional

$$E_0(u) = \int_\Omega W(u(x))dx. \tag{1.2}$$

We seek those density distributions that render the body stable in the sense of Gibbs and hence seek solutions of the variational problem:

(P_0) minimize $E_0(u)$, subject to (1.1), over all u with both u and $W(u)$ in $L^1(\Omega)$.

For this and other variational problems, we will always use the term **solution** to mean global minimizer.

We assume that W is nonconvex, of a form capable of supporting two phases. Precisely, we assume that W consists of two convex sections separated by a concave segment (cf. Figure 1). We also assume that the two minima u_1 and u_2 have equal energy with

$$W(u_1) = W(u_2) = 0. \tag{1.3}$$

The assumption (1.3) involves no loss of generality; indeed, because of the constraint we can always add an affine function of u to the integrand in (1.2) without changing the solution set of P_0.

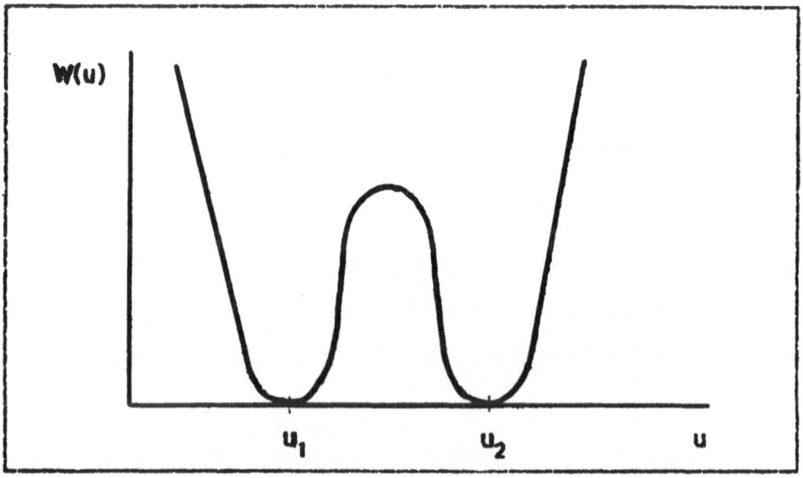

Figure 1. Coarse-grain energy W(u) as a function of density u

Problem P_0 is easily solved. Choose length scale so that[1]

$$\mathrm{vol}\ (\Omega) = 1,$$

and define

$$V = (u_2-m)/(u_2-u_1). \tag{1.4}$$

Then:

(i) for $m < u_1$ or $m > u_2$ the solution u_0 of P_0 is single-phase with $u_0(x) \equiv m$;

(ii) for $u_1 < m < u_2$ all solutions of P_0 are two-phase, the solution set consisting of fields of the form

$$\begin{aligned} u_0(x) &= u_1, \quad x \text{ in } F, \\ u_0(x) &= u_2, \quad x \text{ in } \Omega\backslash F, \end{aligned} \tag{1.5}$$

where F is any (measurable) subset of Ω with volume

$$\mathrm{vol}(F) = V. \tag{1.6}$$

A problem with the two-phase solution (ii) is the drastic <u>lack of uniqueness</u>, as modulo the volume constraint (1.6) the set F is completely <u>arbitrary</u>. This lack of uniqueness arises because <u>interfaces</u> (jumps in u) are allowed to form without a concomitant increase in energy.

One might ask: Which of the infinity of solutions (1.5) are <u>physically preferred</u>? If the physically-preferred solutions are those that arise as limiting cases within a theory which includes interfacial energy, then one might expect the preferred solutions u_0 to be those which minimize the "area"

$$a(u_0) = \mathrm{area}(\Sigma) \tag{1.7}$$

of the interface

$$\Sigma = \partial F \cap \Omega$$

---

[1] We write "vol" and "area", respectively, for n- and $(n-1)$-dimensional Hausdorff measure.

(cf. (1.5) and recall that Ω is open). We are therefore led to the following

Definition. A two-phase solution u_0 of P_0 has **minimal interface** if

$$a(u_0) < a(u)$$

for any solution u (corresponding to the same value of m).

The variational problem associated with this definition, namely finding a subset F of Ω that minimizes area(Σ) subject to vol(F) = V, has a large literature.[1] There is existence, but not uniqueness, and solutions are analytic, at least for $n < 8$. In fact, solutions are surfaces of constant mean-curvature.

2. The Gradient Theory

In a paper [4,5][2] now classic, van der Waals considered fluids whose energy is determined not only by the density, but also by the density gradient[3]. A simple but physically reasonable extension of the classical theory, within van der Waals' framework, is based on the energy[4]

$$E_h(u) = \int_\Omega \{W(u(x)) + h^2|\text{grad}u(x)|^2\}dx. \tag{2.1}$$

Here $W(u)$, still assumed to be of the form shown in Figure 1, represents the coarse-grain energy, that is, the energy, per unit volume, when the density is uniform; while $h>0$ is a small parameter. Note that van der Waals' theory allows for interfacial energy - or more precisely for an increase in energy over regions in which the density undergoes rapid changes.

---

[1] Cf., e.g., Massari and Pepe (1); Giusti [2]; Gonzalez, Massari, and Tamanini [3].

[2] Cahn and Hilliard [6], apparently unaware of van der Waals' paper, rederived what is essentially van der Waal's theory and, using this theory obtained several important results concerning the interfacial energy between phases. Since then gradient theories have been used to analyze phase transitions, spinodal decomposition, and other physical phenomena (cf. [5,7] for selected references).

[3] A theory which directly penalizes a sharp interface is given by Gurtin [8].

[4] We use the following notation: grad is the gradient: Δ is the Laplacian; $\partial/\partial n$ is the outward normal derivative on $\partial\Omega$; $W'(u) = dW(u)/du$.

As before, we seek stable density distributions and hence consider the variational problem:

(P_h) minimize $E_h(u)$ over the set of all u in $H^1(\Omega)$ that satisfy the constraint (1.1).

The **Euler-Lagrange equation** and **natural boundary condition** for this problem are:

$$2h^2 \Delta u = W'(u) - \mu \quad \text{in} \quad \Omega, \tag{2.2}$$

$$\partial u / \partial n = 0 \quad \text{on} \quad \partial\Omega , \tag{2.3}$$

where μ (= constant) is the Lagrange multiplier corresponding to the constraint (1.1). We will refer to μ as the **chemical potential** of u. (This definition is standard for the limiting case $h = 0$.)

Note that for $m < u_1$ or $m > u_2$, Problem P_h has only the single-phase solution $u(x) \equiv m$; for that reason we henceforth restrict our discussion to

$$u_1 < m < u_2 .$$

For W sufficiently regular the direct method of the calculus of variations and elementary regularity theory lead to the conclusion that Problem P_h possesses a (not necessarily unique) solution,[1] so existence is not at issue here. The goal instead is to identify the minimizers of P_h, and, what is more important, to study the asymptotic behavior for small h.

3. Results.

Let $n = 1$ and let Ω be the interval (0,1). Then the solutions of P_0 with minimal interface are the two solutions involving a single transition between phases, namely, the function

$$\begin{aligned} u_0(x) &= u_1, \quad 0 < x < V, \\ u_0(x) &= u_2, \quad V < x < 1 \end{aligned} \tag{3.1}$$

[1] Cf. Morrey [9], Theorems 1.9.1 and 1.10.1.

and its reversal $u_0(1-x)$. (Here V is given by (1.4).) In view of our previous discussion (Section 1), we expect these solutions to be physically preferred in the sense that they, and only they, are limits of solutions u_h of P_h. The following results (a)-(d) of Carr, Gurtin, and Slemrod [10] show, among other things, that this is indeed the case:

(a) All local minimizers of P_h are strictly monotone.[1]

(b) For h small, P_h has exactly two solutions, and one is the reversal of the other.

(c) If u_h denotes the increasing solution, and if u_0 is defined by (3.1), then, for $x \neq V$, $u_h(x)$ approaches $u_0(x)$ as h approaches zero.[2]

(d) For h small,

$$E_h(u_h) = e_h + O(\exp\{-C/h\}), \quad e_h = Kh, \tag{3.2}$$

$$\mu_h = O(\exp\{-C/h\}),$$

where $C,K > 0$ are constants with K the integral of $2\sqrt{W}$ from u_1 to u_2, while μ_h is the chemical potential corresponding to u_h.

One possible definition of interfacial energy is the difference between the actual energy $E_h(u_h)$ and the energy $E_0(u_0)$ which neglects interfacial effects. Since our normalization (1.3) renders $E_0(u_0)$ zero, (3.2) allows us to interpret e_h as **interfacial energy**,[3] at least asymptotically.

For $n>1$, Gurtin and Matano [16] have obtained theorems analogous to (a), one of the simpler results being:

(e) For a (not necessarily circular) cylinder all local minimizers are monotone in the axial direction.

---

[1]Cf. Chafee [11], Casten and Holland [12] and Matano [13]; the first three authors prove that for $n = 1$ all unconstrained local minimizers are constant; Matano generalizes this result to arbitrary n, but convex Ω.

[2]Cf. also Novick-Cohen and Segel [14], Alikakos and Shaing [15].

[3] Cahn and Hilliard [6] show that e_h gives the interfacial energy exactly when Ω is the entire real line.

They also show that:

(f) For $\Omega = D \times G$ (or $\Omega = D$) with D convex and sufficiently small, the solutions of the Euler-Lagrange equation and natural boundary condition are constant on each cross-section parallel to D.

Questions.

(1) What can be said about <u>local</u> minimizers of P_h as h approaches zero.

(2) Suppose the term $h^2|\text{grad}u|^2$ in (2.1) is replaced by the more general[1] "regularization" $g(u,h\text{grad}u)$, $g(u,0) = 0$. To what extent does the asymptotic behavior (for h small) depend on g?

4. Conjectures for $n>1$.

Guided by the results discussed in Section 3, I recently[2] offered the conjectures (C1) - (C3) listed below.

(C1)[3] The limits, as h appoaches zero, of solutions of P_h are exactly the minimal interface solutions of P_0.

To state (C2) and (C3) succinctly:

(i) Let u_0 denote a minimal-interface solution of P_0, let $a(u_0)$ denote the interfacial area (cf. (1.7)), and, $\mathscr{k}$, the sum of principal curvatures of the associated interface, with $\mathscr{k}$ counted positive when the center of curvature lies toward the phase 1 region.

(ii) For each h, let u_h be a solution of P_h that converges to u_0 as h appoaches zero, and let μ_h denote the chemical potential of u_h.

(iii) Let e_h denote the (one-dimensional) interfacial energy (3.2). Then the remaining conjectures are, for h small:

---

[1] Cf. Maddocks and Parry [17], who introduce a regularization of this form for the unconstrained problem.

[2](C1) is contained in [18] and (C2) with (C1) were given at the American Mathematical Society meeting in Minneapolis in November 1984; (C3) was presented at the workshop on Metastability and Incompletely Posed Problems in Minneapolis in May, 1984. Subsequently, Kohn and Sternberg, and Modica, in private communications, have asserted proofs of (C1) and (C2) using as a basis work of Modica and Mortola (cf. [19]). The conjecture (C3) remains open.

[3]Cf. [8] for an analogous result within a theory that directly penalizes a sharp interface.

(C2)[1] $$E_h(u_h) \approx a(u_0)e_h.$$

(C3) $$\mu_h \approx -\kappa e_h/(u_2-u_1). \tag{4.1}$$

The formula (4.1), often referred to as the Gibbs-Thompson relation[2], asserts that the chemical potential differs from the coarse-grain chemical potential (μ=0) by an amount proportional to the mean curvature of the interface.

The motivation behind (C3) is contained in the following strictly formal argument (which hopefully might form the basis of a proof). Consider Ω in R^2. Then granted (C1), for small h the transition from u_1 to u_2 should occur over thin interfacial regions lying between concentric circular arcs. (In R^2 minimial interfaces are arcs of circles.) Near such a region, but away from $\partial\Omega$, the solution should be approximately cylindrically-symmetric. Assuming it is, then the Euler-Lagrange equation (2.2) becomes

$$\mu_h = W'(u) - 2h^2(u_{rr} + r^{-1}u_r), \tag{4.2}$$

where r denotes the radial coordinate and $u_r = \partial u/\partial r$, etc. Let $r = R$ denote the approximate location of the interfacial region. Away from the interface u should be approximately constant. Granted this, choose r_1, r_2 such that

$$u_r(r_i) \approx 0,\ u(r_i) \approx u_i,$$

and assume that $r_1 < R < r_2$. (The analysis for $r_2 < R < r_1$ is similar.) Then, if we multiply (4.2) by u_r and integrate from r_1 to r_2, we conclude, using (1.3), that

$$\mu_h(u_2-u_1) \approx -(2h^2/R)\|u_r\|^2,$$

with $\|\cdot\|$ the $L^2(r_1,r_2)$ norm. We estimate this norm using the corresponding estimate of [10] (p. 350) for Ω an interval:

---

[1]Van der Waals himself asserted that (cf. [5], p. 201): "It will not be without interest to show that the two apparently contradictory hypotheses lead to values of the same order of magnitude for the capillary tension and energy." (The contradictory hypotheses being the classical assumption of an abrupt interface and the smooth transition of van der Waals.)

[2]Cf. e.g. Mullins and Sekerka [20], eqt. (8).

$$\|u_r\|^2 \approx K/(2h).$$

Since $R = k^{-1}$, the last two estimates lead to the desired conclusion (4.1).

5. Inclusion of Boundary Energy.

Problem P_h neglects the contact energy between the fluid and the container walls. For this energy Cahn [21][1] proposes adding the boundary term

$$\int_{\partial\Omega} b(u), \tag{5.1}$$

with $b(z)$ the contact energy, per unit area, between the fluid and the vessel when the density is z. Without contact energy the interfacial energy between phases is[2] $O(h)$ for small h, and this suggests replacing b by hw with w fixed. We are therefore led to the functional

$$T_h(u) = \int_\Omega \{W(u) + h^2|\operatorname{grad} u|^2\} + h \int_{\partial\Omega} w(u), \tag{5.2}$$

and to the problem:

(PB_h) Minimize $T_h(u)$ over the class of all sufficiently regular density distributions u that satisfy the constraint (1.1).

This problem is completely open. One would generally not expect existence,[3] but it would be important to answer the following:[4]

Questions:

(1) What is the lower semi-continuous envelope of the functional T_h over an appropriate space of constrained density fields?

(2) In situations for which PB_h has a solution for all small h, do solutions have limits as h tends to zero, and if so do the limits satisfy an associated variational principle?

---

[1] For $n = 1$.

[2] Cf. (4.2) and Footnote 1 of Section 5.

[3] Cf. Gurtin [22] for a discussion of this issue within a slightly different theory.

[4] Buttazzo (private communication) has investigated (1) for problem PB_h with the term $h^2|\operatorname{grad} u|^2$ in (5.2) replaced by $h|\operatorname{grad} u|$.

I have a conjecture appropriate to Question 2. Given any subset F of Ω, let $A(F)$ and $A_1(F)$, respectively, denote the areas of the sets

$$\partial F \cap \Omega, \quad \partial F \cap \partial\Omega ,$$

let β be a given constant, let

$$G(F) = A(F) + \beta A_1(F),$$

and for each V, $0 < V < 1$, consider the variational problem:

(LD) Minimize $G(F)$ subject to the constraint $\mathrm{vol}(F) = V$.

Remark. LD, known in the literature as the **liquid-drop problem,**[1] may be stated physically as follows: determine the region F occupied by an incompressible liquid drop of volume V in a container Ω with

$$\beta = \frac{\text{contact energy between drop and container walls}}{\text{surface tension of the fluid}} .$$

Definition. Let u_0 be a two-phase solution of P_0 of the form (1.5). Then u_0 **solves the liquid-drop problem** if F is a solution of LD with V given by (1.4) and

$$\beta = \{w(u_1) - w(u_2)\}/K.$$

(Recall that, by (3.2), Kh is the asymptotic form of the interfacial energy between phases.)

Conjecture. Let W and w be such that PB_h has a solution for all small h. Then as h approaches zero the limits of solutions of PB_h are exactly those solutions of P_0 that solve the liquid-drop problem.

Remark. For $|\beta| > 1$, one can show, using results of Massari and Pepe [1], that the liquid-drop problem does not have a solution for general Ω and all

---

[1] Cf. e.g., Massari and Pepe [1], Giusti [2].

V, $0 < V < 1$. This leads me to conjecture that for

$$|w(u_1) - w(u_2)| > K$$

and h sufficiently small Problem PB_h does <u>not</u> have a solution for general Ω and all m.[1]

Acknowledgment. This work was supported by the Army Research Office and the National Science Foundation.

References

[1] Massari, U. and L. Pepe, Su di una impostazione parametrica del problema dei capillari. Ann. Univ. Ferrara **20**, 21-31 (1974).

[2] Giusti, E., The equilibrium configuration of liquid drops. J. reine angew. Math. **321**, 53-63 (1981).

[3] Gonzalez, E. Massari, U., and I. Tamanini, On the regularity of boundaries of sets minimizing perimeter with a volume constraint. Indiana Univ. Math. J. **32**, 25-37 (1983).

[4] van der Waals, J.D., The thermodynamic theory of capillarity under the hypothesis of a continuous variation of density (in Dutch). Verhandel. Konink. Akad. Weten. Amsterdam (Section 1) **1**, No. 8 (1893)..

[5] Rowlinson, J.S., Translation of [4]. J. Stat. Phys. **20**, 197-244 (1979).

[6] Cahn, J.W. and J.E. Hilliard, Free energy of a nonuniform system. I. Interfacial free energy. J. Chem. Phys. **28**, 258-267 (1958).

[7] Davis, H.T. and L.E. Scriven, Stress and structure in fluid interfaces. Advances Chem. Phys. **49**, 357-454 (1982).

[8] Gurtin, M.E., On a theory of phase transitions with interfacial energy. Arch. Rational Mech. Anal. **87**, 187-212 (1985).

[9] Morrey, C.B., <u>Multiple Integrals in the Calculus of Variations.</u> Springer-Verlag, Berlin (1966).

[10] Carr, J., Gurtin, M.E., and M. Slemrod, Structured phase transitions on a finite interval. Arch. Rational Mech. Anal. **86**, 317-351 (1984).

[11] Chafee, N., Asymptotic behavior for solutions of a one-dimensional parabolic equation with homogeneous Neumann boundary conditions. J. Diff. Eqts. 18, 111-134 (1975).

[12] Casten, R.G. and C.J. Holland, Instability results for reaction diffusion equations with Neumann boundary conditions. J. Diff. Eqts. **27**, 266-273 (1978).

[13] Matano, H., Asymptotic behavior and stability of solutions of semilinear diffusion equations. Publ. RIMS, Kyoto Univ. **15**, 401-454 (1979).

---

[2] Cf. [22].

[14] Novick-Cohen, A. and L.A. Segel, Nonlinear aspects of the Cahn-Hilliard equation. Physica **10D**, 278-298 (1984).

[15] Alikakos, N.D. and K.C. Shaing, On the singular limit for a class of problems modelling phase transitions. Forthcoming.

[16] Matano, H. and M.E. Gurtin, On the structure of equilibrium phase transitions within the gradient theory of fluids. Forthcoming.

[17] Maddocks, J.H. and G.P. Parry, A model for twinning. Inst. Math. Appl. Preprint No. 125, Univ. Minnesota (1985).

[18] Gurtin, M.E., Some remarks concerning phase transitions in R^n. Dept. Math., Carnegie-Mellon Univ. (1983).

[19] Modica, L., Gamma convergence to minimal surface problems and global solutions of $\Delta u = 2(u^3-u)$. Proceedings of the International Meeting on Recent Methods in Nonlinear Analysis. (ed. E. De Giorgi, E. Magenes, and U. Mosco) Pitagora Editrice, Bologna (1978).

[20] Mullins, W.W. and R.F. Sekerka, Morphological stability of a particle growing by diffusion or heat flow. J. Appl. Phys. **34**, 323-329 (1963).

[21] Cahn, J.W., Critical point wetting, J. Chem. Phys. **66**, 3667-3672 (1977).

[22] Gurtin, M.E., On phase transitions with bulk, interfacial, and boundary energy. Arch. Rational Mech. Anal. Forthcoming.

THE STABILITY AND METASTABILITY OF QUARTZ

R.D. James

Department of Aerospace Engineering and Mechanics
University of Minnesota
Minneapolis, MN 55455

1. Introduction

In phase transformations, metastability is associated with a transformation temperature on heating which differs from the transformation temperature on cooling. The phenomenon is especially prevalent in solids. From an intuitive point of view, the failure of the body to be "absolutely stable" is a consequence of the observation that at a given temperature and pressure, which would ordinarily cause the body to adopt a well-determined configuration, more than one configuration is observed. If we compare this behavior with, say, a minimum free energy calculation in the spirit of Gibbs, we find that only one configuration minimizes what is thought to be the appropriate free energy.

The intuitive background suggests a way to avoid this dilemma: restrict the class of competitors for the minimum by looking for relative minima of the free energy. This practice has been successful in the analysis of the stability of structures. With so-called material instabilities like phase transformations, workers disagree on what the class of competitors should be. As an extreme example of the lack of such restrictions, some workers ignore conditions (of compatibility) associated with the fact that when a volume element deforms, nearby volume elements also must deform; this approach often leads to the conclusion that a solid cannot support shear stress in equilibrium. In his treatment of solids in contact with fluids, Gibbs [1, eqns. (357)-(374)] enforces conditions of compatibility, but he permits competitors for the minimum to have an arrangement of phase boundaries which can only be obtained from the minimizer by allowing small amounts of mass to cross the phase boundary.

Gibbs' concept of metastability appears to rule out a kind of hysteresis which is observed often in diffusionless solid/solid phase transformations. That is, if we use constitutive equations for a solid of the same form as adopted by Gibbs (summarized by eqn. (355) of [1]), if we introduce mild assumptions on these equations which still allow shear stress, general expressions for the temperature-dependence of elastic moduli and strain-dependence of specific heats (consistent with the appropriate point group symmetry), general transformation strain, etc., and if we calculate metastable states in the sense of Gibbs, then one can effectively predict the absence of hysteresis. This conclusion seems highly insensitive to the details of the theory and to the precise way in which Gibbs' concept of metastability is interpreted. It is a consequence of certain necessary conditions in the calculus of variations. We note that such theories do not model what would be thought of as interfacial energy.

If we adopt a new concept of metastability which places sufficiently severe restrictions on the class of competitors, then configurations observed in experiment can be reconciled with theory. The difficulty with this approach is that a restrictive concept of metastability may easily give rise to a huge variety of metastable configurations in addition to the observed ones. Consequently, the theory loses the ability to predict the onset of transformation*. As far as I know, there does not exist a reasonable concept of metastability which embraces the variety of configurations observed in experiment, but which leads to an acceptable uniqueness theory.

These thoughts lead me to believe that we need new approaches to metastability. With this in mind, we examine here the behavior of two particular transfor-

---

* In some relatively careful experiments, there is so much scatter in measured transformation conditions that it may not be desirable to predict the onset of transformation.

mations in quartz, the α-β transformation and the transformation from one Dauphiné twin to another. Quartz being one of the most common materials, its appearance near transformation (van Tendeloo, van Landuyt and Amelinckx [2]) and on the effect of stress on transformation (Coe and Paterson [3]) have been the subject of extensive observation. Metastability is present in both transformations, but the hysteresis loops act very differently in the two transformations. On the theoretical side, a theory has been given for the Dauphiné twinning by Thomas and Wooster [4] (see also the remarks of Rivlin [5]); the theory predicts with remarkable precision the complicated arrangements of Dauphiné twins produced by specific nonhydrostatic and inhomogeneous stress fields below the α-β transformation temperature. Ericksen [6] gives a nonlinear generalization of this theory. James [7] gives constitutive equations for quartz which cover the behavior above and below the α-β transformation temperature; his equations reduce to those of Ericksen (which in turn linearize to the Thomas-Wooster theory) below the α-β transformation temperature. Promising but incomplete comparisons with the measurements of Coe and Paterson [3] on the effect of nonhydrostatic stress on the α-β transformation are available in reference [7]. All of these studies ignore metastability.

Our starting point is a new theory for the crystallography of complex lattices developed by Pitteri [8]*. This theory is well suited to the study of transformations. We work out the details for quartz and we infer a macroscopic theory, which provides insight into the origins of metastability.

2. Behavior of α- and β-Quartz

The stable phase of quartz at room temperature is α-quartz. If a crystal of α-quartz is heated to about 574°C, it transforms to β-quartz. The transformation involves a change of shape, volume and symmetry. The low temperature α-quartz (point group 32) is piezoelectric, while the more symmetrial β-quartz (point group 622) is not piezoelectric.

---

* The theory has also been developed by Ericksen [9] and Parry [10] and has roots in the early attempts to avoid the Cauchy relations.

The α-β phase transformation involves some hysteresis. The transformation temperature on heating differs from the transformation temperature on cooling by about 2-4°C. Pressure has a substantial effect on the transformation temperature, but almost no effect on the hysteresis. Thus, α-quartz can exist even up to 700°C if a large enough hydrostatic pressure is applied to the specimen. Curiously, the size of the hysteresis loops changes very little as the transformation temperature is changed by the application of pressure. Coe and Paterson [3] studied the effect of both hydrostatic and nonhydrostatic stress on the transformation; their findings are summarized by Figs. 1b and 1c. There is clearly a difference in the amount of hysteresis for different loading devices, but in each case the hysteresis is not sensitive to temperature.

As mentioned above, there is an abrupt change of shape at the α-β transformation. If we consider a reference configuration which corresponds to unstressed β-quartz at the transformation temperature (~ 574°C), the change of shape from this configuration expressed in an orthonormal basis $\{\underset{\sim}{e}_i\}$ with $\underset{\sim}{e}_3$ parallel to the optic axis is given by the transformation strain

$$\underset{\sim}{U}_t = \begin{pmatrix} .9973 & 0 & 0 \\ 0 & .9973 & 0 \\ 0 & 0 & .9988 \end{pmatrix} \qquad (2.1)$$

according to Berger et al. [11,12].

Another striking and well-known consequence of the α-β transformation is that it sometimes cracks the crystal, even if the crystal is unloaded as it passes through the transformation temperature. The cracking can be avoided by choosing a specimen of appropriate shape or by transforming at high pressure. The appearance of a cracked specimen is unusual; occasionally cracks pass all the way through the specimen and a piece breaks off, but much more often many little cracks form in the interior of the specimen and do not reach out to the boundary. These observations are based on our experience with clear "rock shop" specimens in a tube furnace with a long hot zone in which the temperature was very slowly cycled through the α-β transformation.

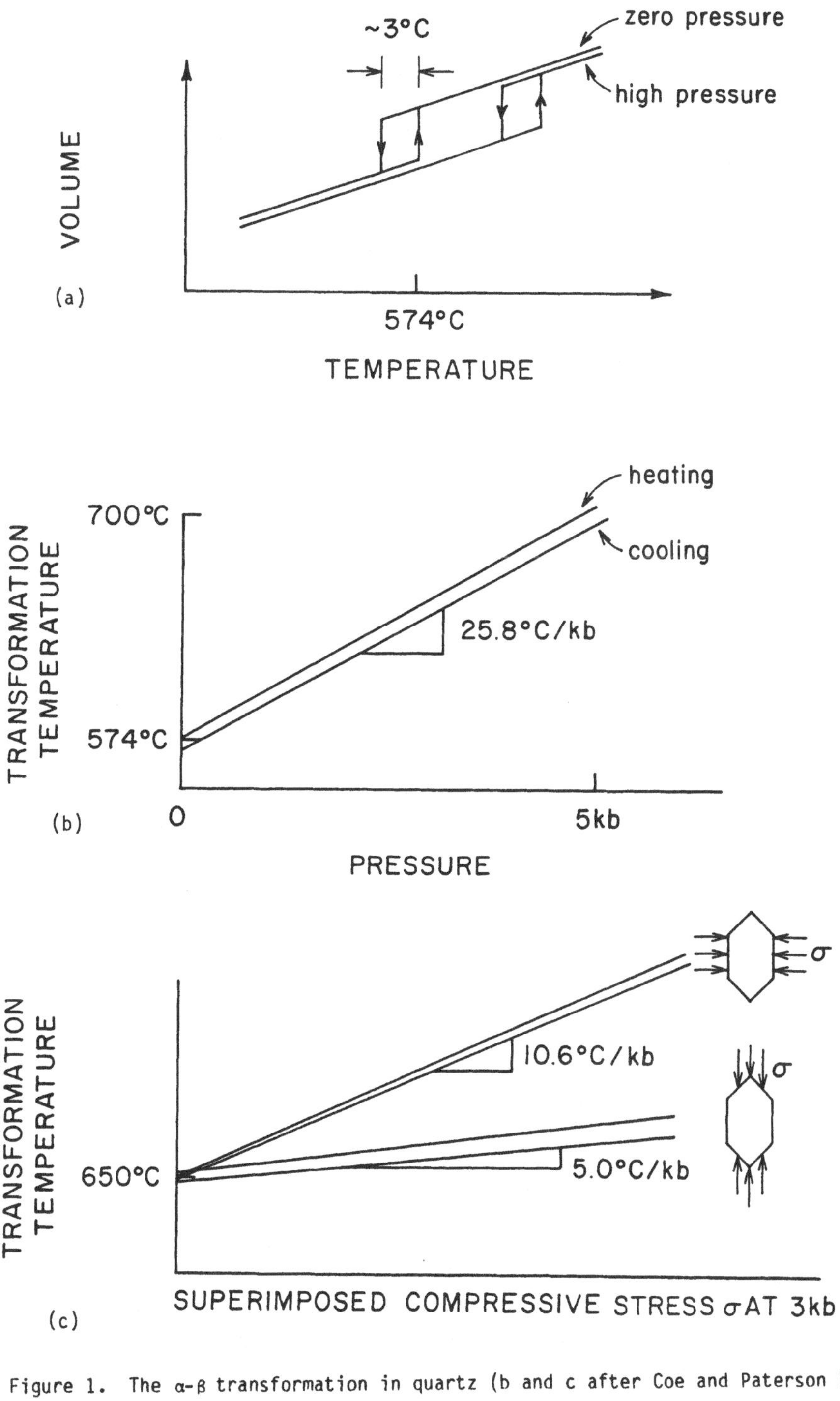

Figure 1. The α-β transformation in quartz (b and c after Coe and Paterson [3]).

α-quartz crystals found in nature almost always contain Dauphiné twins. A typical arrangement is to have a few irregular surfaces parallel to the optic axis (the axis of 3-fold symmetry) which separate the crystal into regions. Adjacent regions are Dauphiné twins. The boundaries between twins are transparent to the eye, but they can be revealed by etching a plane perpendicular to the optic axis; Fig. 2 shows how these boundaries might look after etching. They disappear upon heating through the α-β transformation. In an unstressed crystal, one variety of α-quartz becomes mechanically and piezoelectrically identical with its Dauphiné twin if it is rotated 180° about the optic axis. Wooster and Wooster [13] observed that a Dauphiné twin boundary can be caused to move by the application of a stress. Some particular shear stresses induced by the torsion of specially oriented plates of α-quartz are most effective in causing the Dauphiné twin boundaries to move, and Thomas and Wooster [4] presented a theory for why this is so. They showed that the twin boundaries could be removed by certain mechanical treatments, making the specimens useful for piezoelectric resonators, an important issue at that time.

Unlike the α-β transformation, the hysteresis associated with the conversion of one variety of α-quartz to its Dauphiné twin is extremely sensitive to temperature. As the temperature of a crystal is raised closer to the α-β transformation temperature, the Dauphiné twin boundaries become much more mobile. A few degrees below 574°C, the

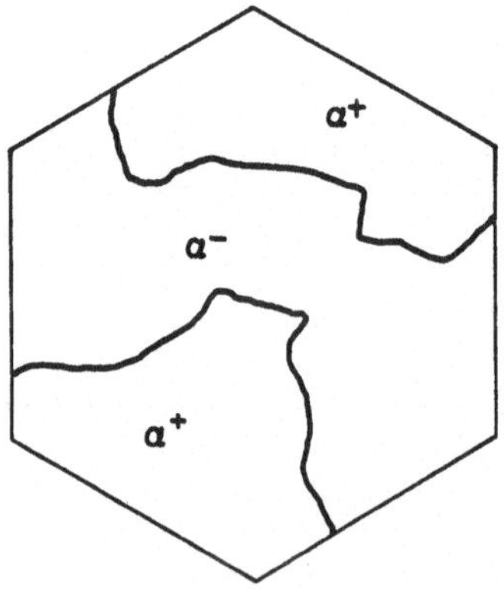

Figure 2. Dauphiné twin boundaries.

positions of the boundaries are highly sensitive to applied stresses and temperature gradients, and beautiful patterns of fluctuating triangular domains are observed in the electron microscope (see van Tendeloo, van Landuyt and Amelinckx [2]). At 300°C, the practical lower limit of the experiments of Thomas and Wooster, the Dauphiné twin boundaries are rather immobile although even at room temperature the severe stresses produced by a sharp indentation can cause Dauphine twinning.

3. Molecular Structure

The macroscopic observations described in the preceding section are known to arise from particular molecular changes. Fig. 3 shows the arrangement of silicon atoms in a perfect crystal of β-quartz. The oxygen atoms are not shown, and the hexagons and circles are there simply to help us explain the locations of the silicon atoms. This is the arrangement of silicon atoms one would see if one's line of sight were parallel to the 6-fold axis (the optic axis) of an unstressed quartz crystal in the high temperature β-phase.

Each of the circles is the projection on a plane perpendicular to the optic axis of a helix. The helices have parallel axes lying on the 3-fold junctions of the honeycomb and all have the same radius, pitch and sense. Each pair of adjacent helices touches at one point and therefore touches at a sequence of regularly spaced points, the distance between neighboring intersections being the pitch c. At each of these intersections lies a silicon atom. Actually, in a crystal at nonzero temperature the atoms are vibrating about these positions, so the atoms pictured represent "average" positions. Later, the mathematical apparatus introduced to describe these positions could be reinterpreted as a description of the periodicity of a probability distribution. In this way we could avoid speaking of atoms as stationary mass points, but we shall not do so.

To see that the conditions introduced so far are not contradictory, we focus attention on the solid atoms in Fig. 3. They all lie in a plane which can be thought of as the plane of the page. If we start at one solid atom and go in a counter-clockwise direction up the helix (I have fixed the sense to be like a right handed screw), we reach another intersection point which is 1/3 c above the page and 120° around the helix. Starting at this intersection point, we continue in a counter-clockwise direction up the same or the adjacent helix and reach an intersection point at 2/3 c above the page. Continuing in this fashion, we can find the positions of all the silicon atoms and these positons are compatible with the assumption that the solid atoms lie in a plane and the

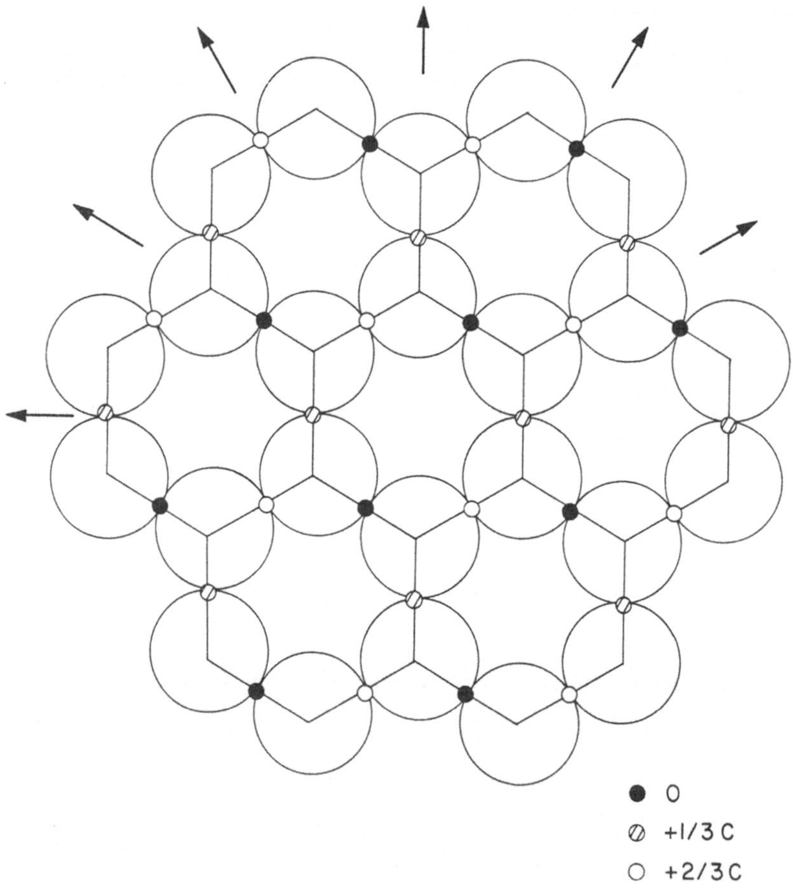

Figure 3. Molecular structure of β-quartz.

helices are parallel translates of one another. However, these assumptions do not restrict the radius or pitch of the helices.

Each silicon atom is surrounded by four oxygen atoms (not pictured) such that the oxygen atoms lie at the corners of a regular tetrahedron with the silicon atom at the center. Each oxygen atom lies near but not on a helix in a position which is equidistant from two adjacent silicon atoms on the same helix.

At about 574°C, the high temperature β form transforms to α-quartz, pictured in Fig. 4. The transformation involves small but co-operative movements of the silicon atoms and rotation and slight compression of the tetrahedron of oxygens. Each silicon moves to the nearby position shown in Fig. 4, but the silicons still lie on perfect helices which are translates of one another. However, in α-quartz the helices overlap a little bit, the same amount for each adjacent pair. We have exaggerated the amount of overlap to make the pictures clear. It is a matter of geometry that if two parallel helices (of the same radius, pitch and sense) intersect at one point, then they intersect at a sequence of regularly spaced pairs of points, spaced according to the scheme: $x^+ + nc$, $x^- + nc$, n = integer, c = pitch of the helices. In α-quartz, adjacent helices intersect and the silicon atoms on one helix lie at either the $x^+ + nc$ positions or the $x^- + nc$ positions but not at both. Furthermore, adjacent silicons on a helix are always 120° apart.

Given an array of these parallel, self-intersecting helices, there are exactly two ways to arrange the silicon atoms consistent with the restrictions given above, which give rise to the two Dauphiné twins. In fact, if we place one atom at the x^+ position, all the other atomic positions are uniquely determined by these conditions. If we place it at the x^- position, we get another set of atomic positions related to the first set by a rigid rotation of 60° about the optic axis and a translation. But this is not the only transformation which takes one configuration of α-quartz into its twin, because it can be composed with transformations which take a configuration of α-quartz into itself. For example, a rotation of 120° about the optic axis takes a configuration of α-quartz into a translate of itself, so a rotation of 120° + 60° = 180° takes one configuration into its twin, which justifies the use of the word twin.

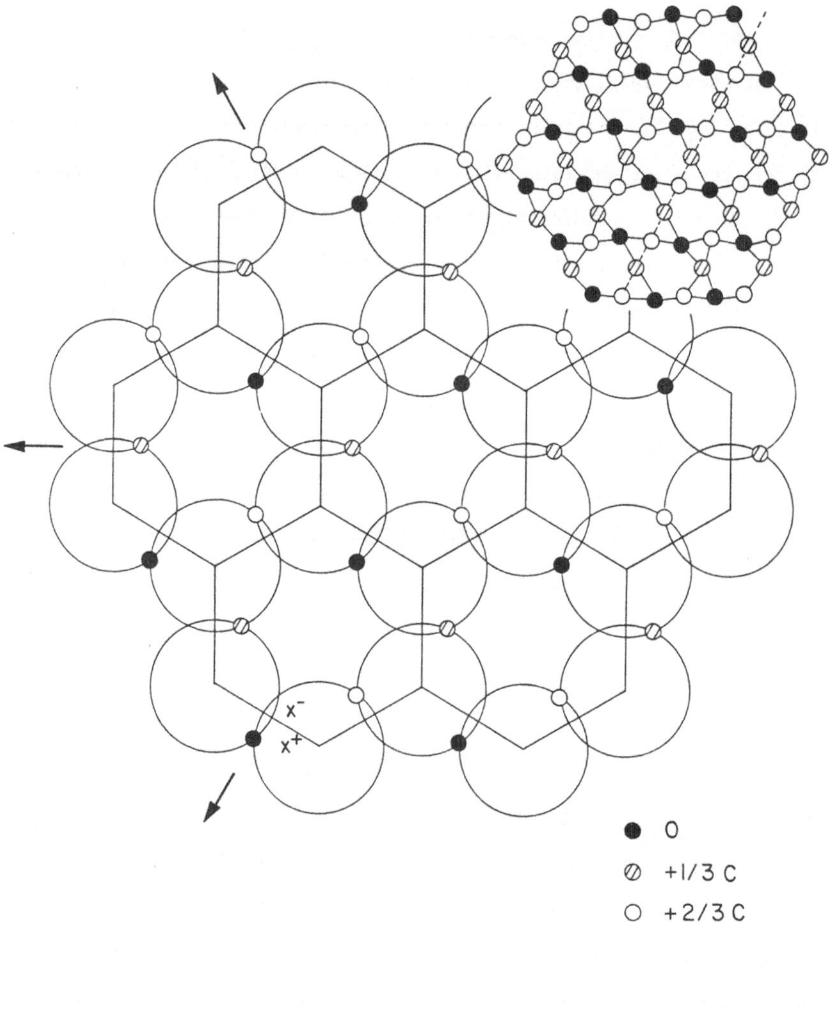

Figure 4. Molecular structure of α-quartz. The inset shows the approximate arrangement of silicon atoms near a Dauphiné twin boundary (dashed).

It can be seen from Fig. 3 that the rotations which map a configuration of β-quartz into a translate of itself are the 6, 3 or 2-fold rotations about the optic axis and the 2-fold rotations about the axes indicated by arrows. In α-quartz the 2-fold axes are indicated by arrows and the only other rotation is the 3-fold rotation about the optic axis. These sets of rotations form the point groups 622 (β-quartz) and 32 (α-quartz) respectively. It is evident that (622) = (32) × (any rotation mapping a configuration of α-quartz into its Dauphiné twin).

A piece of quartz found in nature usually contains several Dauphiné twins. The arrangement of atoms near the twin boundary is shown roughly by the inset of Fig. 4. It also may contain so-called optical twins, which are two adjacent regions in the crystal having opposite sense of the helices.

In α-quartz, a regular tetrahedron of oxygen atoms surrounds each silicon atom. The transformation from one Dauphiné twin to another is accomplished by co-operative movement of the silicon atoms to the nearest intersection point on the helix $(x^+ \to x^-)$ and by rigid rotation of the tetrahedron of oxygen atoms.

Futher information on the positions of the oxygen atoms and a molecular calculation of the piezoelectric constant can be found in Vigoureux [14].

4. Constitutive Relations Inferred From the Molecular Structure

In the terminology of Pitteri [8], an infinite set of points in a three dimensional vector space is a $(\nu + 1)$-lattice if there are a set of three linearly independent <u>lattice vectors</u> $\underset{\sim}{e}_1, \underset{\sim}{e}_2, \underset{\sim}{e}_3$ and a set of ν vectors called <u>shifts</u> $\underset{\sim}{p}_0, \underset{\sim}{p}_1, \ldots, \underset{\sim}{p}_\nu$ such that the position vector $\underset{\sim}{x}$ of each point in the infinite set can be expressed in the form

$$\underset{\sim}{x} = n^j \underset{\sim}{e}_j + \underset{\sim}{p}_k, \tag{4.1}$$

n^1, n^2, n^3 being integers and k being an integer in the set $(o, \ldots, \nu)$. Furthermore, the definition of a $(\nu + 1)$-lattice requires every expression of the form (4.1) to give a point of infinite set. A $(\nu + 1)$-lattice can be thought of as a set of ν interpenetrating congruent Bravais lattices.

The positions of the silicon atoms in unstressed α- or β-quartz can be regarded as points in the $(\nu + 1)$-lattice in an infinite number of ways. This is easily visualized for a two-dimensional simple cubic lattice defined by orthonormal vectors $\underset{\sim}{e}_1, \underset{\sim}{e}_2$; it can be regarded as a 1-lattice with say $\underset{\sim}{p}_o = \underset{\sim}{0}$, as a 2-lattice with lattice vectors $\underset{\sim}{e}_1, 2\underset{\sim}{e}_2$ and shifts $\underset{\sim}{p}_o = \underset{\sim}{0}$, $\underset{\sim}{p}_1 = \underset{\sim}{e}_2$, etc. For deformable crystals, it is not really clear where we should stop. On one hand, it is desirable to have a complex description in order to cover the variety of ways the individual atoms move when the crystal is subject to a stress, but on the other hand very complicated constitutive relations will be useless. Already we have simplified the crystal by only considering the positions of the silicon atoms.

In the absence of a physical description of an arbitrarily deformed crystal, we shall choose a description of positions of the silicon atoms as a $(\nu + 1)$-lattice which (with appropriate choices of lattice vectors and shifts but with ν fixed) applies to both β-quartz and to the Dauphiné twins of α-quartz in unstressed configurations. It is easy to see from Fig. 4 that the minimum value of ν which suffices is 2 so according to Pitteri's definition, the silicon atoms of α- or β-quartz lie on a 3-lattice.

A convenient choice of lattice vectors and shifts for unstressed α-quartz is shown in Fig. 5. The vector $\hat{\underset{\sim}{e}}_3$ is perpendicular to the page and its length equals the pitch of the helices. It is important to note that these vectors define only the special configurations associated with unstressed α- (or β-) quartz, hence the carets. To describe a deformed crystal we admit (at this stage) any atomic positions compatible with the definition of a 3-lattice.

We really have to be more specific about the special lattice vectors shown in Fig. 5 because of ordinary thermal expansion. Thus, let θ_0 denote the α-β transformation temperature and let $\hat{\underset{\sim}{e}}_i(\theta)$, $\underset{\sim}{p}^{\pm}(\theta)$ be assigned temperature dependent lattice vectors and shifts with the properties $\hat{\underset{\sim}{e}}_1 \cdot \hat{\underset{\sim}{e}}_3 = 0$, $\hat{\underset{\sim}{e}}_2 \cdot \hat{\underset{\sim}{e}}_3 = 0$, $\langle \hat{\underset{\sim}{e}}_1, \hat{\underset{\sim}{e}}_2 \rangle = 60°$, $|\underset{\sim}{e}_1| = |\underset{\sim}{e}_2|$, and

$$\text{for } \theta > \theta_0 , \quad \hat{p}_1^+ = \hat{p}_1^- = (1/2)\,\hat{e}_1 + (2/3)\,\hat{e}_3,$$

$$\hat{p}_2^+ = \hat{p}_2^- = (1/2)(\hat{e}_1 + \hat{e}_2) + (1/3)\,\hat{e}_3,$$

$$(4.2)$$

$$\text{for } \theta < \theta_0 , \quad \hat{p}_1^\pm = (1/2)\,\hat{e}_1 \pm \lambda(2\hat{e}_2 - \hat{e}_1) + (2/3)\,\hat{e}_3,$$

$$\hat{p}_2^\pm = (1/2)(\hat{e}_1 + \hat{e}_2) \pm \lambda(\hat{e}_1 + \hat{e}_2) + (1/3)\hat{e}_3,$$

$\lambda(\theta)$ being an assigned function of temperature. (All temperature dependence is suppressed in (4.2)). The notation $\pm$ is associated with the Dauphiné twins. The description (4.2) is appropriate for right-handed quartz, as pictured in Figs. 3 and 4. Obviously, the positions of the silicon atoms in left-handed quartz are points of a 3-lattice, so the description can cope with transformations between left and right-handed quartz, which apparently have not been found to occur in the solid state. The special lattice vectors and shifts $\hat{e}_i(\theta)$ and $\hat{p}^\pm(\theta)$ suffer jump discontinuities at θ_0.

Our basic assumption is that at a given temperature, the free energy is determined by the positions of the silicon atoms, which are any positions compatible with the definition of a 3-lattice. Specifically we assume the existence of a function

$$\phi(e_1, e_2, e_3, p_1, p_2, \theta) \tag{4.3}$$

which is Galilean invariant: the condition

$$\phi(Re_i, Rp_k, \theta) = \phi(e_i, p_k, \theta) \tag{4.4}$$

holds for all rotations R and all e_i, p_k in the domain of ϕ. (The domain of ϕ is assumed to be Galilean invariant, too). ϕ represents the free energy of a large part of the lattice divided by the number of silicon atoms in that part.

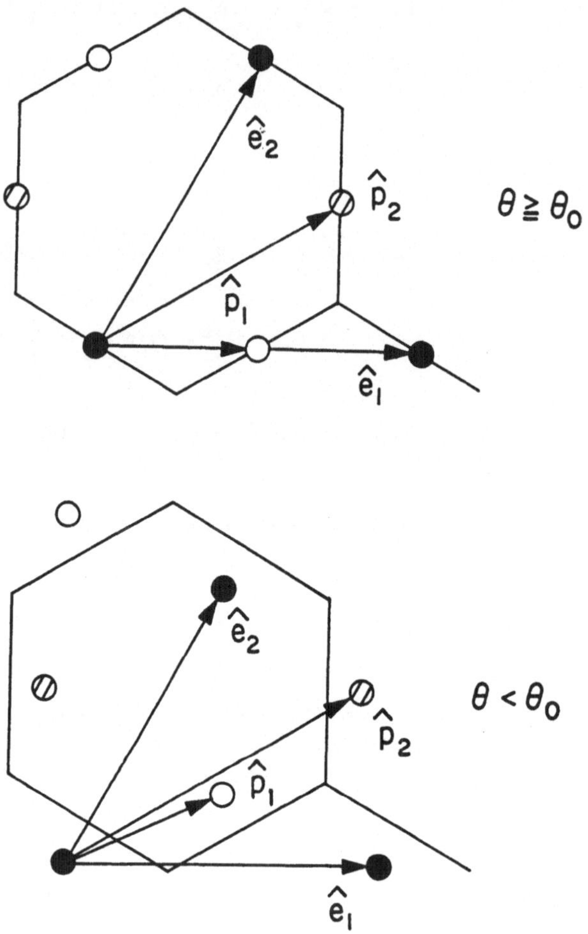

Figure 5. Choice of lattice vectors and shifts for unstressed α and β quartz.

It can happen that two sets of lattice vectors and shifts generate the same 3-lattice. Pitteri [8] has worked out the most general transformation between pairs of such "equivalent" lattice vectors and shifts. Let $\underset{\sim}{e}_i$, $\underset{\sim}{p}_k$ and $\bar{\underset{\sim}{e}}_i, \bar{\underset{\sim}{p}}_k$ generate the same 3-lattice according to the definition (4.1). Then there is a 3×3 matrix of integers M_i^j with determinant ±1, a 2×2 matrix A_k^{ℓ} which belongs to the group

$$\begin{pmatrix}-1 & 1\\ -1 & 0\end{pmatrix}, \begin{pmatrix}0 & -1\\ 1 & -1\end{pmatrix}, \begin{pmatrix}1 & -1\\ 0 & -1\end{pmatrix}, \begin{pmatrix}-1 & 0\\ -1 & 1\end{pmatrix}, \begin{pmatrix}0 & 1\\ 1 & 0\end{pmatrix}, \begin{pmatrix}1 & 0\\ 0 & 1\end{pmatrix}, \tag{4.5}$$

and a 2×3 matrix of integers L_k^j such that

$$\bar{e}_i = M_i^j e_j,$$
$$\bar{p}_k = A_k^\ell p_\ell + L_k^j e_j. \tag{4.6}$$

A key observation is that the transformations in (4.6) represented by the three matrices do not depend on any particular choice of lattice vectors or shifts. Therefore, they embody one precise interpretation of the inherent symmetry of the lattice. If ϕ is to represent the free energy of the lattice itself, it is natural to assume that whenever $\bar{e}_i, \bar{p}_k$ and e_i, p_k are related as in (4.6), then

$$\phi(e_i, p_k, \theta) = \phi(\bar{e}_i, \bar{p}_k, \theta). \tag{4.7}$$

Some restrictions on the domain of ϕ must be introduced. For example, it is not possible to load a quartz crystal to the point where the silicon atoms are arbitrarily far apart. We shall adopt restrictions which essentially confine attention to the α-β transformation and the Dauphiné twinning, including the effects of small distortion. Obviously, the special lattice vectors and shifts (4.2) should belong to the domain of ϕ, as should nearby lattice vectors and shifts. Thus, we assume that for each temperature θ, there is a bounded open set $N(\theta)$ in the vector space of objects $(e_1, e_2, e_3, p_1, p_2)$ which contains the special lattice vectors $(\hat{e}_1(\theta), \hat{e}_2(\theta), \hat{e}_3(\theta), \hat{p}_1^\pm(\theta), \hat{p}_2^\pm(\theta))$. We assume $N(\theta)$ is mapped into itself by Galilean transformations. From now on we confine attention to lattice vectors and shifts in $N(\theta)$.

Since (4.6) involves matrices of integers, it is plausible that only a finite number of transformations of the form (4.6) map $N(\theta)$ into $N(\theta)$. Of special interest are those particular transformations given by (4.2) which eventually yield members of the point group. Arguments of Pitteri can be adapted to prove that there is a sufficiently small neighborhood $N(\theta)$ of

$$\{(R\hat{e}_i(\theta), R\hat{p}_k^\pm(\theta)) \mid R \text{ is a rotation}\}$$

such that the particular transformations which map $N(\theta)$ into itself must satisfy

$$\begin{aligned} \underset{\sim}{R}\hat{\underset{\sim}{e}}_i(\theta) &= M_i^j\, \hat{\underset{\sim}{e}}_j(\theta), \\ \underset{\sim}{R}\hat{\underset{\sim}{p}}_k^{\pm}(\theta) &= A_{k\ell}\, \hat{\underset{\sim}{p}}_\ell^{\pm}(\theta) + L_{ki}\hat{\underset{\sim}{e}}_i(\theta), \end{aligned} \tag{4.8}$$

$\underset{\sim}{R}$ being a rotation and the matrices M, A and L having the properties stated just before (4.6). The rotations which satisfy (4.8) are precisely the members of the point groups of α- and β- quartz for $\theta < \theta_0$ and $\theta > \theta_0$, respectively. For example, we have the following solutions of (4.8) which are the generators of these groups:

	$\underset{\sim}{R}$	M	A	L
$\theta > \theta_0$	$\underset{\sim}{R}^{\pi/3}_{\hat{e}_3}$	$\begin{pmatrix} 0 & 1 & 0 \\ -1 & 1 & 0 \\ 0 & 0 & 1 \end{pmatrix}$	$\begin{pmatrix} -1 & 1 \\ -1 & 0 \end{pmatrix}$	$\begin{pmatrix} 0 & 0 & 1 \\ 0 & 1 & 1 \end{pmatrix}$
	$\underset{\sim}{R}^{\pi}_{\hat{e}_1}$	$\begin{pmatrix} 1 & 0 & 0 \\ 1 & -1 & 0 \\ 0 & 0 & -1 \end{pmatrix}$	$\begin{pmatrix} -1 & 0 \\ -1 & 1 \end{pmatrix}$	$\begin{pmatrix} 1 & 0 & 0 \\ 1 & -1 & 0 \end{pmatrix}$
$\theta < \theta_0$	$\underset{\sim}{R}^{2\pi/3}_{\hat{e}_3}$	$\begin{pmatrix} -1 & 1 & 0 \\ -1 & 0 & 0 \\ 0 & 0 & 1 \end{pmatrix}$	$\begin{pmatrix} 0 & -1 \\ 1 & -1 \end{pmatrix}$	$\begin{pmatrix} 0 & 1 & 1 \\ -1 & 1 & 0 \end{pmatrix}$
	$\underset{\sim}{R}^{\pi}_{\hat{e}_1}$	$\begin{pmatrix} 1 & 0 & 0 \\ 1 & -1 & 0 \\ 0 & 0 & 1 \end{pmatrix}$	$\begin{pmatrix} -1 & 0 \\ -1 & 1 \end{pmatrix}$	$\begin{pmatrix} 1 & 0 & 0 \\ 1 & -1 & 0 \end{pmatrix}$

At some point in the transition from molecular to macroscopic theory, we have to decide how to relate atomic and macroscopic deformation. In most macroscopic theories, homogeneous deformations play a central role. For example, in elasticity theory the strain energy of an arbitrarily deformed body is determined by a knowledge of the strain energies of all homogeneously deformed configurations of the body. So a basic question is: when we apply loads so as to homogeneously deform a relatively large quartz crystal, what is a good assumption about the way the silicon atoms move? Of course, we have already assumed that they move to positions on some 3-lattice, but this still leaves a lot of freedom in relating macroscopic and molecular theory. The classical rule due to Born is to select some Bravais sublattice of the 3-lattice, and to require that this sublattice experience exactly the same deformation as the assigned macroscopic homogeneous deformation. An interpretation of the Born rule in the case of quartz is that a macroscopic deformation of the form

$$\underset{\sim}{y} = \underset{\sim}{F}\underset{\sim}{x}, \quad \underset{\sim}{F} = \text{const.}, \quad \underset{\sim}{x} \in \hat{R} \tag{4.9}$$

is analogous to a change of lattice vectors of the form

$$\underset{\sim}{e}_i = \underset{\sim}{F}\hat{\underset{\sim}{e}}_i, \tag{4.10}$$

where for convenience we have imagined the special lattice vectors $\hat{\underset{\sim}{e}}_i$ being analogous to the reference configuration $\hat{R}$.

In its favor, the Born rule works for the special lattice vectors given in equation (4.2) associated with the undistorted configuration of quartz at various temperatures, since these have been obtained from x-ray measurements and (4.10) is closely satisfied. While we feel a little uneasy about it, we adopt the Born rule here, since it seems to lead to a reasonable theory which in a special case agrees with a successful theory for Dauphiné twinning, which appears to model correctly the effect of stress and loading device on the α-β transformation, and which is quite suggestive about metastability.

Also in its favor, the Born rule says nothing about the shifts. Born regarded these as being adjusted so as to satisfy equilibrium of the lattice, according to a central force calculation. The central force calculation is not appropriate for quartz whose bonds have both ionic and covalent character, but the basic idea seems sound. In the absence of any obvious forces* which do work on the shifts, we shall minimize them out of the problem to obtain a free energy depending only on the deformation gradient which is appropriate for equilibrium calculations. Thus, assume that there exist functions

$$\underset{\sim}{p}_1^{\pm}(\underset{\sim}{e}_k,\theta),\ \underset{\sim}{p}_2^{\pm}(\underset{\sim}{e}_k,\theta) \tag{4.11}$$

such that for all $(\underset{\sim}{e}_1,\underset{\sim}{e}_2,\underset{\sim}{e}_3,\underset{\sim}{p}_1,\underset{\sim}{p}_2)$ belonging to $N(\theta)$, we have

$$\phi(\underset{\sim}{e}_i,\underset{\sim}{p}_k,\theta) \geq \phi(\underset{\sim}{e}_i,\underset{\sim}{p}_k^{\pm}(\underset{\sim}{e}_j,\theta),\theta) \ . \tag{4.12}$$

We shall assume that the inequality in (4.12) is strict when $\underset{\sim}{p}_k \neq \underset{\sim}{p}_k^{+}(\underset{\sim}{e}_j,\theta)$ and $\underset{\sim}{p}_k \neq \underset{\sim}{p}_k^{-}(\underset{\sim}{e}_j,\theta)$. Furthermore, it is natural to assume that the functions (4.11) can be continued beyond $N(\theta)$ as relative minima of ϕ. Special values of the functions (4.11) are given by (4.2).

Any such strict minima of an invariant function are themselves invariant in a certain sense. Schematically, if $f(x,y)$ is a function of two sets of variables represented by x and y, and if $f(x,y)$ satisfies (for all x,y in its domain)

$$f(x,y) = f(Tx,\bar{T}y), \tag{4.13}$$

for transformations T and $\bar{T}$ which belong to a group under which the domain of f is invariant, and if there is a function $g(x)$ with a suitable domain and co-domain such that for each fixed x

$$f(x,g(x)) < f(x,y) \tag{4.14}$$

holds for all $y \neq g(x)$, then

$$\begin{aligned} f(x,g(x)) &= f(Tx,\bar{T}g(x)) \geq f(Tx,g(Tx)) \\ &= f(T^{-1}Tx,\bar{T}^{-1}g(tx)) \geq f(x,g(x)). \end{aligned} \tag{4.15}$$

But (4.15) can only be satisfied if the inequalities are not strict, which by

---

* We are ignoring electric fields, which would apparently be the natural conjugate forces for the shifts.

(4.14) occurs if and only if

$$g(Tx) = \overline{T}g(x). \tag{4.16}$$

This argument can be adapted to the functions given in (4.11) if allowance is made for the fact that both + and - shifts minimize the energy. With strong restrictions on $N(\theta)$ such as those mentioned above, (4.16) translates into a condition on the functions introduced in (4.11) of the form

$$\underset{\sim}{p}_k^s(M^j \underset{\sim}{Q}\underset{\sim}{e}_j, \theta) = A_k^\ell Q \underset{\sim}{p}_\ell^{s'}(\underset{\sim}{e}_j, \theta) + L_k^j \underset{\sim}{Q}\underset{\sim}{e}_j, \tag{4.17}$$

which must hold for all rotations Q, for all matrices M, A and L satisfying (4.8), and for <u>some</u> choice of s as + or - and s' as + or -. It follows from (4.17) with $A_k^\ell = \delta_k^\ell$ and $L_k^j = 0$, that $\underset{\sim}{p}_k^\pm(\underset{\sim}{e}_i, \theta)$ is Galilean invariant: For all rotations $\underset{\sim}{Q}$,

$$\underset{\sim}{p}_k^\pm(\underset{\sim}{Q}\underset{\sim}{e}_i, \theta) = \underset{\sim}{Q}\underset{\sim}{p}_k^\pm(\underset{\sim}{e}_i, \theta). \tag{4.18}$$

Further restrictions follow from (4.17) with Q omitted. For example, it is clear that s and s' are opposite signs when $\underset{\sim}{e}_i = \hat{\underset{\sim}{e}}_i(\theta)$, $\theta < \theta_0$ and M, A and L correspond to members of the point group 622 which are not members of the point group 32.

If we now use the Born rule, we arrive at a sequence of free energy functions of increasing simplicity and decreasing ability to describe various molecular changes. To describe these, consider a reference configuration $\hat{R}_0$ which is analogous to the special lattice vectors evaluated at $\theta = \theta_0$, <u>i.e.</u> β-quartz at the transformation temperature. A rather general free energy function is defined by

$$\Phi(\underset{\sim}{F}, \underset{\sim}{p}_1, \underset{\sim}{p}_2, \theta) = \phi(\underset{\sim}{F}\hat{\underset{\sim}{e}}_i(\theta_0), \underset{\sim}{p}_1, \underset{\sim}{p}_2, \theta) \tag{4.19}$$

with conditions of invariance and domain inherited from (4.4) and (4.7). We view (4.19) as providing a free-energy function Φ for continuum theory; in this context $\underset{\sim}{F}$ is the deformation gradient and the shifts $\underset{\sim}{p}_1$ and $\underset{\sim}{p}_2$ play the role of so-called internal variables. It follows in particular from (4.4), (4.7) and (4.8) that if $\underset{\sim}{Q}$ is a rotation and $\underset{\sim}{R}$ is a rotation in the point group associated with matrices M, L and A via (4.8) with $\theta = \theta_0$, then

$$\phi(\underset{\sim}{QFR}, A_k^{\ell}\underset{\sim}{Q}\underset{\sim}{p}_{\ell} + L_k^i \underset{\sim}{QFR}\hat{\underset{\sim}{e}}_i(\theta_0), \theta)$$
$$= \hat{\phi}(\underset{\sim}{F}, \underset{\sim}{p}_k, \theta). \tag{4.20}$$

To obtain simpler but more restrictive theories, we can restrict the shifts in various ways. A natural but severe restriction is suggested by the form of the special lattice vectors (4.2). Imagine taking the carets off $(4.2)_{3,4}$ and replacing the assigned function $\lambda(\theta)$ by an independent variable μ. The effect of this restriction is to enforce the Born rule for the shifts when μ is fixed, but with different values of μ we can describe configurations associated with both α- and β- quartz as well as some intermediate configurations. Specifically, this theory is based on the assumption,

$$\hat{\phi}(\underset{\sim}{F}, \mu, \theta) =$$
$$\phi(\underset{\sim}{e}_i, 1/2\ \underset{\sim}{e}_1 + \mu(2\underset{\sim}{e}_2 - \underset{\sim}{e}_1) - 1/3\ \underset{\sim}{e}_3, (1/2 + \mu)(\underset{\sim}{e}_1 + \underset{\sim}{e}_2) + 1/3\ \underset{\sim}{e}_3, \theta)\Big|_{\underset{\sim}{e}_i = \underset{\sim}{F}\hat{\underset{\sim}{e}}_i(\theta_0)} \tag{4.21}$$

the minus sign occuring if and only if R does <u>not</u> belong to the point group 32. This seems to be the simplest free energy which may be appropriate for dynamic calculations.

For equilibrium calculations, we follow the reasoning leading up to the inequality (4.12). The natural choice of an equilibrium energy function is obtained by assuming the shifts are determined by the functions given in (4.11), <u>viz.</u>

$$\hat{\phi}^{\pm}(\underset{\sim}{F}, \theta) = \phi(\underset{\sim}{F}\underset{\sim}{e}_i(\theta_0), \underset{\sim}{p}_k^{\pm}(\underset{\sim}{F}\hat{\underset{\sim}{e}}_i(\theta_0), \theta), \theta). \tag{4.23}$$

This energy function is double-valued and defined on a domain determined by $N(\theta)$ in the obvious way.

The free energy function $\hat{\phi}^{\pm}$ inherits invariance from the general theory. A little calculation shows that for any rotation $\underset{\sim}{Q}$ and for any rotation $\underset{\sim}{R}$ in the point group 32,

$$\hat{\phi}^{\pm}(\underset{\sim}{QFR}, \theta) = \hat{\phi}^{\pm}(\underset{\sim}{F}, \theta). \tag{4.24}$$

However, if $\underset{\sim}{Q}$ is a rotation and $\underset{\sim}{R}$ is in the point group 622 but not in the point group 32,

$$\phi^{+}(\underset{\sim}{QFR},\theta) = \phi^{-}(\underset{\sim}{F},\theta). \tag{4.25}$$

Of course

$$\phi^{+}(\underset{\sim}{1},\theta_0) = \phi^{-}(\underset{\sim}{1},\theta_0), \tag{4.26}$$

and

$$\phi^{+}(\underset{\sim}{U}_t,\theta_0) = \phi^{-}(\underset{\sim}{U}_t,\theta_0), \tag{4.27}$$

where $\underset{\sim}{U}_t$ is given by

$$\underset{\sim}{U}_t \hat{\underset{\sim}{e}}_i(\theta_0^+) = \hat{\underset{\sim}{e}}_i(\theta_0^-). \tag{4.28}$$

We identify $\underset{\sim}{U}_t$ with the macroscopic transformation strain given in (2.1).

Let $U_\beta(\theta)$, $\theta > \theta_0$, $\underset{\sim}{U}_\alpha(\theta)$, $\theta < \theta_0$ be tensors satisfying*

$$\begin{aligned} \hat{\underset{\sim}{e}}_i(\theta) &= \underset{\sim}{U}_\beta(\theta)\hat{\underset{\sim}{e}}_i(\theta_0) \qquad && \theta > \theta_0, \\ \hat{\underset{\sim}{e}}_i(\theta) &= \underset{\sim}{U}_\alpha(\theta)\hat{\underset{\sim}{e}}_i(\theta_0) \qquad && \theta < \theta_0. \end{aligned} \tag{4.29}$$

Without loss of generality we arrange the special lattice vectors so that $\underset{\sim}{U}_\beta$ and $\underset{\sim}{U}_\alpha$ are positive-definite and symmetric. Also,

$$\underset{\sim}{U}_\beta(\theta_0) = \underset{\sim}{1}, \quad \underset{\sim}{U}_\alpha(\theta_0) = \underset{\sim}{U}_t. \tag{4.30}$$

The theory suggests that we should have an exchange of stability at θ_0. We assume that

$$\begin{aligned} &\text{for} \quad \theta > \theta_0 \qquad && \phi^{\pm}(\underset{\sim}{U}_\beta(\theta),\theta) < \phi^{\pm}(\underset{\sim}{F},\theta), \\ &\text{for} \quad \theta < \theta_0 \qquad && \phi^{\pm}(\underset{\sim}{U}_\alpha(\theta),\theta) < \phi^{\pm}(\underset{\sim}{F},\theta) \end{aligned} \tag{4.31}$$

hold for all $\underset{\sim}{F}$ in the domain of $\phi^{\pm}$.

---

* We have extended the definition $\hat{e}_i(\theta)$ to θ_0 from both side by continuity.

5. Effect of Stress on the Transformation Temperature

The free energy function $\phi^{\pm}(\underset{\sim}{F},\theta)$ found in the preceding section with the properties listed in (4.24)-(4.31) is of the same form as the one we gave in reference [7]. We arrived at it on intuitive grounds and by analogy to the theory of Dauphiné twinning. Using this free energy, we studied a class of deformations $\underset{\sim}{y}(\underset{\sim}{x})$ whch minimize the total free energy

$$E = \int_{\hat{R}} \rho\phi^{\pm}(\nabla\underset{\sim}{y}(\underset{\sim}{x}),\theta)d\underset{\sim}{x} + L. \tag{5.1}$$

In (5.1) L represents the potential of the loading device and ρ is the number of silicon atoms per unit volume of the reference configuration. The temperature θ was held constant in the minimization and both $+$ and $-$ branches of the free energy were allowed to compete for the minimum. Here, we briefly describe the minimizers of (5.1) for some choices of L.

If we confine attention to the minimizers of E, there is no loss of generality in replacing $\phi^{\pm}$ by the function ϕ given for each $\underset{\sim}{F}$ by

$$\phi(\underset{\sim}{F},\theta) = \min_{+-} \phi^{\pm}(\underset{\sim}{F},\theta). \tag{5.2}$$

At points F in the common domain of ϕ^{+} and ϕ^{-}, ϕ remains continuous, although typically ϕ will fail to be continuously differentiable on this domain.

First consider an unloaded body. In this case, we put $L = 0$ in (5.1). At a fixed temperature θ, a homogeneous deformation

$$\underset{\sim}{y}(\underset{\sim}{x}) = \hat{\underset{\sim}{F}}\underset{\sim}{x} \tag{5.3}$$

minimizes the total energy E (in an appropriate function space; see Ball [15]) if and only if $\hat{\underset{\sim}{F}}$ minimizes the integrand, that is, if and only if

$$\phi(\hat{\underset{\sim}{F}},\theta) < \phi(\underset{\sim}{F},\theta) \quad \text{for all } \underset{\sim}{F}. \tag{5.4}$$

By (4.31) and the Galilean invariance of ϕ, some minima of the integrand are given by

$$\underset{\sim}{f} = \begin{cases} \underset{\sim\alpha\sim\alpha}{R} \underset{\alpha}{U}(\theta) & \theta < \theta_0, \\ \\ \underset{\sim\beta\sim\beta}{R} \underset{\beta}{U}(\theta) & \theta > \theta_0. \end{cases} \tag{5.5}$$

$\underset{\sim\alpha}{R}$ and $\underset{\sim\beta}{R}$ being arbitrary rotations. A known theorem of continuum mechanics [16] states that under fairly mild conditions of continuity on deformations, no inhomogeneous deformations can be obtained from (5.5) by allowing $\underset{\sim\alpha}{R}$ and $\underset{\sim\beta}{R}$ to depend on x if $\theta \neq \theta_0$. Thus, it is plausible to assume that for $\theta \neq \theta_0$ the only minimizers of E are homogeneous deformations of the form (5.3) with $\underset{\sim}{f}$ given by (5.5).

There remains the possibility that α- and β- quartz can co-exist in a stable deformation at $\theta = \theta_0$. However, if $\underset{\sim}{y}(\underset{\sim}{x})$ is a continuous, piecewise differentiable deformation and if $\nabla\underset{\sim}{y}(\underset{\sim}{x})$ is given by the right hand side of (2.3) at $\theta = \theta_0$ (with $\underset{\sim\alpha}{R}$ and $\underset{\sim\beta}{R}$ possibly depending on $\underset{\sim}{x}$), then it can be shown that there are vectors $\underset{\sim}{a}$ and $\underset{\sim}{n}$ and a rotation $\underset{\sim}{R}$ such that

$$\underset{\sim\sim t}{RU} - \underset{\sim}{1} = \underset{\sim}{a} \times \underset{\sim}{n}. \tag{5.6}$$

In particular, (5.6) implies that one eigenvalue of $\underset{\sim t}{U}$ equals 1, which is <u>not</u> true of the measured transformation strain (2.1). This already gives a hint about the metastability associated with the α-β transformation.

Now consider the case of loading by a hydrostatic pressure:

$$L = p \text{ vol}(\underset{\sim}{y}(\hat{R})) = p \int_R \det \nabla\underset{\sim}{y}(\underset{\sim}{x})d\underset{\sim}{x}. \tag{5.7}$$

Similar reasoning as above leads us to study the problem

$$\min_{\underset{\sim}{F}} [\phi(\underset{\sim}{F},\theta) + p \det \underset{\sim}{F}]. \tag{5.8}$$

with p and θ positive constants. Let us assume that the functions $\underset{\sim\alpha}{U}(\theta)$ and $\underset{\sim\beta}{U}(\theta)$ can be extended smoothly as relative minima of both ϕ^+ and ϕ^- beyond θ_0; assume that each of these functions continue to give equal values of ϕ^+ and ϕ^-. In view of the exchange of stability at θ_0 assumed in (4.31), we must have

$$\frac{d}{d\theta} [\phi(\underset{\sim\beta}{U}(\theta),\theta) - \phi(\underset{\sim\alpha}{U}(\theta)),\theta)]_{\theta=\theta_0} < 0, \tag{5.9}$$

which implies that

$$\phi_\theta(\underset{\sim}{U}_\beta(\theta_0),\theta_0) - \phi_\theta(\underset{\sim}{U}_\alpha(\theta_0),\theta_0) < 0. \qquad (5.10)$$

A detailed treatment of the existence of minima of (5.8) is made a little complicated by the invariance groups. It is plausible that there exist smooth functions $\overline{\underset{\sim}{U}}_\beta(\theta)$, $\overline{\underset{\sim}{U}}_\alpha(\theta)$ which agree with the functions $\underset{\sim}{U}_\alpha(\theta)$, $\underset{\sim}{U}_\beta(\theta)$ at $\theta = \theta_0$ and which jointly minimize $\phi^{\pm}(\underset{\sim}{F},\theta) + \overline{p}(\theta)\det \underset{\sim}{F}$, $\overline{p}(\theta)$ being a smooth function of θ, for an open interval of values of θ containing θ_0. If so, we interpret $\overline{p}(\theta)$ as the hydrostatic pressure on a specimen whose transformation temperature is θ. Under these conditions, a routine calculation [7] yields a form of the Clausius-Clapeyron equation:

$$\frac{d\overline{p}}{d\theta}\,[\det \overline{\underset{\sim}{U}}_\beta(\theta) - \det \overline{\underset{\sim}{U}}_\alpha(\theta)] = \phi_\theta(\overline{\underset{\sim}{U}}_\alpha(\theta),\theta) - \phi_\theta(\overline{\underset{\sim}{U}}_\beta(\theta),\theta)\ . \qquad (5.11)$$

If we evaluate this equation at $\theta = \theta_0$, use $\overline{\underset{\sim}{U}}_\beta(\theta_0) = \underset{\sim}{1}$ and $\overline{\underset{\sim}{U}}_\alpha(\theta_0) = \underset{\sim}{U}_t$ and make use of (5.10), we conclude that if the pressure goes up, then the transformation temperature also goes up, in agreement with observations (Fig. 1b), if we overlook the metastability.

We can also use Fig. 1b to get a measured value for $d\overline{p}/d\theta$ at $\theta = \theta_0$; then this value can be put in equation (5.11) to yield a value for the right-hand side of (5.11), the "latent heat". Once this is known, we can quantitatively study some of the nonhydrostatic loading devices.

This was done in reference [7] for certain loading devices which produced the same stress field as intended by Coe and Paterson in their nonhydrostatic experiments summarized in Fig. 1c. The agreement was within experimental error, even though the actual energy of the loading device (its "hardness") was only roughly modelled.

To get an idea of how these nonhydrostatic loading devices affect the transformation temperature, we consider the <u>points of convexity</u> of the free energy. These are tensors $\hat{\underset{\sim}{F}}$ such that the surface $\phi(\underset{\sim}{F},\theta)$ <u>vs.</u> $\underset{\sim}{F}$ lies above the tangent plane at $\hat{\underset{\sim}{F}}$, the slope of this tangent plane $\phi_{\underset{\sim}{F}}(\hat{\underset{\sim}{F}},\theta)$ being interpreted as the

Piola-Kirchhoff stress. Points of convexity $\hat{\underset{\sim}{F}}$ yield homogeneous deformations $\underset{\sim}{y}(\underset{\sim}{x}) = \hat{\underset{\sim}{F}}\underset{\sim}{x}$ which are minimizers of E when L is given the expression for an appropriate dead loading device. It is geometrically plausible (and it can be proved) that if a number of points of convexity correspond to the same Piola-Kirchhoff stress, then no point of convexity corresponding to a different Piola-Kirchhoff stress lies in the convex hull of these points. To apply this result to quartz at the transformation temperature θ_0, we observe that there are an infinite number of points of convexity corresponding to zero Piola-Kirchhoff stress; these are given by (5.5) evaluated at $\theta = \theta_0$, which reduces to

$$\hat{\underset{\sim}{F}} = \begin{cases} \underset{\sim}{R}_\alpha \underset{\sim}{U}_t, \\ \underset{\sim}{R}_\beta , \end{cases} \tag{5.12}$$

R_α and R_β being any rotation tensors. In [7], it is shown that the $\underset{\sim}{U}_t$ for quartz is in the interior of the convex hull of the rotation tensors. Hence, there are no points of convexity near $\underset{\sim}{U}_t$ corresponding to a nonzero Piola-Kirchhoff stress. Physically speaking, if we apply dead loads so as to bring a quartz crystal at θ_0 to a homogeneously deformed absolutely stable configuration, then it will always end up in the β- phase.*

The study of which Dauphiné twin is present in a minimizer $\underset{\sim}{y}(\underset{\sim}{x}) = \hat{\underset{\sim}{F}}\underset{\sim}{x}$ of E is simply a study of which branch of the free energy function (ϕ^+ or ϕ^-) actually agrees with ϕ at $\hat{\underset{\sim}{F}}$. Unlike the α-β transformation, the difference between ϕ^+ and ϕ^- is evident even when $\phi^\pm$ are approximated by quadratic forms. Thus, Thomas and Wooster obtained a detailed picture of the arrangement of Dauphiné twins in a loaded body by studying two simultaneous problems in linear elasticity.

---

* This prediction is not tested by the experiments of Coe and Paterson (Fig. 1c), since their stress fields cannot be set up in a dead loading device (see [7] for a full explanation). Note that the α- phase always was stable in their loading devices at θ_0.

6. Speculative Remarks on Metastability

The remarks of section 5 ignore the fact that the α-β transformation temperature really different on heating than cooling. Nor do they address the fact that even at deformation gradients where ϕ^+ is much less than ϕ^-, a body with a - twin does not transform to the + twin at temperatures sufficiently less than 574°C, in the experiments of Thomas and Wooster [4].

As for the hysteresis in the α-β transformation, we lean toward an explanation based on the failure of the transformation strain to satisfy an equation of the form (5.6), which is a failure of conditions of compatibility. In physical terms, imagine cooling an unloaded crystal in the β-phase. At just a little below θ_0, it is really energetically favorable for the crystal to be in the α-phase. Suppose by some accident a small part of the crystal transforms to the α-phase. Because (5.6) fails to have a solution, the boundary of this part will necessarily be a stressed transition layer which probably contains a family of strains intermediate between $\underset{\sim}{U}_t$ and $\underset{\sim}{1}$. Associated with the layer is an extra free energy. If this free energy exceeds the free energy gained by having the interior of this part in the α-phase, it seems likely that the part will spontaneously transform back to the β-phase. If we cool the body enough, it will reach a point where the difference in free energy between α and β is sufficient to compensate for the free energy built up in the transition layer; at this point a wave of transformation passes through the crystal.

These remarks suggest how cracking could occur. Suppose we reach the stage at which the necessarily stressed transition layer is moving through the crystal. As a guess, we would say that on one side of this layer the principal stresses are tensile while on the other side they are compressive, gradually fading to zero away from the layer on either side. It seems reasonable that on the tensile side, these stresses could be large enough to cause cracks at some of the larger defects. As the crystal boundary is approached from inside, one of these principal stresses would have to tend to zero since the crystal is not loaded at its boundary.

This could suppress cracking near the boundary, as observed. Crystal shape* would clearly have an influence on the stress field, and high pressure could suppress the cracking by converting tensile to compressive stresses.

The question is: how would we analyze this phenomenon? The free energy ϕ does in fact sense the energy associated with $\underset{\sim}{F}$ between $\underset{\sim}{1}$ and $\underset{\sim}{U}_t$ so we might consider the following scheme. Formally, the equations of balance of mass, momentum, energy together with a statement of the second law of thermodynamics such as the Clausius-Duhem inequality and boundary conditions appropriate to an unloaded body in a constant temperature heat bath have a Lyapunov functon of the form

$$\int_{\hat{R}} \rho \, [\varepsilon - \hat{\theta}\eta] d\underset{\sim}{x}, \qquad (6.1)$$

ε being the internal energy density, $\hat{\theta}$ the temperature of the heat bath and η the entropy density. The integrand of (6.1) is not quite the free energy because $\hat{\theta}$ is the temperature of the heat bath, not the body. If we overlook this distinction, the total free energy serves as a Lyapunov function. We could start the equations at $t = 0$ with the body in the β-phase, with sufficiently small disturbances. Because ϕ does sense the energy of intermediate configurations, it seems likely that with appropriate conditions of continuity, there would not exist a dynamic solution of the equations of motion at $\theta = \theta_0$ which would allow (6.1) to decrease in time. These remarks are necessarily quite vague, because of the lack of precise knowledge about solutions of the equations of motion of nonlinear thermoelasticity. If a failure of existence is found, it would seem interesting to study in a dynamic setting whether there is a temperature of the heat bath which will cause the transformation with <u>any</u> sufficiently small disturbances in a certain class. At fixed temperature, it would be interesting to know what disturbances are effective in causing transformation.

The only study related to these ideas which I know of is the recent work by Pego [17]. His study is in one space dimension where the conditions of continuity

---

* but according to a purely elastic theory, crystal size (at constant shape) would have no influence on the stress field, due to the basic invariance of the equations of motion under the transformation $\underset{\sim}{x} \to \lambda\underset{\sim}{x}$, $\underset{\sim}{y} \to \lambda\underset{\sim}{y}$, $\underset{\sim}{t} \to \lambda\underset{\sim}{t}$.

must be somewhat different from those indicated above, since the analog of (5.6) can be satisfied in one dimension with any (scalar) transformation strain.

The hysteresis associated with Dauphiné twinning is obviously much different. To discuss metastability in this case, we first need a free-energy function that senses "intermediate configurations" between the twins of α-quartz. $\phi^{\pm}$ appears too special in that it only assigns two energies when $\underset{\sim}{F} = \underset{\sim}{U}_t$ and $\theta = \theta_0$. The simplest choice which might be realistic is the energy function $\hat{\phi}(\underset{\sim}{F},\mu,\theta)$ defined by (4.21), while a much more general choice is provided by (4.19). The case where $\underset{\sim}{F}$ is absent from $\hat{\phi}$ is essentially Landau's theory of phase transitions, with μ as the order parameter; however, the presence of $\underset{\sim}{F}$ might be considered the most interesting aspect of quartz. Landau's theory, the macroscopic observations and the molecular model all suggest that we adopt a theory associated with the notion of a thermally activated process. Treatments of aspects of this kind of theory are given by Landau [18], Müller [19] and Wiener [20]. Often, these theories lead to an ordinary differential equation which in the simplest case is

$$\dot{\mu} = f(\underset{\sim}{F},\mu,\theta), \tag{6.2}$$

which is to be solved along side of the dynamic equations of motion.

Acknowledgement

The research reported here was supported by the Materials Research Laboratory at Brown University and by the National Science Foundation through the grant MEA 8209303.

References

1. J.W. Gibbs, "On the equilibrium of heterogeneous substances", Trans. Conn. Acad. III (1875-1878), p. 108 and p.343.

2. G. van Tendeloo, J. van Landuyt and S. Amelinckx, "The α-β phase transition in quartz and $AIPO_4$ as studied by electron microscopy and diffraction", Phys. Stat. Sol. a 33 (1976), p. 723.

3. R.S. Coe and M.S. Paterson, "The α-β inversion in quartz: a coherent phase transition under nonhydrostatic stress", J. Geophys. Res. 74 (1969), p. 4921.

4. L.A. Thomas and W.A. Wooster, "Piezocrescence - the growth of Dauphiné twinning under stress", Proc R. Soc. A208 (1951), p. 43.

5. R.S. Rivlin, "Some thoughts on material stability" in Finite Elasticity (ed. D.E. Carlson and R.T. Shield), Martinus Nijhoff (1982).

6. J.L. Ericksen, "Multi-valued strain energy functions for crystals", Int. J. Solids Struct. 18, (1982), p. 913.

7. R.D. James, "Displacive phase transformations in solids", to appear.

8. M. Pitteri, "On $(\nu + 1)$-lattices", J. Elasticity 15 (1985), p. 3.

9. J.L. Ericksen, "Nonlinear elasticity of diatomic crystals", Int. J. Solids Struct. 6 (1970), p. 951.

10. G.P. Parry, "On phase transitions involving internal strain", Int. J. Solids Struct. 17 (1981), p. 361.

11. C. Berger, L. Eyraud, M. Richard and R. Rivière, "Étude radiocristallograpique de variation de volume pour quelques materiaux subissant des transformations de phase solide-solide", Bull. Soc. Chim. France (1966), p. 628.

12. C. Berger, M. Richard and L. Eyraud, "Applications de la microcalorimetrie a la détermination précise des variations d'enthalpie de quelques transformations solide-solide", Bull. Soc. Chim. France (1965), p. 1491.

13. W.A. Wooster and N. Wooster, Nature 157 (1946), p. 405.

14. P. Vigoureux, Quartz Resonators and Oscillators, H.M. Stationery Office (1931).

15. J.M. Ball, "Constitutive inequalities and existence theorems in nonlinear elastostatics", in Nonlinear Analysis and Mechanics, Vol. 1 (ed. R.J. Knops), Pitman (1977).

16. M.E. Gurtin, An Introduction to Continuum Mechanics, Academic Press (1981).

17. R. Pego, "Phase Transitions in one Dimensional Nonlinear Viscoelasticity: Admissibility and Stability", to appear in IMA Volumes and its Applications, Dynamical Problems in Continuum Physics, Springer Verlag.

18. L.D. Landau and E.M. Lifshits, Statistical Physics, 3rd edition, (Trans. J.B. Stykes and M.J. Kearsley), Pergamon (1980).

19. M. Achenbach and I. Müller, "Creep and yield in martensitic transformations", Ingenieur Archiv (in press).

20. J.H. Wiener, Statistical Mechanics of Elasticity, John Wiley and Sons (1983).

CONTINUATION THEOREMS FOR SCHRODINGER OPERATORS

by

Carlos E. Kenig

Department of Mathematics
University of Chicago
Chicago, Ill.

The purpose of this note is to describe the connection between quantum physics, the absence of positive eigenvalues for Schrödinger operators, and unique continuation theorems for Schrödinger operators. We shall also describe some recent work on unique continuation theorems.

A Schrödinger operator is a partial differential operator on R^n of the form $H = -\Delta + V = H_0 + V$, where $H_0 = -\Delta = -\sum_{j=1}^{n} \frac{\partial^2}{\partial x_j^2}$, and V is a function on R^n. For a recent survey of many of the properties of Schrödinger operators, see [18]. Their name comes from the form of Schrödinger's equation

$$(1) \qquad i\frac{\partial\psi}{\partial t} = H\psi .$$

H is thus the Hamiltonian operator of a nonrelativistic particle, H_0 is the kinetic energy, and V the potential energy. The function V is not supposed to be smooth, continuous, or even locally bounded. In fact the Coulomb potentials $(V(x) = \frac{1}{|x|})$ are unbounded, and appear in models for hydrogen atoms. The goal is to study very general V.

The solutions to (1) are given by $e^{-itH}\psi_0$ for some $\psi_0 \in L^2(\mathbb{R}^n)$, and so we are interested in studying the spectral properties of H. The eigenfunctions of H, i.e. solutions of $H\psi = E\psi$, $\psi \in L^2(\mathbb{R}^n)$, $E \in \mathbb{R}$ are called stationary or bound states. We consider $H = -\Delta + V$ as a self-adjoint operator on the Hilbert space $L^2(\mathbb{R}^n)$. The spectral theorem assigns a unique spectral family, or projection-valued measure, on the real line to any self-adjoint operator on a Hilbert space. Von Neumann (1932) placed the spectral theorem at the heart of his axiomatic formulation of quantum mechanics. Ever since then, spectral analysis of Schrödinger operators has been a central part of mathematical physics.

A very useful division of the spectrum (see for example [13] for the precise definitions), distinguishes among σ_p, the point spectrum, consisting of eigenvalues, isolated or not, and $\sigma_{cont} = \sigma_{ac} \cup \sigma_{sc}$, the continuous spectrum, which is associated to the restriction of H to the part of $L^2(\mathbb{R}^n)$ orthogonal to eigenvectors. σ_{ac} is the spectrum of the restriction of H to the part of $L^2(\mathbb{R}^n)$ which corresponds to the part of the spectrum which is absolutely continuous with respect to Lebesgue measure, and σ_{sc}, the spectrum of the restriction of H to the part of $L^2(\mathbb{R}^n)$ which corresponds to the singular continuous spectrum. The definitions allow for the possibility that σ_{cont} and σ_p intersect.

In one-body physics, the potential V tends to 0 at ∞, and typically,

$$\sigma_{cont} = \sigma_{ac} = [0,\infty) , \quad \sigma_{sc} = \emptyset ,$$

(2)

$$\sigma_p \cap \sigma_{cont} = \{0\} \text{ or } \emptyset .$$

This is expected on physical grounds. Usually, it is easy to prove that $\sigma_{sc} = \emptyset$. The relationship of this decomposition of the spectrum to quantum physics is that σ_p comprises the energy of bound states, and the spectral subspace associated with σ_{ac} consists of dynamical states that may participate in scattering.

The reason why one expects on physical grounds that $\sigma_p \cap \sigma_{cont} = \{0\}$ or $\emptyset$ is that if the potential goes to 0 at ∞, and the energy of a particle is positive, one might expect that quantum fluctuations would eventually propel the particle to a place where its motion would not be confined, and this would make the bound state large, and hence not in $L^2(\mathbb{R}^n)$.

Oppenheimer assumed this in his Göttingen dissertation (1927), to calculate the ionization rate of hydrogen in an electric field. In response to this, Von Neumann and Wigner (1929) ([21]) constructed an example of a one-dimensional potential $V(x)$, which goes to 0 at ∞, but with positive eigenvalues.

<u>Example 1</u>: (Wigner-Von Neumann ([21] and also [13]).

$$V(x) = -\frac{8 \sin 2|x|}{|x|} + O(1/|x|^2) \text{ at } \infty, \quad V(x) \in L^{\infty}_{loc},$$

and yet we have positive eigenvalues. The reason that physical intuition goes wrong is that there are violent oscillations in the potential at ∞, which 'trap' the particle.

Example 2. (Reed-Simon, Ex2, XIII. 13 Vol. 4 ([13])).

$V(x) = ||x|-1|^{-1}$ on $\mathbb{R}^3$ has positive eigenvalues. The reason for this is the local behavior of $V(x)$, which is not in $L^1_{loc}(\mathbb{R}^3)$.

Thus, we see that the problem of absence of positive eigenvalues is a difficult one. The most successful philosophy for eliminating them in dimensions greater than one has been developed by Kato [11], Agmon [1], [2], Simon [17] and others. To illustrate this philosophy, let us assume that V has compact support, supp $V \subset \{|x|<R\}$. Suppose that $H = -\Delta + V$ has a positive eigenvalue E. We therefore have $[-\Delta + V]u = Eu$ in $\mathbb{R}^n$, with $E>0$, $u \in L^2(\mathbb{R}^n)$. By the support assumption on V, we have

$$-\Delta u - Eu = 0 \quad \text{on} \quad |x| > R, \ E>0, \text{ and } \ u \in L^2(|x| > R).$$

A classical theorem of Rellich ([14]) now shows that $u \equiv 0$ for $|x|>R$. From this we would like to conclude that $u \equiv 0$ on R^n. We have $[-\Delta + (V-E)]u = 0$ in $|x|<2R$, and u vanishes on an open subset of $|x|<2R$. This leads us to unique continuation theorems. Recall that if u is real analytic, and it vanishes together with all its derivatives at a point, then it must be identically zero. If we had the same conclusion for solutions of $[-\Delta+V]u = 0$, we would be done (replacing V by $V-E$). Such results are known as unique continuation theorems. If the potential V is such that solutions of $[-\Delta+V]u = 0$ which vanish of infinite order at a point are identically 0, we shall say that V has the strong unique continuation property (s.u.c.p.). Unique continuation theorems for solutions of elliptic partial differential equations have been studied for many years. The first results for non-analytic V go back to T. Carleman (1939) [5]. He proved that if $V \in L^{\infty}_{loc}(\mathbb{R}^2)$, then V has the s.u.c.p. The method introduced by

Carleman (the so-called Carleman estimates) has been used to attack all problems of this kind until very recently, when N. Garofalo and F.H. Lin ([7]) developed a new method to prove Carleman's result. In 1954, C. Muller ([12]) extended Carleman's result to any number of dimensions. However, as we explained at the outset of this note, we are interested, from physical considerations, in potentials V which are not locally bounded. Therefore, in recent years, people have been interested in establishing the s.u.c.p when $V \in L^p_{loc}(\mathbb{R}^n)$, $p < \infty$. The question becom For what p's does the s.u.c.p hold? The first results in this direction seem to be due to A.M. Berthier ([4]) and V. Georgescu ([8)], in 1979. They show the s.u.c.p. for $V \in L^p_{loc}(\mathbb{R}^n)$, $p = 2$ when $n = 1,2,3$, $p > n-2$ for $n \geq 4$. Shortly afte M. Schechter and B. Simon (1980)[16], established the s.u.c.p. for $p > 1$, $n=1,2$, $p > \frac{2n-1}{3}$, $n = 3,4,5$, $p > n-2$, $n \geq 6$. Saut and Scheurer (1981)([15]), proved the s.u.c.p. for $p > \frac{2n}{3}$ all $n > 2$, while Amrein, Berthier and Georgescu (1981) [3] proved the s.u.c.p for $p > \frac{n}{2}$ for $n = 2,3,4$. Hörmander (1983) [9] was able to establish the s.u.c.p. for $p > \frac{4n-2}{7}$, $n > 4$. One is naturally curious about the meaning of all these numbers. Let us clarify them with an example.

Example 3 ([10]): Let $u(x) = e^{-(\log 1/|x|)^{1+\varepsilon}}$, $\varepsilon > 0$. Then, u vanishes at 0 of infinite order. Let $V(X) = -\frac{\Delta u}{u}$. Then, $V(X) \sim (\log 1/|x|)^{2\varepsilon}/|x|^2 \in L^p_{loc}(\mathbb{R}^n)$ for all $p < \frac{n}{2}$, and so the s.u.c.p. cannot hold in this range of p's. As $n\to\infty$, the best result of the ones mentioned above is Hörmander's ($\frac{4n}{7} > \frac{n}{2}$). We now have:

Theorem 1 (David Jerison and Carlos Kenig [10]): The s.u.c.p. holds for $V \in L^{n/2}_{loc}(R^n)$, $n \geq 3$.

We see that this improves all previous results, and is the best possible result in the scale of L^p spaces.

Remark 1: In the application to absence of positive eigenvalues we sketched above, what is needed is to show that if $-\Delta u + Vu = 0$ and u vanishes in an open set, then $u = 0$. This is called the unique continuation property (u.c.p.).

It is not known what the best L^p class on V is for the u.c.p. to hold. The best result known today is the one given by Theorem 1. However, we conjecture that $V \in L^1_{loc}(\mathbb{R}^n)$ suffices for u.c.p. to hold.

Remark 2: Recently ([20]) E.M. Stein was able to simplify one of the steps in the proof of Theorem 1, and give an extension to Lorentz spaces. He shows that if V is in the closure in $L^{n/2,\infty}(R^n)$ of the continuous functions, then V satisfies the s.u.c.p.

I will now try to give an idea of how one proves Theorem 1. The main estimate that is needed is

Theorem 2: (David Jerison and Carlos Kenig [10]). There exists a sequence $\{\tau_j\}$, $\tau_j \to +\infty$, such that if $\frac{1}{q} - \frac{1}{p} = \frac{2}{n}$, $\frac{1}{p} + \frac{1}{q} = 1$, $n \geq 3$ (i.e. $q = \frac{2n}{n+2}$, $p = \frac{2n}{n-2}$), and $f \in C^\infty(\mathbb{R}^n)$ and vanishes of infinite order at 0, we have

$$(3) \qquad \| |x|^{-\tau_j} f \|_{L^p(\mathbb{R}^n)} \leq C \| |x|^{-\tau_j} \Delta f \|_{L^q(\mathbb{R}^n)},$$

where C is independent of f and j.

Remark 3: If f is not 0 of infinite order at 0, there is no hope for such an inequality. The main point is that C is independent of j. This is a strengthening of the Sobolev inequality, which forces $\frac{1}{q} - \frac{1}{p} = \frac{2}{n}$. The idea of considering the special case $\frac{1}{p} + \frac{1}{q} = 1$ comes from the application to the absence of positive eigenvalues (see [10]).

Let us now show how Theorem 2 implies Theorem 1. Suppose $0 \in \Omega$, Ω connected, and $|\Delta u| \leq |V||u|$, $V \in L^{n/2}_{loc}(\Omega)$. Assume also that u vanishes of ∞ order at 0. We can assume, without loss of generality with that $B = \{|x| < 1\} \subset \Omega$. We want to show that $u \equiv 0$ in Ω. Pick $\phi \in C^\infty(R^n)$ with $\phi \equiv 1$ in $|x| < 1/2$, supp $\phi \subset B$, and let $f = \phi u$. We apply Theorem 2 to f, to obtain $\| |x|^{-\tau_j} f \|_{L^p} \leq C \| |x|^{-\tau_j} \Delta f \|_q$. Choose $\rho < 1/2$ so that $C \|V\|_{L^{n/2}(B_\rho)} < 1/2$, where $B_\rho = \{ |x| < \rho \}$. We will show $u \equiv 0$ in B_ρ. A connectivity argument finishes the proof.

$$\| |x|^{-\tau_j} u\|_{L^p(B_\rho)} \le \| |x|^{-\tau_j} f\|_{L^p} \le C\| |x|^{-\tau_j}\Delta f\|_{L^q}$$

$$\le C\| |x|^{-\tau_j} \Delta u\|_{L^q(B_\rho)} + C\| |x|^{-\tau_j}\Delta f\|_{L^q({}^cB_\rho)}$$

$$\le C\| |x|^{-\tau_j} Vu\|_{L^q(B_\rho)} + C\| |x|^{-\tau_j}\Delta f\|_{L^q({}^cB_\rho)}$$

$$\le C\|V\|_{L^{n/2}(B_\rho)} \| |x|^{-\tau_j} u\|_{L^p(B_\rho)} + C\| |x|^{-\tau_j} \Delta f\|_{L^q({}^cB_\rho)}$$

by Hölder's inequality and the relationship $\frac{1}{q} - \frac{1}{p} = \frac{2}{n}$. Therefore,

$$\|(\frac{|x|}{\rho})^{-\tau_j} u\|_{L^p(B_\rho)} \le 2c\|\Delta f\|_{L^q({}^cB_\rho)},$$

and thus, letting $j \to \infty$, we see that $u \equiv 0$ in B_ρ.

We will now sketch an idea of the proof of Theorem 2. We want to show, for $f \in C^\infty(\mathbb{R}^n/\{0\})$ that, for a certain discrete set $\{\tau_j\}$, if $\tau \notin \{\tau_j\}$, we have

(4) $$\| |x|^{-\tau} f\|_{L^p} \le C\| |x|^{-\tau}\Delta f\|_{L^q},$$

when $\frac{1}{q} - \frac{1}{p} = \frac{2}{n}$, $\frac{1}{p} + \frac{1}{q} = 1$.

Note that $1 < q < 2 < p < \infty$, $\frac{1}{p} + \frac{1}{q} = 1$. A classical theorem in this situation is the Hausdorff-Young Theorem: If $1 < q < 2$, $\frac{1}{p} + \frac{1}{q} = 1$, then

$$\|\hat{f}\|_{L^p} \le \|f\|_{L^q}.$$

This is proved by noting the following inequalities:

$$\|\hat{f}\|_{L^2} \le \|f\|_{L^2} \qquad \text{(Plancherel)} \quad \text{and}$$

$$\|\hat{f}\|_{L^\infty} \le \|f\|_{L^1}.$$

The result follows from the inequalities by the M. Riesz interpolation theorem. We are thus led to try to prove (4) by interpolation, first proving an $L^2 \to L^2$ result, and then an $L^1 \to L^\infty$ result. Because of the homogeneities of the problem (coming from Sobolev's inequalities), we cannot use the M. Riesz interpolation theorem, but we must use an interpolation procedure where not only the function space, but also the operator changes analytically with a complex parameter. This is the so-called analytic interpolation procedure, due to E.M. Stein ([19)]. We fix τ and introduce an analytic family of operators T_z given by

$$T_z g = |x|^{-\tau}(-\Delta)^{-z/2}(|x|^{\tau-z}g), \tag{5}$$

modifed by a Taylor series of order the integer part of τ. Estimates for T_z, uniform in τ, when $z = 2$ are easily seen to be equivalent to (4).

We need to show that $L^2 \to L^2$ when $\mathrm{Re}z = 0$, which is reasonable, since T_z then has the character of a singular integral operator. Because of rotation and dilation invariance, this is accomplished by using spherical harmonics, and Mellin analysis. The result ends up being a consequence of Stirling's formula.
At the other end point of the interpolation, $\mathrm{Re}z = n$, and T_z then has the character of a logarithmic potential. This maps L^1 not into L^∞, but into BMO, which by a theorem of C. Fefferman and E.M. Stein ([6]) has the same interpolation properties as L^∞. The actual proof relies on the calculation of certain Mellin transforms, and on certain delicate uniform asymptotic estimates for the hypergeometric function. (See [10] for the full details). It is this last step which was later simplified by E.M. Stein [20].

References

[1] S. Agmon, Lower bounds for solutions of Schrödinger type equations in unbounded domains, Proc. of International Conference on Functional Analysis and related topics, (1969), Univ. Tokyo Press, Tokyo.

[2] S. Agmon, Lower bounds for solutions of Schrödinger equations, J. Analyse Math. 23 (1970), 1-25.

[3] W. Amrein, A. Berthier and V. Georgescu, L^p inequalities for the Laplacian and unique continuation, Ann. Inst. Fourier (Grenoble) 31 (1981), 153-168.

[4] A.M. Berthier, Sur le spectre ponctuel de l'opérateur de Schrödinger, C.R. Acad. Sci. Paris, Ser A. 290 (1980), 393-395.

[5] T. Carleman, Sur un problème d'unicité pour les systèmes d'équations aux derivées partielles a deux variables independantes, Ark. Mat. 26B (1939), 1-9.

[6] C. Fefferman and E.M. Stein, H^p-spaces of several variables, Acta Math. 129 (1972), 137-193.

[7] N. Garofalo and F.H. Lin, Monotonicity properties of variational integrals, Ap weights and unique continuation, to appear Indiana U. Math. J.

[8] V. Georgescu, On the unique continuation property for Schrödinger Hamiltonians, Helv. Phys. Acta 52 (1979), 655-670.

[9] L. Hörmander, Uniqueness theorems for second order elliptic differential equations, Comm in P.D.E. 8(1) (1983), 21-64.

[10] D. Jerison and C. Kenig, Unique continuation and absence of positive eigenvalues for Schrodinger operators, to appear, Annal of Math. 121, (1985).

[11] T. Kato, Growth properties of solutions of the reduced wave equation with variable coefficients, Comm. Pure and Appl. Math. 12 (1959), 403-425.

[12] C. Muller, On the behavior of the solution of the differential equation $\Delta u = f(x,u)$ in the neighborhood of a point, Comm. Pure and Appl. Math. 1 (1954), 505-515.

[13] M. Reed and B. Simon, Methods of Modern Mathematical Physics, Vol. 1-4, Academic Press, New York, 1978.

[14] F. Rellich, Über das asymptotische Verhalten der Lösungen von $\Delta u + \lambda u = 0$ in unendlichen Gebieten, Uber. Deutsch. Math. Verein 53 (1943), 57-65.

[15] J. Saut and B. Scheurer, Un théoreme de prolongement unique pour des opérateurs elliptiques dont les coefficients ne sont pas localement bornés, C.R. Acad. Sci. Paris Ser A 290 (1980), 598-599.

[16] M. Schechter and B. Simon, Unique continuation for Schrödinger operators with unbounded potential, J. Math. Anal. Appl. 77 (1980), 482-492.

[17] B. Simon, On positive eigenvalues of one-body Schrödinger operators, Comm. Pure and Appl. Math. 22 (1969), 531-538.

[18] B. Simon, Schrödinger semigroups, Bull. AMS 7(3) (1982), 447-526.

[19] E.M. Stein, Interpolation of linear operators, Trans. Amer. Math. Soc. 83, (1956), 482-492.

[20] E.M. Stein, Appendix to [10], to appear, Annals of Math. 121, (1985)

[21] J. Von Neumann and E. Wigner, Über merkwürdige diskrete Eigenwerte, Z. Phys. 30 (1929), 465-467.

TWINNING OF CRYSTALS (II)

David Kinderlehrer

School of Mathematics
University of Minnesota

1. Introduction

Certain properties of a crystalline substance may appear only below, for instance, a certain critical temperature and are frequently accompanied by equilibrium states exhibiting a marked decrease in symmetry. One such example of this is the appearance of twinned crystals in what sometimes may be regarded as an austenite/martensite transition. In the higher temperature austenite the crystal is cubic, while in the lower temperature martensite it is tetragonal. Another example is the appearance of spontaneous polarization in a ferroelectric, like Rochelle salt. Also in this instance, the crystal structure is more symmetric in the absence of spontaneous polarization.

It is tempting to regard the onset of these phenomena, the transitions in phase, as consequences in part of a decrease in symmetry and to attempt to predict some of their properties by the methods of thermoelasticity. It would be especially agreeable were this theory amenable to study by differential equations. The objective here is to examine the equations at the critical temperature. How does the material know what is about to happen next? If the transition is sufficiently gradual, it is legitimate to assume that it is of second order and that a smooth stored energy density governs behavior near this critical temperature. It may then be feasible to provide some insight about post-transition configurations by examining an appropriate boundary value problem. In the case of Rochelle salt, we are able to give a derivation of Müller's interaction theory pertinent to the transition, cf. §5.

One scope of this discussion is to illustrate the remarkable instability possessed by this boundary value problem. It turns out that the linearized system at the critical temperature has an infinite dimensional kernel and cokernel; the Agmon-Douglis-Nirenberg [1] complementing condition necessarily fails. The equations defining the kernel have real characteristics which indicate incipient twinning planes.

The plan of these remarks is to give a brief summary of the thermoelastic crystal and then to analyze the cubic/tetragonal and Rochelle salt transitions. We close with some observations about recent work.

2. Recapitulation of the Thermoelasticity of Crystals.

Here we follow Part I and Ericksen [11]. Consider the simplest crystal configuration, which we identify with a three-dimensional lattice described in terms of three linearly independent lattice vectors

$$a_1\ ,\ a_2\ ,\ a_3\ \varepsilon\ R^3\ .$$

A point $x \ \varepsilon\ R^3$ is on the lattice if and only if

$$x = n^1 a_1 + n^2 a_2 + n^3 a_3\ ,\quad n^i\ \varepsilon\ Z\ ,\ \text{or}$$
$$x = An\ ,\ A = \text{matrix with columns}\ \ a_1, a_2, a_3.$$
$$n = (n^1,\ n^2,\ n^3)\ \varepsilon\ R^3$$

The lattice vectors are not unique; in fact, selecting instead $\{a_1', a_2', a_3'\}$, or A' = matrix with columns a_1', a_2', a_3', we may write

$$x = A'n' \quad \text{for some}\ \ n'^i\ \varepsilon\ Z\ .$$

Thus the change of coordinates mapping

$$n = A^{-1}A'n' = Mn'\ ,$$

a bijection of the integer lattice Z^3, is of the form

$$M = (m_{ij})\ ,\ m_{ij}\ \varepsilon\ Z\ ,\ \det M = \pm 1 \qquad (2.1)$$

We denote by G this group of matrices.

Owing to its frame indifference, the Helmholtz free energy density may be written as

$$\Phi = \Phi(A^T A, \theta)\ ,$$

where θ denotes the temperature. Since Φ does not distinguish among choices of lattice vectors,

$$\Phi(A^TA,\theta) = \Phi(A'^TA',\theta) = \Phi(M^TA^TAM,\theta) , \quad M \in G .$$

Thus the energy density of the lattice exhibits the invariance property

$$\Phi(C,\theta) = \Phi(M^TCM,\theta) \quad \text{for } M \in G \text{ and } C = C^T , \det C > 0 . \tag{2.2}$$

Thermoelasticity is introduced into this framework via the Cauchy-Born hypothesis. Let $\{a_1,a_2,a_3\}$ be a fixed reference basis of lattice vectors and let $y(x)$, $F = \nabla y$, be a deformation of the lattice. Then the energy density (per unit reference volume) is given by

$$W(F,\theta) = \Phi((FA)^TFA,\theta) = \Phi(A^TCA,\theta) , \quad C = F^TF . \tag{2.3}$$

The Piola stress at F is given, as usual, by

$$S(F,\theta) = \frac{\partial W}{\partial F}(F,\theta) \tag{2.4}$$

and the stiffness tensor, which defines the linearized equations there, is denoted by

$$S'(F,\theta)[\xi] = \lim_{t\to 0} \frac{1}{t}(S(F + t\xi,\theta) - S(F,\theta)) , \quad \xi \text{ a } 3 \times 3 \text{ matrix.} \tag{2.5}$$

When $F = 1$, we write $S'(\theta)[\xi] = S'(1,\theta)[\xi]$.

We suppose the lattice vectors $A(\theta)$ to be chosen so that $W(1,\theta)$ is a local minimum of $W(F,\theta)$ so

$$S(1,\theta) = 0 \tag{2.6$_1$}$$

and

$$S'(\theta)[\xi] \cdot \xi > 0 \quad \text{for all } 3 \times 3 \text{ matrices } \xi . \tag{2.6$_2$}$$

Note that by (2.6_1), $S'(\theta)[\xi] = S'(\theta)[\frac{1}{2}(\xi + \xi^T)] = S'(\theta)[\xi]^T$. In our formulation, the reference lattice vectors at a temperature θ correspond to a natural equilibrium state.

So far, no mention has been made of equations; we bring them up now. An essential feature of this treatment of phase transition is that the process is sufficiently slow that temperature is to be regarded as a control parameter. A typical problem is to find the equilibrium configuration of a body occupying $\Omega \subset R^3$ with smooth boundary Γ, subjected to an appropriate distribution of forces and displacements, just as in an ordinary problem in finite elasticity. As an instance, consider a class Y of deformations of Ω constrained in some way, perhaps, on all or part of Γ and lying in a favorite function space. One formulation may be given in terms of energy:

$$\begin{aligned} \text{Find } \ y \in Y\text{: } E(y) &= \int_\Omega W(\nabla y,\theta)dx - \int_\Omega f\cdot y \; dx - \int_\Gamma g\cdot y \; dS \\ &= \inf_{\eta \in Y} E(\eta) \; . \end{aligned}$$

Another may be to ask for a solution

$$\begin{aligned} y \in Y\text{:} \quad - \operatorname{div} S(\nabla y,\theta) &= f \quad \text{in } \Omega \; , \\ S(\nabla y,\theta)\nu &= g \quad \text{on } \Gamma \end{aligned}$$

with ν the outer normal to Γ. A solution may be sought at a given temperature θ_1 or a family may be sought varying with θ. Other boundary conditions are also of interest, or to be more precise, the nature of the boundary conditions relative to a particular environment is not always clear, cf. Podio-Guidugli and Vergara-Caffarelli [24]. A minimum of the variational principle may or may not correspond to a solution of the system of differential equations, and vice-versa.

It might be worthwhile to recall from Part I that since

$$W(M,\theta) = W(1,\theta) \quad \text{for } \ M \in G \; ,$$

cf. (2.2), W does not have the growth properties as $|F| \to \infty$ usually associated with direct methods in the calculus of variations, Ball [2].

In certain circumstances, for example, if $\{a_1,a_2,a_3\}$ are mutually orthogonal and

$$|a_1| = |a_2| \neq |a_3|, \tag{2.7}$$

we have seen in Part 1 that there are simple shearing motions $y(x)$, $F = \nabla y = 1 + \xi \otimes n$, with

$$W(F) = W(1) = \text{local minimum},$$

where temperature dependence is momentarily suppressed. However the entire path of deformations $y_t(x)$, $F_t = 1 + t\xi \otimes n$, $0 < t < 1$, cannot be experienced by the body since for some $t \in (0,1)$,

$$S'(F_t)[\xi \otimes n] \cdot (\xi \otimes n) = \frac{\partial^2 W}{\partial F_{ij}\, \partial F_{hk}} (F_t) \xi_i \xi_h n_j n_k < 0 ,$$

violating the local condition of material stability.

Moreover, let $E \subset R$ be a given set of positive measure, χ its characteristic function, and set

$$h(x) = \int_0^{n \cdot x} \chi(t)dt ,$$

$$y(x) = x + h(x)\xi .$$

Then $\nabla y = 1 + \chi(n \cdot x)\xi \otimes n$ satisfies $S(\nabla y) = 0$, so uniqueness and regularity for solutions of the traction boundary value problem fail in a rather striking way.

3. Decrease in Symmetry and the Onset of Transition.

A given configuration A has a symmetry group, for $C = A^T A$,

$$L(C) = \{M \in G: M^T C M = C\} \tag{3.1}$$

Generally speaking, at a stable minimum, (2.6) will hold in a strict sense, so at $\theta = \overline{\theta}$,

$$S'(\overline{\theta})[\varepsilon] \cdot \varepsilon > \text{const.}\ |\varepsilon|^2 , \quad \varepsilon = \varepsilon^T . \tag{3.2}$$

Interpreted in terms of Φ,

$$\nabla\Phi(C,\overline{\theta}) = 0 \quad \text{and} \quad \nabla^2\Phi(C,\overline{\theta}) > 0 \text{ on symmetric matrices.}$$

So near $\overline{\theta}$, we may find a unique family $A(\theta)$, $|A(\theta) - A(\overline{\theta})|$ small, so that

$$W(1,\theta) = \Phi(A(\theta)^T A(\theta), \theta)$$

is again a local minimum.

Suppose now that we are given a continuous family of local minima $A(\theta)$, $C(\theta) = A(\theta)^T A(\theta)$, for $|\theta - \theta_0|$ small such that

$$L(C(\theta)) = L(C(\theta_0)) \quad \text{for} \quad \theta > \theta_0 \quad \text{and}$$
$$L(C(\theta)) \subset L(C(\theta_0)) \quad \text{but} \quad L(C(\theta)) \neq L(C(\theta_0)) \quad \text{for} \quad \theta < \theta_0.$$

For example the crystal might be cubic, $A(\theta) = \alpha(\theta)1$, for $\theta > \theta_0$, with $A(\theta)$ tetragonal, namely satisfying (2.7), for $\theta < \theta_0$. Consequently there is an $M \in L(C(\theta_0))$ such that

$$M^T C(\theta) M \neq C(\theta) \qquad \text{for} \quad \theta < \theta_0 .$$

Since

$$\Phi(M^T C(\theta)M, \theta) = \Phi(C(\theta), \theta) = \min ,$$

there is no unique path of minima passing through $A(\theta_0)$ and thus $\nabla^2 \Phi(C(\theta_0), \theta_0)$ is not positive definite on symmetric matrices.

In this way we see that a decrease in symmetry of the minimum energy configuration induces some kind of degenerate behavior in the energy density as well as a multiplicity of configurations. Our concern is the nature of this degeneracy and its relationship to the phase transition.

4. The Cubic/Tetragonal Transition.

Here again, our principal references are Ericksen [11] and Part I. Other papers connected to the subject are listed in the references, as well as the references of Part I. The articles [10], [11] include a discussion of the nature of the physical hypotheses which lead us to think that an analysis in this spirit is justified.

Suppose that the body occupies a (reference) region $\Omega \subset R^3$ with $\Gamma = \partial\Omega$ smooth and that for $\theta > \theta_0$,

$$a_i(\theta) = \alpha(\theta)e_i \ , \ e_i \ \text{the Euclidean basis, } i = 1,2,3, \ \ \alpha(\theta_0) = 1,$$

where the a_i are lattice vectors realizing a local minimum of Φ. So $A(\theta) = \alpha(\theta)1$ and

$$\Phi(\alpha(\theta)^2 1, \theta) = \min \quad \text{for} \quad \theta > \theta_0.$$

Suppose that for $\theta < \theta_0$, the lattice vectors realizing a local minimum of Φ are given by

$$a_1(\theta) = \beta(\theta)e_1 \; , \; a_2(\theta) = \gamma(\theta)e_2, \; a_3(\theta) = \gamma(\theta)e_3$$

with $\beta(\theta) - \gamma(\theta) \neq 0$ for $\theta < \theta_0$, but, of course, $\alpha(\theta_0) = \beta(\theta_0) = \gamma(\theta_0) = 1$. So, for $C(\theta) = A(\theta)^T A(\theta)$,

$$L(C(\theta)) \subsetneq L(C(\theta_0)) \quad \text{for} \quad \theta < \theta_0$$

and in particular

$$C_1(\theta) = \text{diag}(\beta(\theta)^2, \gamma(\theta)^2, \gamma(\theta)^2), \quad C_2(\theta) = (\gamma(\theta)^2, \beta(\theta)^2, \gamma(\theta)^2),$$
$$C_3(\theta) = \text{diag}(\gamma(\theta)^2, \gamma(\theta)^2, \beta(\theta)^2)$$

are all minima of $\Phi(C,\theta)$.

Before continuing, let us note that the form of the linearized equations at a cubic natural state $A = \alpha 1$ is well established. It follows from the invariance of C under $L(C)$, here all orthogonal matrices with integer coefficients, that there are three constants c_{11}, c_{12}, c_{44} such that

$$S'(\theta)[\varepsilon] = 4\alpha^4 \nabla^2 \Phi(\alpha^2 1, \theta)[\varepsilon],$$

$$S'(\theta)[\varepsilon] = c_{11}\begin{pmatrix} \varepsilon_{11} & & 0 \\ & \varepsilon_{22} & \\ 0 & & \varepsilon_{33} \end{pmatrix} + c_{12}\begin{pmatrix} \varepsilon_{22}+\varepsilon_{33} & & 0 \\ & \varepsilon_{11}+\varepsilon_{33} & \\ 0 & & \varepsilon_{11}+\varepsilon_{22} \end{pmatrix}$$
$$+ c_{44}\begin{pmatrix} 0 & \varepsilon_{12} & \varepsilon_{13} \\ \varepsilon_{12} & 0 & \varepsilon_{23} \\ \varepsilon_{13} & \varepsilon_{23} & 0 \end{pmatrix}.$$

In fact,

$$c_{11} = 4\alpha^4 \frac{\partial^2\Phi}{\partial C_{ii}^2} \;, \; c_{12} = 4\alpha^4 \frac{\partial^2\Phi}{\partial C_{ii}\partial C_{jj}} \;, \; c_{44} = 4\alpha^4 \frac{\partial^2\Phi}{\partial C_{ij}^2} \;, \; i \neq j \;.$$

Details of such calculations may be found in Love [20].

Now for $\theta < \theta_0$,

$$\begin{aligned} 0 &= \nabla\Phi(C_2(\theta),\theta) - \nabla\Phi(C_1(\theta),\theta) \\ &= \int_0^1 \nabla^2\Phi(C_1(\theta) + t(C_2(\theta) - C_1(\theta)),\theta)[C_2(\theta) - C_1(\theta)]dt \end{aligned}$$

and since $\beta^2 - \gamma^2 \neq 0$,

$$0 = \int_0^1 \nabla^2\Phi(C_1(\theta) + t(C_2(\theta) - C_1(\theta)),\theta)[\xi]dt$$

where $\xi = \mathrm{diag}(1,-1,0)$. Now let $\theta \to \theta_0$ where $C_1(\theta_0) = C_2(\theta_0) = 1$. Thus

$$S'(\theta_0)[\xi] = \nabla^2\Phi(1,\theta_0)[\xi] = 0$$

or $c_{11} = c_{12}$. We assume that this is the only degeneracy of $S'(\theta_0)$ so that $c_{11} + 2c_{12} = 3c_{11} > 0$ and $c_{44} > 0$. We then have that

$$S'(\theta_0)[\xi] = c_{11}\mathrm{tr}\ \varepsilon\ 1 + c_{44}\begin{pmatrix} 0 & \varepsilon_{12} & \varepsilon_{13} \\ \varepsilon_{12} & 0 & \varepsilon_{23} \\ \varepsilon_{13} & \varepsilon_{23} & 0 \end{pmatrix}. \tag{4.2}$$

Let $v(x)$ be a solution of the linearized equations

$$\begin{aligned} -\mathrm{div}\ S'(\theta_0)[\nabla v] &= 0 \quad \text{in } \Omega\ , \\ S'(\theta_0)[\nabla v]\nu &= 0 \quad \text{in } \Gamma. \end{aligned} \tag{4.3}$$

Multiplying the equation (4.3) by v and integrating by parts implies that

$$\int_\Omega\{c_{11}(\mathrm{div}\ v)^2 + 2c_{44}(\varepsilon_{12}(v)^2 + \varepsilon_{23}(v)^2 + \varepsilon_{13}(v)^2)\}dx = 0.$$

thus v is in the kernel of the mapping (*)

$$v \to (-\mathrm{div}\ S'(\theta_0)[\nabla v],\ S'(\theta_0)[\nabla v]\nu) \tag{4.4}$$

(*) In order to maintain an elementary level of exposition we suppress naming the technical spaces.

if and only if it satisfies the four equations

$$\left.\begin{aligned} \operatorname{div} v &= 0 \\ \varepsilon_{12}(v) &= 0 \\ \varepsilon_{23}(v) &= 0 \\ \varepsilon_{13}(v) &= 0 \end{aligned}\right\} \quad \text{in } \Omega \tag{4.5}$$

The principal symbol matrix of this system,

$$\begin{pmatrix} \xi_1 & \xi_2 & \xi_3 \\ \xi_2 & \xi_1 & 0 \\ 0 & \xi_3 & \xi_2 \\ \xi_3 & 0 & \xi_1 \end{pmatrix}$$

is suggestive of a symmetric hyperbolic system. In fact, solutions of (4.5) may be characterized in terms of six arbitrary real valued functions $f_1(t),\ldots,f_6(t)$ of a real variable t by

$$\begin{aligned} v(x) = &\begin{pmatrix}1\\1\\0\end{pmatrix} f_1(x_1 - x_2) + \begin{pmatrix}1\\-1\\0\end{pmatrix} f_2(x_1 + x_2) + \begin{pmatrix}0\\1\\1\end{pmatrix} f_3(x_2 - x_3) + \\ &\begin{pmatrix}0\\1\\-1\end{pmatrix} f_4(x_2 + x_3) + \begin{pmatrix}1\\0\\1\end{pmatrix} f_5(x_1 - x_3) + \begin{pmatrix}1\\0\\-1\end{pmatrix} f_6(x_1 + x_3) \end{aligned} \tag{4.6}$$

Thus the kernel of (4.4) is infinite dimensional. Its cokernel is also. Possible planes to serve as twin boundaries start, at $\theta = \theta_0$, as the planes normal to the six vectors appearing in (4.6). They are not all kinematically compatible, cf. James [16], [17].

The system defined by (4.2) is not completely degenerate. The problem

$$\begin{aligned} -\operatorname{div} S'(\theta_0)[\nabla v] &= 0 \quad \text{in } \Omega , \\ v &= 0 \quad \text{on } \Gamma \end{aligned} \tag{4.7}$$

has only the trivial solution $v = 0$, and the corresponding operator is a Fredholm operator. This suggests that perfect, or sufficient, control over the deformation at the boundary may lead to control or even elimination of twinning. We shall attempt to illuminate this issue in our discussion of Rochelle salt.

A different system, related to isotropic elasticity, with an infinite-dimensional kernel was found in Simpson and Spector [25].

5. A Ferroelectric Transition: Motivation

Ferroelectric crystals are distinguished by the presence of a spontaneous electric dipole moment, or polarization, which can be reversed, with no change in magnitude, by an applied electric field. This spontaneous polarization may be limited to a certain temperature range. Rochelle salt, $NaKC_4H_4O_6 \cdot 4H_2O$, a ferroelectric between -18°C and 24°C, is a suitable example to take up here. The evidence for this is primarily that

- it is always piezoelectric,
- there is no spontaneous polarization at the critical temperatures (or Curie points),
- the spontaneous polarization may reasonably be taken to be a continuous function,
- the specific heat at the Curie point has a very small discontinuity.

This information may be found in [6], [18], [27], [19], [21] as well as other places. There seems to be general agreement that the transition in question is of second order, among workers who believe in second-order transitions. Not all ferroelectrics enjoy these properties; the well known example of $BaTiO_3$ is one.

The first macroscopic theory describing Rochelle salt was the "interaction theory" of Müller dating to 1940. Tisza [27] used it as an example of his quasi thermo-dynamics. Here we show that the relations of Müller's theory may be seen as a result of the decrease in symmetry in passing to the state which admits spontaneous polarization. One consequence of this is that the domain structure is accompanied by mechanical twinning. This seems to be recognized. The results of the analysis, in so far as we are concerned with a continuum theory, seem to be in general agreement with the experimental facts.

6. Description of the Elastic Dielectric.

The elastic dielectric is presumed to admit a stored energy density of deformation F, temperture θ, and polarization density $p \in R^3$, $W(F,p,\theta)$. Perhaps the clearest explanation of this theory at the continuum level remains Toupin [28]. We wish to discuss a dielectric crystal, so by analogy with the preceding, assume $\{a_1,a_2,a_3\}$ are lattice vectors and an energy density is given of the form

$$\Psi(A,p,\theta), \quad A = \text{matrix with columns } a_1, a_2, a_3 . \qquad (6.1)$$

Now frame-indifference demands that

$$\Psi(A,p,\theta) = \Psi(QA,Qp,\theta) , \; Q^TQ = 1 ,$$

so after a little manipulation we find a Φ such that

$$\Psi(A,p,\theta) = \Phi(A^TA,A^{-1}p,\theta).$$

The invariance of Φ under change in lattice vectors may be determined as in the previous situation when there was no polarization. Let us briefly review this argument, which is due to J. Ericksen. Note that the coordinates of p in the basis $\{a_1,a_2,a_3\}$, $[p]_A$, is just given by the formula

$$p = A[p]_A$$

so

$$\Psi(A,p,\theta) = \Phi(A^TA,[p]_A,\theta) .$$

If A' is the matrix of another set of lattice vectors, then Ψ does not distinguish between A and A'. So with $A' = AM$, $M \in G$,

$$\Psi(A',p,\theta) = \Psi(A,p,\theta) , \text{ and}$$

$$p = A'[p]_{A'} = AM[p]_{A'} = A[p]_A ,$$

so

$$\begin{aligned}\Phi(A^TA,[p]_A,\theta) &= \Phi(M^T(A^TA)M, [p]_{A'},\theta) \\ &= \Phi(M^T(A^TA)M, M^{-1}[p]_A,\theta)\end{aligned}$$

Reviving our prejudice for Euclidean coordinates, we may express this as

$$\Phi(A^T A, A^{-1}p, \theta) = \Phi(M^T(A^T A)M, M^{-1}A^{-1}p, \theta), \quad M \in G \tag{6.2}$$

To set ourselves in the framework of elasticity theory, define

$$W(F,p,\theta) = \Phi(A^T CA, A^{-1}F^{-1}p, \theta), \quad \text{for } 3 \times 3 \text{ matrices } F \tag{6.3}$$

with $\det F > 0$.

The stress tensor in the dielectric and the field equations satisfied by the deformed body are derived in Toupin [28] by means of a principle of virtual work and are noted here in the briefest fashion. Another treatment, perhaps less formal but emphasizing the transitions we have in mind, may be found in Grindlay [14].

Suppose that the material occupies a reference region $\Omega \subset R^3$ with smooth boundary Γ and experiences a motion $y = y(x)$ and a polarization $P = \rho p$, where $\rho = \rho_0/\det F$ is the density. Here $\rho_0 > 0$, a constant, is the reference density. In the region $y(\Omega)$ the stress tensor of the body is given by

$$T = \rho\, S_0\, F^T + T_{MS} \tag{6.4}$$

with

$$S_0(F,p,\theta) = \frac{\partial W}{\partial F}(F,p,\theta)$$

and

$$T_{MS} = E_{MS} \otimes E_{MS} + E_{MS} \otimes P - \frac{1}{2}|E_{MS}|^2\, 1,$$

the Maxwell self stress defined in terms of the self-field E_{MS} determined by the polarization, in a convenient system of units. Note that

$$\begin{aligned} E_{MS}(\eta) &= -\nabla\left\{\frac{1}{4\pi}\int_{y(\Omega)} P \cdot \nabla_{\eta'}\left(\frac{1}{|\eta - \eta'|}\right)d\eta'\right\}, \quad \eta \in R^3 \\ &= -\nabla\phi(\eta). \end{aligned}$$

In summary, the stress must reflect that the comportment of the body is influenced by the electric field induced by the polarization as well as the polarization.

Let us briefly take note of the field equations. Let div_y denote the differential operator defined on vectors or matrices depending on the deformation $y(x)$, $F = \nabla y$,

$$\mathrm{div}_y C = \sum \frac{\partial C^i}{\partial y^i} = \sum \frac{\partial C^i}{\partial x_k} F^{-1}_{ki} \ .$$

in the presence of an electric field E_0 applied to the system,

$$\begin{aligned} &-\mathrm{div}_y T = (\nabla_y E_0)P, \\ &\frac{\partial W}{\partial p} + E_{MS} + E_0 = 0 \quad \text{in} \quad \Omega, \\ &-\Delta_y \phi + \mathrm{div}_y\ P = 0. \end{aligned} \tag{6.6}$$

In terms of the reference configuration, the Piola-Kirchoff stress is

$$S(F,p,\theta) = \frac{1}{\rho} TF^{-T}$$

or

$$S(F,p,\theta) = S_0(F,p,\theta) + T_{MS} \det F\ F^{-T} \tag{6.7}$$

and the first equation of (6.6) may be written

$$-\mathrm{div}\ S = \nabla_y\ E_0 p \ . \tag{6.8}$$

Boundary conditions are also important. Here our tale assumes greater interest, because I was unable to determine to my satisfaction the kernel of the linearized system we are to discuss employing the common ones, based on the dielectric crystal and an isolated external conductor. However, were the crystal grounded instead, there was some promise. I then learned from Kanzig ([18], p. 33) that this seems to be the condition under which the anomalies in the behavior of Rochelle salt and KH_2PO_4 are observed. Such a boundary condition amounts to specifying the potential ϕ on $\partial\Omega$, say ϕ vanishes there.

Indeed the discussion of [12] suggests that no anamolous behavior ought to be observed in an insulated crystal, whose boundary conditions are very similar to the crystal/conductor system. We examine this possibility later. It is also an example of live boundary conditions, which have been studied more carefully by Gurtin and Spector, Podio-Guidugli and Vergara-Caffarelli, Podio-Guidugli, Vergara-Caffarelli, and Virga, and Spector, [15], [23], [24], [26].

By the natural state $y(x) = Fx$, $p \in R^3$ fixed at a temperature θ, we intend that for F and p,

$$\left.\begin{aligned} T &= 0, \\ -\frac{\partial W}{\partial p}(F,p,\theta) + E &= 0, \end{aligned}\right\} \quad \text{in } \Omega \tag{6.9}$$

the last equation of (6.6) being satisfied automatically. So a natural state may be thought of as merely a homogeneous state for which $T = 0$, since E may be determined from the second relation. We further stipulate that at a natural state $y(x) = x$, $p = 0$,

$$\frac{\partial W}{\partial p}(1,0,\theta) = 0,$$

so

$$T_{MS} = 0 \quad \text{and} \quad S_o = 0 . \tag{6.10}$$

Homogeneous natural states may be thought of as critical points of the energy density

$$\rho W(F,p,\theta) \quad - \frac{1}{2}|\nabla_y \phi|^2 + \nabla_y \phi \cdot p$$

when ϕ is linear in y. Agreeing to think of vectors as columns, so

$$\nabla_y \phi = F^{-T} \nabla \phi$$

and choosing $\eta = -\nabla\phi$, or $E = F^{-T}\eta$, we may write this density in the reference configuration as

$$\psi = W - \frac{1}{2}|F^{-T}\eta|^2 \det F - F^{-T}\eta \cdot p . \tag{6.11}$$

Now

$$d\psi = W_F \cdot dF + F^{-T}\eta \cdot F^{-T}\, dF^T\, F^{-T}\eta \det F - \frac{1}{2}\,|E|^2 \det F\, F^{-T} \cdot dF$$

$$+ F^{-T}\, dF^T\, F^{-T}\eta \cdot p + W_p \cdot dp - F^{-T}\eta \cdot dp\ .$$

Since $F^{-T}dF^T E \cdot p = E \otimes p\, F^{-T} \cdot dF$,

$$d\psi = (W_F + E \otimes E \det FF^{-T} + E \otimes pF^{-T} - \frac{1}{2}\,|E|^2 \det FF^{-T}) \cdot dF$$

$$+ (W_p \cdot dp - E \cdot dp)$$

or

$$d\psi = S \cdot dF + (W_p - E) \cdot dp. \tag{6.12}$$

Consequently, $y(x) = Fx$ and p is a natural state if $d\psi = 0$.

The conditions (6.9) are two equations in three unknowns (or 6 equations in 9 unknowns); we may think of holding one set fixed and solving for the rest. We have thereby many natural states.

In the "experiment" to be discussed, we postulate the existence of $F = F(\theta)$ and $p = p(\theta) \neq 0$, for θ near the critical temperature, such that

$$\left.\begin{aligned} T &= 0 \\ \frac{\partial W}{\partial p} &= 0 \end{aligned}\right\} \quad \text{in } \Omega. \tag{6.13}$$

This corresponds to imposing the boundary conditions

$$\left.\begin{aligned} S\nu &= 0 \\ \phi &= 0 \end{aligned}\right\} \quad \text{on } \Gamma \tag{6.14}$$

and then assuming the resulting configuration to be homogeneous. More precisely, we choose a potential ϕ for η, cf. (6.11), with

$$\phi = 0 \quad \text{on} \quad \Gamma.$$

For then, since $p(\theta)$ and $F(\theta)$ are constants,

$$\Delta\phi = \operatorname{div} P = 0 \quad \text{in} \quad \Omega,$$

so $\phi = 0$ by the maximum principle.

7. The Transition in Rochelle Salt.

Briefly stated, Rochelle salt is orthorhombic in its paraelectric phases and monoclinic in its ferroelectric one. Near its upper Curie point θ_0, for example, it has lattice vectors

$$a_i(\theta) = \alpha_i(\theta)e_i \;, \quad i = 1,2,3 \quad \text{(no sum)}, \quad \theta > \theta_0 \tag{7.1}$$

with the $\alpha_i(\theta)$ all different and it has no spontaneous polarization([1]). Below θ_0, but above the lower Curie point, it has lattice vectors

$$\begin{aligned} a_1(\theta) &= \alpha_1(\theta)e_1 \\ a_2(\theta) &= \alpha_2(\theta)e_2 \\ a_3(\theta) &= \gamma(\theta) = \gamma^2(\theta)e_2 + \gamma^3(\theta)e_3 \end{aligned} \qquad \theta < \theta_0 \tag{7.2}$$

and exhibits the spontaneous polarization

$$p(\theta) = g(\theta)e_1 \qquad \theta < \theta_0 \tag{7.3}$$

and Maxwell self-field

$$E_{MS} = f(\theta)e_1 \qquad \theta < \theta_0 \,. \tag{7.4}$$

We shall take (7.1) and (7.2), (7.3) to be natural states.

The linearized equations of (6.6) at θ_0, $F = 1$, $p = 0$ are simply, for variations v, q, ϕ,

$$\begin{aligned} -\operatorname{div}\{S_0'(\theta_0)[\nabla v] + Hq\} &= 0 \\ \nabla\phi + H\varepsilon + bq &= 0 \qquad \text{in} \quad \Omega \\ -\Delta\phi + \operatorname{div} q &= 0 \end{aligned} \tag{7.5}$$

where by H we mean the tensor

$$H = (h_{ij\ell}) = \left(\frac{\partial^2 W}{\partial F_{ij} \partial p_\ell} (1,0,\theta_0) \right) ,$$

$$Hq = (h_{ij\ell} q_\ell) \qquad 3 \times 3 \text{ matrix},$$

$$H\varepsilon = (h_{ij\ell} \varepsilon_{ij}) \qquad \text{a vector in } R^3,$$

and

$$bq = (b_{ij} q_j) = \left(\frac{\partial^2 W}{\partial p_i \partial p_j} (1,0,\theta_0) q_j \right).$$

These are nothing more than the classical equations for linear piezoelectricity. The boundary conditions (6.7) become

$$\left. \begin{aligned} (S_0'(\theta_0)[\nabla v] + Hq)\nu &= 0 \\ \phi &= 0 \end{aligned} \right\} \text{ on } \Gamma . \tag{7.6}$$

The orthorhombic symmetry of Rochelle salt implies that

$$S_0'(\theta_0)[\varepsilon] = \begin{pmatrix} a_{11kk}\,\varepsilon_{kk} & 0 & 0 \\ 0 & a_{22kk}\,\varepsilon_{kk} & 0 \\ 0 & 0 & a_{33kk}\,\varepsilon_{kk} \end{pmatrix}$$

$$+ \begin{pmatrix} 0 & a_{1212}\varepsilon_{12} & a_{1313}\varepsilon_{13} \\ a_{1212}\varepsilon_{12} & 0 & a_{2323}\varepsilon_{23} \\ a_{1313}\varepsilon_{13} & a_{2323}\varepsilon_{23} & 0 \end{pmatrix} \tag{7.7}$$

and it is known that

$$bq = \begin{pmatrix} b_1\, q_1 \\ b_2\, q_2 \\ b_3\, q_3 \end{pmatrix} , \quad b_i > 0 , \ i = 1,2,3. \tag{7.8}$$

The nonzero piezoelectric coefficients $h_{ij\ell} = h_{ij\ell}$ are

$$h_{231}, \ h_{312}, \text{ and } h_{123} .$$

We now employ the material symmetry available from (6.2). For example, with

$$M = \begin{pmatrix} -1 & 0 & 0 \\ 0 & -1 & 0 \\ 0 & 0 & 1 \end{pmatrix} , \tag{7.9}$$

$$\Phi(M^T CM, M^{-1}A^{-1}p, \theta) = \Phi(C, A^{-1}p, \theta) .$$

Thus, with the obvious notations,

$$M \in L(C(\theta), 0) \quad \text{for} \quad \theta > \theta_0 \quad \text{but}$$

$$M \notin L(C(\theta), p(\theta)) \quad \text{for} \quad \theta < \theta_0$$

Indeed, we may calculate that for $\theta_0 - \delta < \theta < \theta_0$,

$$C(\theta) = \begin{pmatrix} \alpha_1^2 & 0 & 0 \\ 0 & \alpha_2^2 & \alpha_2\gamma^2 \\ 0 & \alpha_2\gamma^2 & |\gamma|^2 \end{pmatrix} , \quad A^{-1}p = \frac{g}{\alpha_1} e_1 ,$$

and (7.10)

$$M^T C(\theta) M = \begin{pmatrix} \alpha_1^2 & 0 & 0 \\ 0 & \alpha_2 & -\alpha_2\gamma^2 \\ 0 & -\alpha_2\gamma^2 & |\gamma|^2 \end{pmatrix} , \quad M^{-1}A^{-1}p = -\frac{g}{\alpha_1} e_1$$

At this point we may argue as in §5. Since $S = 0$ for each natural state when $\theta < \theta_0$,

$$\nabla_F \Phi(C(\theta), A(\theta)^{-1}p, \theta) + T_{MS} = 0$$

and

$$\nabla_F \Phi(M^T C(\theta) M, (MA(\theta))^{-1}p, \theta) + T_{MS} = 0 .$$

Recalling that T_{MS} vanishes quadratically with $|p|$ as $\theta \to \theta_0$, we calculate that

$$\nabla_F^2 \Phi(C(\theta_0), 0, \theta_0)[\xi] + \frac{\partial}{\partial p} \nabla_F \Phi (C(\theta_0), 0, \theta_0) q = 0$$

for

$$\xi = \begin{pmatrix} 0 & 0 & 0 \\ 0 & 0 & \lambda \\ 0 & \lambda & 0 \end{pmatrix} \qquad \text{and} \quad q = \begin{pmatrix} \mu \\ 0 \\ 0 \end{pmatrix}$$

where λ and μ are nonzero coefficients. From the second equation in (6.13),

$$\nabla_F \Phi_p(C(\theta_0), 0, \theta_0)[\xi] + \frac{\partial^2}{\partial p^2} \phi(C(\theta_0), 0, \theta_0) q = 0.$$

Thus after some checking, we find

$$S_o'(\theta_o)[\xi] + Hq = 0 \,, \tag{7.11}$$

$$H\xi + bq = 0 \,.$$

Given the form of the linearized equations (7.7) and (7.8),

$$\begin{aligned} a_{2323}\lambda + h_{231}\mu &= 0 \,, \\ h_{231}\lambda + b_1\mu &= 0 \,. \end{aligned} \tag{7.12}$$

Thus

$$a_{2323}b_1 - (h_{231})^2 = 0 \tag{7.13}$$

which is Müller's result, cf. Tisza [27], for example.

Now it is easy to check by a local stability argument, based on the energy density (6.12), that the form

$$U(\varepsilon, q) = \frac{1}{2} S_o'(\theta_o)[\varepsilon] \cdot \varepsilon + h_{ij\ell}\, \varepsilon_{ij} q_\ell + \frac{1}{2} b_{ij} q_i q_j \geq 0 \,. \tag{7.14}$$

So if $U(\varepsilon, q) = 0$ for some $\varepsilon = \varepsilon^T$ and q, then

$$\begin{aligned} S_0'(\theta_o)[\varepsilon] + Hq &= 0 \,, \\ H\varepsilon + bq &= 0 \,. \end{aligned} \tag{7.15}$$

In the present circumstances this means that

$$a_{2323}\varepsilon_{23} + h_{231}q_1 = 0 ,$$
$$h_{231}\varepsilon_{23} + b_1 q_1 = 0 . \tag{7.16}$$

There are analogous relations between ε_{12} and q_3 and ε_{13} and q_2, and

$$\varepsilon_{11} = \varepsilon_{22} = \varepsilon_{33} = 0$$

Now the least anomolous situtation is that (7.13) is the only degeneracy of the form $U(\varepsilon,q)$. Assuming this, (7.13) implies that (7.14) holds and

$$\varepsilon_{11} = \varepsilon_{22} = \varepsilon_{33} = \varepsilon_{12} = \varepsilon_{13} = q_2 = q_3 = 0 \tag{7.17}$$

We now inspect the kernel of the linearized equations (7.5) and (7.6)

Multiplying the first equation of (7.5) by v and integrating by parts gives

$$\int_\Omega \{S_0'(\theta_0)[\varepsilon(v)] + Hq\} \cdot \varepsilon(v)dx = 0 . \tag{7.18}$$

Multiplying the second equation by q and integrating, we see that

$$\begin{aligned} \int_\Omega (H\varepsilon + q) \cdot q\, dx &= -\int_\Omega \nabla\phi \cdot q\, dx \\ &= \int_\Omega \phi \operatorname{div} q\, dx - \int_\Gamma \phi\, q \cdot \nu dS \\ &= \int_\Omega \phi \operatorname{div} q\, dx \\ &= \int_\Omega \phi\, \Delta\phi\, dx \\ &= -\int_\Omega |\nabla\phi|^2 dx + \int_\Gamma \phi_\nu \phi dS \\ &= -\int_\Omega |\nabla\phi|^2 dx . \end{aligned} \tag{7.19}$$

From (7.18) and (7.19) we have that for a pair v,q in the kernel,

$$\int_\Omega U(\varepsilon(v),q)dx = -\int_\Omega |\nabla\phi|^2 dx \leq 0 .$$

So

$$\left.\begin{aligned} \phi &= 0 \\ U(\varepsilon(v),q) &= 0 \end{aligned}\right\} \text{ in } \Omega , \tag{7.20}$$

from which we infer that (7.16) and (7.17) are satisfied. It remains to identify this space; an elementary calculation indicated here so the author does not forget it.

First of all, the conditions of (7.17) give that

$$q = (q_1, 0, 0)$$

and

$$v^1 = v^1(x_2,x_3), \quad v^2 = v^2(x_1,x_3), \quad v^3 = v^3(x_1,x_2).$$

Moreover, div $q = \Delta\phi = 0$, so $q_1 = q_1(x_2,x_3)$. Now $(v^1,v^2,0)$ is an affine rigid motion in x_1,x_2 and $(v^1,0,v^3)$ is one in x_1,x_3, thus there are scalar functions $\mu(x_3)$ and $\lambda(x_2)$ and vector functions $f(x_3)$ and $g(x_2)$ such that

$$\begin{pmatrix} v^1 \\ v^2 \\ 0 \end{pmatrix} = \begin{pmatrix} 0 & \mu(x_3) & 0 \\ -\mu(x_3) & 0 & 0 \\ 0 & 0 & 0 \end{pmatrix} \begin{pmatrix} x_1 \\ x_2 \\ 0 \end{pmatrix} + f(x_3) \quad \text{and}$$

$$\begin{pmatrix} v^1 \\ 0 \\ v^3 \end{pmatrix} = \begin{pmatrix} 0 & 0 & \lambda(x_2) \\ 0 & 0 & 0 \\ -\lambda(x_2) & 0 & 0 \end{pmatrix} \begin{pmatrix} x_1 \\ 0 \\ x_3 \end{pmatrix} + g(x_2).$$

Comparing the representations for v^1 gives that $\mu(x_3)$, $\lambda(x_2)$ are linear and $f^1(x_3)$ and $g^1(x_2)$ are linear so that

$$v^1(x_2,x_3) = \mu_1 x_3 x_2 + \mu_0 x_2 + \lambda_1 x_1 x_3 + \lambda_0 x_3 + \kappa, \quad \kappa \text{ constant.}$$

Now

$$v^2(x_1,x_3) = -(\mu_1 x_3 + \mu_0)x_1 + f^2(x_3) ,$$
$$v^3(x_1,x_2) = -(\lambda_1 x_2 + \lambda_0)x_1 + g^3(x_2).$$

From (7.13),

$$q_1(x_2,x_3) = -\frac{h_{231}}{2b_1}\left(f^{2\prime}(x_3) + g^{3\prime}(x_2)\right) .$$

In summary, to within an affine rigid motion and some torsions, pairs (v,q) in the kernel are given by

$$v = \begin{pmatrix} 0 \\ \psi(x_3) \\ \eta(x_2) \end{pmatrix} \quad \text{and} \quad q = -\frac{h_{231}}{2b_1} \begin{pmatrix} \psi'(x_3) + \eta'(x_2) \\ 0 \\ 0 \end{pmatrix} \tag{7.21}$$

for arbitrary ψ, η.

Observed twin regimes, or domains, as they are called in the theory of ferroelectricity or ferromagnetism, are obtained by setting either $\psi = 0$ or $\eta = 0$. Both occur, although the more stable are those with walls parallel to the x_1, x_2 - plane, which is the case $\eta = 0$. The analysis of stability obviously demands analysis of the nonlinear system. One configuration is not a $\pi/2$ rotation in the x_2, x_3 - plane of the other because the crystal structure starts at orthorhombic, with $\alpha_2 \neq \alpha_3$. In fact, $\alpha_2/\alpha_3 \sim 2.291$.

The two expressions for $C(\theta)$ are kinematically compatible and mechanical twinning is a necessary companion of the domain structure. As we have mentioned, this fact seems to be recognized in the literature.

8. Alternative Boundary Conditions.

Thinking about the crystal/conductor system, but in the absence of an applied field for simplicity, we are led to the field equations

$$\begin{aligned} -\operatorname{div} S &= 0 \\ -\frac{\partial W}{\partial p} + E_{MS} &= 0 \qquad \text{in } \Omega, \\ -\Delta_y \phi + \operatorname{div}_y P &= 0 \end{aligned} \tag{8.1}$$

$$-\Delta_y \phi = 0 \qquad \text{in } R^3 - \Omega, \tag{8.2}$$

and the boundary conditions

$$\begin{aligned} S\nu &= 0 \\ [\nabla\phi \cdot F^{-T}\nu] + q \cdot F^{-T}\nu &= 0 \end{aligned} \qquad \text{on } \Gamma, \tag{8.3}$$

where $[\nabla\phi \cdot F^{-T}\nu] = (\nabla\phi|_{R^3-\Omega} - \nabla\phi|_{\Omega}) \cdot F^{-T}\nu$ is the jump in the normal derivative across $y(\Omega)$, ν the outward pointing normal to Ω. In (8.2) one may imagine y as defined in all of R^3 on the equations (8.1), (8.2) given in spatial coordinates.

Assuming that this situation is achieved by homogeneous deformations of a crystal occupying the reference configuration Ω, that is, by natural states in the sense of (6.10), the boundary conditions (8.3) gives that p and E_{MS} are proportional. At the critical temperature one discovers $(7.13)_1$ but not $(7.13)_2$. Instead, a relation obtains between the linearized polarization density q and the linearized electric field intensity. Thus the fact that the dielectric response is "hard" in the crystal/conductor system may also be predicted.

We may inquire about the linearized equation. Linearized at the critical temperature θ_0, together with $y(x) = x$ and $p = 0$ gives the system (2.5), as before, in Ω,

$$\Delta\phi = 0 \quad \text{in} \quad R^3 - \Omega , \tag{8.4}$$

and

$$\left.\begin{aligned} (S_0'(\theta_0)[\nabla v] + Hq)\nu &= 0 \\ [\phi_\nu] + q \cdot \nu &= 0 \end{aligned}\right\} \quad \text{on} \quad \Gamma . \tag{8.5}$$

Reasoning as before, we are led to the formula

$$\int_\Omega U(\varepsilon(v),q)dx + \int_\Omega |\nabla\phi|^2 dx + \int_{R^3-\Omega} |\nabla\phi|^2 dx = 0 ,$$

where we assume ϕ decays sufficiently at ∞ to permit this to make sense. Thus

$$\begin{aligned} U(\varepsilon(v),q) &= 0 \quad \text{in} \quad \Omega , \\ \phi &= 0 \quad \text{in} \quad R^3, \end{aligned} \tag{8.6}$$

which implies that (v,q) satisfies (7.13) and (7.14). Thus (v,q) is given by (2.18), but must now also obey

$$q_1 \nu_1 = 0 \quad \text{on} \quad \Gamma .$$

Thus, in any reasonable domain, $q_1 \equiv 0$, so $v = \text{const}$, or more precisely, the kernel for this problem consists of pairs $(v,0)$ with v an affine rigid motion, the same as the ordinary traction system in linear elasticity.

This problem thus exhibits much less degeneracy then its predecessor. It would be interesting to understand if the nonlinear equations may be solved, perhaps not uniquely, but perhaps not with the extreme degeneracy present in twinning.

9. Some Observations.

The larger issue we bring up here is a theory to accomodate the multiplicity of solutions which may conceivably emanate at the Curie point. Were we to proceed in a manner standard to analysis, we would seek a family of solutions $y(x,\theta;\eta)$ for each η in the null space such that

$$\frac{dy}{d\theta}\Big|_{\theta = \theta_0} = \eta$$

and since there are many η we would be tempted to regard the collection $\{y(x,\theta;\eta)\}$, $\theta < \theta_0$, as fluctuations. The ensuing question of what statistics to apply to the collection to extract meaningful solutions does not seem to have been treated in this context.

This discussion has been limited to the analysis of equilibrium configurations whose morphology is altered with variations in its thermal environment. Investigations of the mechanical behavior of a crystal, generally away from Curie points, are also under way.

The dead load problem is especially unstable. I. Fonseca [12] has shown that if

$$\inf \{ \int_\Omega W(\nabla v)dx - \int_\Omega f.vdx - \int_{\partial\Omega} g\cdot vds \} > -\infty,$$

where the infimum is taken over, say, Lipschitz functions and if

$$\int_\Omega fdx + \int_{\partial\Omega} gdS = 0,$$

then $f = g = 0$. She has also studied the question of local minima.

The pure displacement problem is very sensitive to boundary conditions even at normal temperatures. Generally, the least energy configuration is determined by its subenergy. Following Ericksen [8], the subenergy density of W is

$$\phi(\det F) = \inf_{\det F = \det A} W(A).$$

Consider the simplest case where one seeks

$$\inf_V \int_\Omega W(\nabla v)dx,$$

V = the set of Lipschitz v with $v(x) = Fx$ on $\partial\Omega$ where F is a given constant matrix.

and ϕ is convex.

Then

$$\inf_V \int_\Omega W(\nabla v)dx = \inf_V \int_\Omega \phi(\det \nabla v)dx = |\Omega|\phi(\det F).$$

It is not clear that there is a $y(x)$ which realizes this minimum. On the other hand given any smooth $y(x) \in V$ with the property $\det \nabla y = \det F$ in Ω, there is a sequence $(u^k) \subset V$ such that

$$u^k \to y \text{ uniformly and}$$

$$\int_\Omega W(\nabla u^k)dx \to |\Omega|\phi(\det F).$$

Such a sequence (u^k) determines a Young measure, or parametrized measure, at least in some restricted sense.

It would seem that the crystal seeks the lowest energy configuration available to it by undergoing small kinematically admissible shears [4], [13]. Possible behavior of the linearized equation in such a state has been discussed in [5].

Ball and James [3] have begun a study of fine twinning related to indium-thallium alloys. Pitteri [22] has shown how the theory studied here accomodates the classical theory. Additional discussion may be found in the articles in this volume by Ericksen, Goldenfeld, James, Pitteri, and Wright.

To actually determine twinned solutions in the martensite, or tetragonal, configuration considered in §§2-4 is not difficult: It has been described at length, of the references in Part I as well as [17], and consists in subdividing the domain Ω by a suitable family of parallel planes, assigning two different deformation gradients alternately in the subregions so determined. Since the stress vanished identically, the boundary conditions are automatically fulfilled. This is not so in the ferroelectric transition, and, except for some special regions Ω it does not seem that homogeneous deformations will play the same role. A mathematical proof that twinning occurs is not know to this author.

(1) In equilibrium in its paraelectric phase, Rochelle salt admits a primitive orthorhombic Bravais lattice with pointgroup 222. This justifies the choice of M in (7.9).

The author acknowledges with pleasure many hours of discussion with J.L. Ericksen. He also wishes to thank M. Chipot and I. Fonseca for their collaboration.

This research was partially supported by the N.S.F.

References

1. Agmon, S., Douglis, A., and Nirenberg, L., Estimates near the boundary for solutions of elliptic partial differential equations satisfying general boundary conditions, II., CPAM 22 (1964), 35-92.

2. Ball, J.M., Constitutive inequalities and existence theorems in nonlinear elastostatics, Heriot-Watt Symp. vol. I, (R. Knops, ed.) Pitman (1976).

3. Ball, J.M. and James, R., Fine phase mixtures as minimizers of energy

4. Chipot, M. and Kinderlehrer, Equilibrium configurations of crystals, to appear.

5. Chipot, M., Kinderlehrer, D., and Vergara-Caffarelli, G., The smoothness of linear laminates, IMA Preprint 199, Arch. Rat. Mech. and Anal.

6. Devonshire, A.F., Theory of Ferroelectrics, Adv. in Physics, 3 (1954), 85-130.

7. Ericksen, J.L., Special topics in elastostatics, Adv. in Appl. Mech. (C.-S. Yih, ed.), Academic Press (7 (1977), 189-243).

8. ——————, Some simpler cases of the Gibbs phenomenon for thermoelastic solids, J. of thermal stresses, 4 (1981), 13-30.

9. ——————, Twinning of crystals (1), these proceedings, 1986.

10. ——————, Special topics in elastostatics, Adv. in appl. mechanics, (C.-S. Yih, ed.) Academic Press 7, (1977), 189-243.

11. ——————, Some phase transitions in crystals, Arch. Rat. Mech. Anal. 73 (1980), 99-124.

12. Fonseca, I. Thesis, University of Minnesota, Arch. for Rat. Mech. and Anal.

13. —————' The lower quasiconvex envelope of the stored energy function for an elastic crystal.

14. Grindlay, J. An introduction to the phenomenological theory of ferroelectrics, Pergamon (1970).

15. Gurtin, M.E., and Spector, S. On stability and uniqueness in finite elasticity, Arch. Rat. Mech. and Anal. 70 (1979), 153-165.

16. James, R., Finite deformation by mechanical twinning, Arch. Rat. Mech. Anal. 22 (1981), 143-176.

17. ________, Diplacive phase transitions in solids, J. of Mech. and Physics of Solids (1986).

18. Kanzig, W., Ferroelectrics and antiferroelectrics, Academic Press (1957).

19. Lines, M.E. and Glass, A.M., Principles and applications of ferroelectrics and related materials, Oxford (1977).

20. Love, A.E.H., A treatise on the mathematical theory of elasticity, Dover (1948).

21. Mitsui, M., Tatsuzaki, I., and Nakamura, E., An introduciton to the physics of ferroelectrics, Gordon and Breach (1976).

22. Pitteri, M., Reconciliation of local and global symmetries of crystals, J. Elasticity 14 (1984), 175-190.

23. Podio-Guidugli, P. and Vergara-Caffarelli, G., On a class of traction problems in elasticity (preprint).

24. Podio-Guidguli, P., Vergara-Caffarelli, G., and Virga, E., The role of ellipticity and normality assumptions in formulating live boundary conditions in elasticity, (preprint).

25. Simpson, H. and Spector, S., On Hadamard stability in finite elasticity, IMA preprint 171, Arch. Rat. Mech. and Anal.

26. Spector, S., On uniqueness in finite elasticity with general loading, J. Elasticity 10 (1980), 145-161.

27. Tisza, L. On the general theory of phase transformations in solids, (Smoluchowski, R. Mayer, J.E., and Weyl, W.A. eds.) Wiley (1951), 1-35.

28. Toupin, R. The elastic dielectric, J. Rat. Mech. Anal. 5, (1956), 849-915.

SIMULATION OF PSEUDO-ELASTIC BEHAVIOUR IN A SYSTEM OF RUBBER BALLOONS

by

Wolfgang Kitsche, Ingo Müller, Peter Strehlow

FB9-Hermann Föttinger Institut
Technical University, Berlin

1. Introduction

The pressure-radius relation of a spherical rubber balloon is non-monotone, a feature that suggests a nontrivial stability problem whose implications we investigate here from several viewpoints. After the formulation of the stability criterion we show that the balloon can be stabilized at any radius by loading it with a piston under an elastic spring, if only the spring is hard enough. It will then be shown that the measurement of internal pressure of the balloon by the height of a water pressure head is equivalent to the action of a spring and, in fact, that this may suffice to stabilize the balloon.

If two connected balloons are subject to an inflation-deflation cycle, the pressure-volume curve exhibits a simple hysteresis loop and several other interesting features due to the fact that one balloon inflates first while the other remains behind to follow later. We then study increasing numbers of connected balloons and observe the gradual emergence of a pseudo-elastic hysteresis much like the one that occurs in materials with shape memory. The lesson to be learned from that analogy is not entirely clear, but as it stands the paper offers an interesting study of stability, metastability and instability.

The arguments presented here are extracted from the bachelor's thesis and the diploma thesis of W. Kitsche (see [1] and [2]).

2. Pressure-Radius Relation and Pressure-Particle Number Relations in a Rubber Balloon

2.1. Pressure-Radius Relation.

Rubber is an incompressible isotropic material whose stress-strain relation we take to be that of a Mooney-Rivlin material. Thus we have

$$t_{ij} = s_{-1}B^{-1}_{ij} + P\delta_{ij} + s_1 B_{ij} \tag{2.1}$$

where t_{ij} is the Cauchy stress and B_{ij} is the left Cauchy-Green tensor. The coefficients s_{-1} and s_1 are constants.

Let D,R and d,r be the thickness and the radius in the natural configuration and in the inflated configuration of a spherical balloon. And let us focus our attention on a tiny slab of material inside the rubber membrane which is cut out by the coordinate lines δ, $\delta + \Delta\delta$, ϕ, $\phi + \Delta\phi$ and ρ, $\rho + \Delta\rho$ of the natural spherical coordinate system. The tensor B for this state has the form

$$\underset{\sim}{B} = \begin{bmatrix} B\langle rr\rangle & 0 & 0 \\ 0 & B\langle\delta\delta\rangle & \\ 0 & 0 & B\langle\phi\phi\rangle \end{bmatrix} = \begin{bmatrix} (\frac{d}{D})^2 & & 0 \\ 0 & (\frac{r}{R})^2 & 0 \\ 0 & 0 & (\frac{r}{R})^2 \end{bmatrix}$$

and the stress is therefore given by

$$\underset{\sim}{t} - \underset{\sim}{P1} = s_{-1}\begin{bmatrix} (\frac{D}{d})^2 & 0 & 0 \\ 0 & (\frac{R}{r})^2 & 0 \\ 0 & 0 & (\frac{R}{r})^2 \end{bmatrix} + s_1\begin{bmatrix} (\frac{d}{D})^2 & 0 & 0 \\ 0 & (\frac{r}{R})^2 & 0 \\ 0 & 0 & (\frac{r}{R})^2 \end{bmatrix} . \tag{2.3}$$

Incompressibility requires $d \cdot r^2 = D \cdot R^2$ so that $\frac{d}{D}$ can be eliminated from (2.3). Elimination of P between the three equations (2.3) gives

$$t\langle\delta\delta\rangle = t\langle\phi\phi\rangle = -\frac{1}{2}(p_e + p_i) + s_1(\frac{r}{R})^2)(1 - \frac{s_{-1}}{s_1}(\frac{r}{R})^2)(1-(\frac{R}{r})^6) \tag{2.4}$$

where $t\langle rr\rangle$ has been assumed equal to $-\frac{1}{2}(p_e + p_i)$, the arithmetic mean of the external and internal pressure.

Figure 1 shows the tangential and radial forces on a piece of the balloon. In equilibrium these forces must be balanced. After a little calculation we obtain from this requirement that

$$[p] = p_i - p_e = 2s_1 \frac{D}{R} (\frac{R}{r} - (\frac{R}{r})^7) - \frac{s_{-1}}{s_1} (\frac{r}{R})^2) . \qquad (2.5)$$

In this calculation we have dropped powers of $\frac{D}{R}$ as seems appropriate for thin balloons.

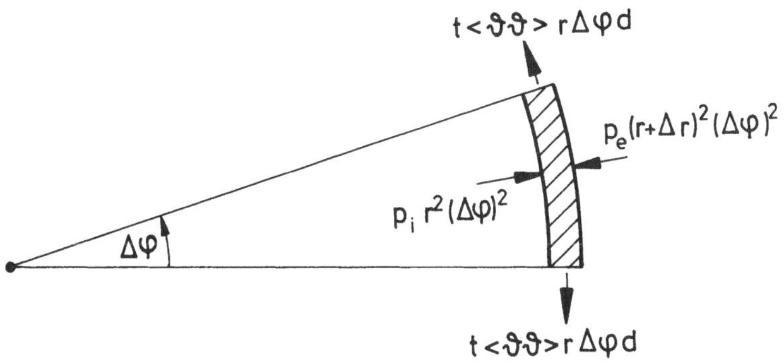

Figure 1: Force equilibrium on the balloon.

A proper derivation of (2.5) can be found in [3]; a shorter version - but more extensive than the above one - is presented in [4].

Figure 2 shows graphically how [p] depends on $\frac{r}{R}$. That graph is drawn for the values

$$D = 0.17 \text{ mm} \;\Big|\; R = 25 \text{ mm} \;\Big|\; s_1 = 353.7 \cdot 10^3 \frac{N}{m^2}, \;\Big|\; s_{-1} = -15.9 \cdot 10^3 \frac{N}{m^2}, \qquad (2.6)$$

which have been measured in our laboratory for a particular balloon at a temperature of 20°C.

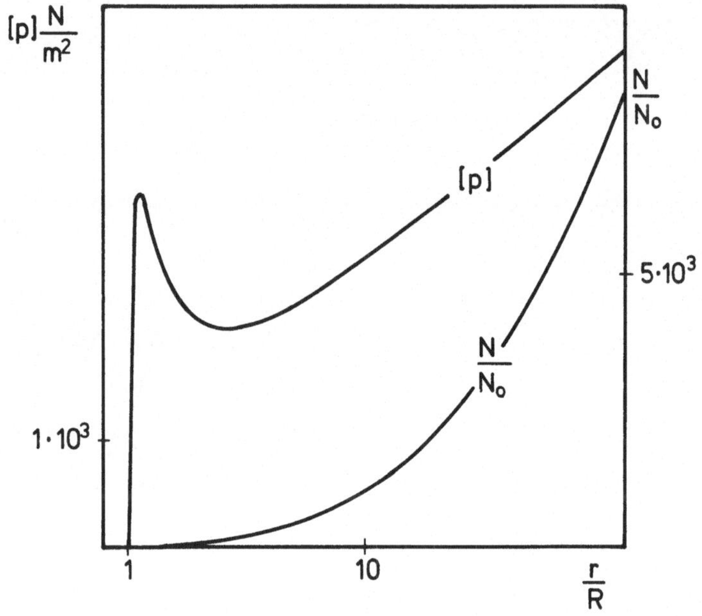

Figure 2: Pressure-radius relation and pressure-particle number relation of a balloon.

2.2 Pressure-Particle Number Relation

The non-monotone character of the ([p], $\frac{r}{R}$) - curve is suggestive of unstable behaviour and indeed, this behaviour is our main concern in this paper. A curve of equal interest to the ([p], $\frac{r}{R}$) - curve in the investigation of stability properties is the ([p], N) - curve, where N is the number of gas molecules inside the balloon. This curve is constructed as follows: First we calculate how N depends on $\frac{r}{R}$. We have

$$N = \frac{4\pi}{3} r^3 n, \tag{2.7}$$

where n is the particle number density, which is related to [p] by the thermal equation of state of an ideal gas

$$n = \frac{p_i}{kT} = \frac{[p] + p_e}{kT} . \tag{2.8}$$

k is the Boltzmann constant and T the absolute temperature. We combine (2.7) and (2.8) to obtain

$$\frac{N}{N_0} = \left(\frac{r}{R}\right)^3 \left(1 + \frac{[p]}{p_e}\right) \qquad (2.9)$$

where N_0 is the number of molecules in the uninflated balloon. The monotone curve of Figure 2 represents the (N,r) - dependence for a pressure $p_e = 10^5 \frac{N}{m^2}$, which is normal atmospheric pressure.

The two curves of Figure 2 allow us to construct the dependence of $[p]$ on $\frac{N}{N_0}$ and Figure 3 shows that curve. Unfortunately, because of the initial slow growth of $\frac{N}{N_0}$ with $\frac{r}{R}$ the interesting region around the maximum of $[p]$ is compressed into a simple very sharp spike as shown in Figure 3. A curve like that is useless for the graphical method of determining stability of states which will be described below. Therefore, for the purpose of explanation we shall replace it by the curve of Figure 4 which, while artificially constructed, preserves all qualitative features of the correct curve of Figure 3.

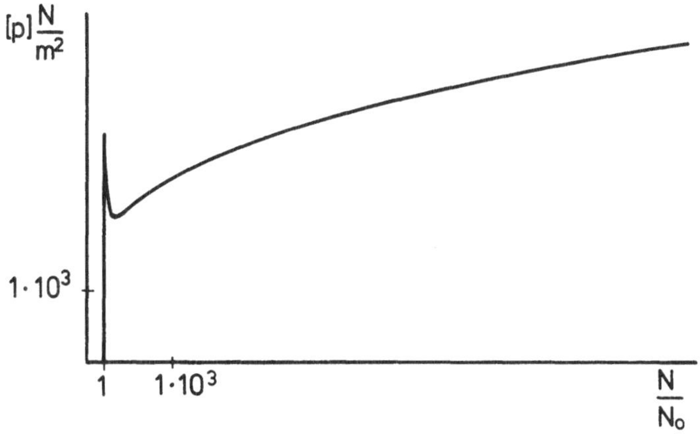

Figure 3: Pressure-particle number relation

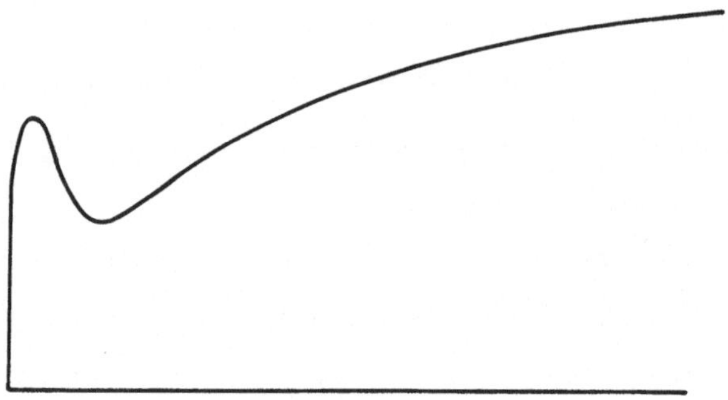

Figure 4: Qualitative pressure-particle number relation (left side of curve stretched considerably by arbitrary factor. To be used only for qualitative arguments).

3. Stability of a Simple Balloon

3.1. Stability Criterion.

Thermodynamic stability criteria are usually based upon the entropy inequality and the energy balance

$$\frac{dS}{dt} + \int_{\partial V} \frac{q_i}{T} da_i \geq 0 ,$$

$$\frac{dE}{dt} + \int_{\partial V} (q_i - t_{ij}v_j) da_i - \int_V f_i v_i dV = 0 . \tag{3.1}$$

Under certain conditions we can extract a Lyapunov function from these relations. Conditions under which this is possible are as follows:

i). $T = T_0$ = const holds throughout V. In that case T_0 can be pulled out from $(3.1)_1$ and the integral $\int q_i da_i$ can be eliminated between $(3.1)_1$ and $(3.1)_2$ with the following result:

$$\frac{d(E - T_0 S)}{dt} - \int_{\partial V} t_{ij}v_j da_i - \int_V f_i v_i dV \leq 0. \tag{3.2}$$

ii). Either $t_{ij} = - p_0 \delta_{ij}$ = const. holds on ∂V or $v_i da_i = 0$. Therefore the surface integral in (2) can be rewritten as follows

$$\int_{\partial V} t_{ij} v_j da_i = - p_0 \int_{\partial V} v_i da_i = - p_0 \frac{dV}{dt} = - \frac{dp_0 V}{dt}$$

and we obtain

$$\frac{d(E - T_0 S + p_0 V)}{dt} - \int_V f_i v_i \, dV < 0. \tag{3.3}$$

iii). The power of the body force can be derived from a potential energy so that

$$\int_V \rho f_i v_i dv = \frac{dE_{pot}}{dt} . \tag{3.4}$$

Combining (3) and (4) we obtain

$$\frac{d}{dt} (E - T_0 S + p_0 V + E_{pot}) < 0 , \tag{3.5}$$

which we express by saying that the <u>availability</u>

$$A = E - T_0 S + p_0 V + E_{pot} \tag{3.6}$$

assumes a minimum in equilibrium.

E is the sum of the kinetic and internal energy, but we shall neglect kinetic energy here and consider E as the internal energy.

3.2. Free Energies of Air and Balloons.

Since we are going to evaluate the condition that A has a minimum in equilibrium for various systems, all involving air-filled rubber balloons, we must know the free energies $E - T_0 S$ of air and rubber balloons as functions of volume. The easiest way to obtain this knowledge is by starting from the Gibbs equation in the form

$$dE = TdS + Ldt \tag{3.7}$$

which states that the change of internal energy in a time dt is due to heating TdS and to working Ldt. The equation (7) may be written as

$$d(E - TS) = - SdT + Ldt, \tag{3.8}$$

and it is quite general in the sense that it holds for different materials; in particular, it holds for ideal gases and for balloons. What is different in those different materials is the form of the working term.

In gases, the work done on the body by the pressure is

$$Ldt = - pdV = -NkT \frac{1}{V} dV \tag{3.9}$$

where the thermal equation of state has been used. N is the number of molecules. Then we obtain by combining (3.8) and (3.9)

$$d(E - TS)_{Air} = - SdT - NkT \frac{1}{V} dV \tag{3.10}$$

and we conclude that

$$(E - TS)_{Air} = - NTk \ln V + C(T) . \tag{3.11}$$

Since $T = T_0$ = const in our case, $E - TS$ is a function of N and V.

The work done on a rubber balloon is

$$Ldt = [p] \, 4\pi r^2 \, dr, \tag{3.12}$$

if we only consider radial deformation. With (8) we thus have

$$d(E - TS)_{Balloon} = - SdT + [p] \, 4\pi r^2 \, dr . \tag{3.13}$$

Of course $[p]$ is given by (2.5); it is only a function of r. Since T is constant for us, we conclude that $(E\text{-}TS)_{Balloon}$ is a function of r, or in fact of the volume of the balloon.

3.3. Example: A Balloon loaded by an Elastic Spring.

To fix the ideas let us look at the setup shown in Figure 5 where a balloon is connected with a cylinder of cross section F that is closed off by a piston. The motion of the piston will be resisted by a linear elastic spring with spring constant λ. $V_{1,2}$ are the volumes of the cylinder and the balloon respectively. When the pressure throughout the system is equal to the surrounding pressure p_0 , the spring is unloaded and the cylinder has volume V_{10} . The mass of the piston is ignored. The system is loaded by pumping air into V_1.

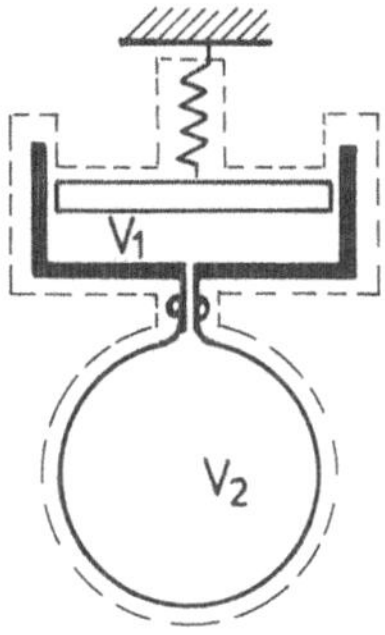

Figure 5. A balloon connected with a cylinder loaded by a spring.

First of all we choose a surface ∂V for the system and this is indicated by the dashed line in Figure 5. It is easy to see that the surface satisfies the conditions i) through iii). of Section 3.1 provided that $T = T_0$ = const. is assumed. Therefore the availability from A is the relevant Lyapunov function and we have

$$A = (E-T_0S)_{Air1} + (E - T_0S)_{Air2} + (E - T_0S)_{Balloon} + p_0(V_1+V_2) + \frac{1}{2}\lambda \frac{(V_1-V_{10})^2}{F_1^2} . \tag{3.14}$$

As we have seen, the three free energies in this formula are functions of N_1,V_1, of N_2,V_2, and of V_2, respectively. Therefore A is a function of the three variables N_1,V_1,V_2, since of course $N_1 + N_2 = N$ is constant.

A necessary condition for A to be minimal is the vanishing of the derivatives of A with respect to N_1,V_1 and V_2:

$$\frac{\partial A}{\partial N_1} = (e_A^1 - T_0 s_A^1 + p_1 v_1) - (e_A^2 - T_0 s_A^2 + p_2 v_2),$$

$$\frac{\partial A}{\partial V_1} = -p_1 + p_0 + \lambda \frac{V_1 - V_{10}}{F_1^2} , \tag{3.14}$$

$$\frac{\partial A}{\partial V_2} = - p_2 + p_0 + [p] .$$

e,s and v are specific values of E, S and V respectively. The first condition states that the two specific free enthalpies in V_1 and V_2 are equal. But since T is constant throughout, this statement is equivalent to $p_1 = p_2$, i.e. equal pressures. Therefore $(3.14)_{2,3}$ can be combined to give one condition, namely

$$[p]_{\text{equilibrium}} = \lambda \frac{V_1 - V_{10}}{F_1^2} \tag{3.15}$$

which states that the jump of pressure across the piston or the balloon is equal to the pressure exerted by the spring.

A sufficient condition for equilibrium is that the matrix of second derivatives of A with respect to its variables $X_i = \{N_1, V_1, V_2\}$ be positive definite. If we use (10) and (13) a little calculation will show that this sufficient condition assumes the specific form

$$\left\| \begin{matrix} \frac{N_1 + N_2}{N_1 N_2} & -\frac{1}{V_1} & \frac{1}{V_2} \\ -\frac{1}{V_1} & \frac{N_1}{V_1^2} + \frac{\lambda}{kTF_1^2} & 0 \\ \frac{1}{V_2} & 0 & \frac{N_2}{V_2^2} + \frac{\partial[p]}{\partial V_2} \end{matrix} \right\| \text{ is positive-definite} \tag{3.16}$$

These are three conditions of which two are trivially satisfied, since N_1, N_2 and λ are all positive. The third condition however is non-trivial. It reads

$$\frac{\partial[p]}{\partial r} > -\frac{1}{kT} \frac{4\pi r^2}{\frac{F_1^2}{\lambda} + \frac{(N_1+N_2)kT}{p_2^2}} \quad . \tag{3.17}$$

We shall refer to (3.15) as the condition for equilibrium, and if (3.17) is satisfied as well as (3.15) we shall say that the equilibrium is stable.

From (3.17) we conclude that points on the descending part of the ([p], r) - curve may well be stable. In fact, if only λ is large enough, all those points will be stable. But as λ is made smaller, the steeper parts of the curve lose their stability and eventually for very small λ's the whole branch is unstable. [Note that $\lambda = 0$ must not be considered because, by (3.15), the necessary condition for equilibrium allows only the trivial solution $[p] = 0$ in that case.]

3.4 A Simple Method of Measuring the ([p],r) - Curve.

Figure 6 shows a simple device for loading a balloon by a water pressure head in two connected tubes of cross-sections F_0 and F_1. The dashed line represents the height of the water level in both tubes when the balloon is not attached and in that case the value of V_1 is called V_{10}. The device of Figure 6 is dynamically equivalent to the device of Figure 5 with the water surface in tube one assuming the role of the piston. Indeed, an increase of V_1 from its initial value V_{10} is resisted by the force

$$\rho g\left(1 + \frac{F_1}{F_0}\right) F_1 \frac{V_1 - V_{10}}{F_1}$$

which is linear in $V_1 - V_{10}$ just as the force in the spring of Figure 5. The "spring constant" of the new device is

$$\lambda = \rho g F_1\left(1 + \frac{F_1}{F_0}\right) \tag{3.18}$$

and all formulae of Section 3.3 remain valid, as do, in particular, the stability conditions (15) and (17). Note that λ can be made large by choosing tube 0 with a small cross-section F_0 .

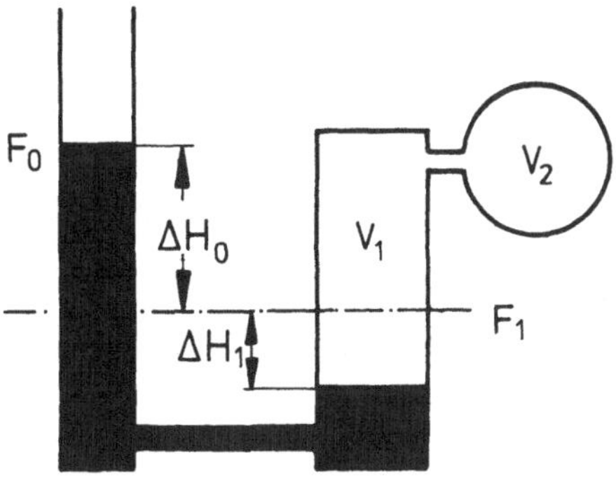

Figure 6: Loading and measuring device for a balloon

Knowing this we find it trivially easy to measure the $([p],r)$ - curve of a balloon: We just make F_0 small enough to have a big enough λ that we can expect all radii to be stable. Then we inflate the balloon by adding water to tube 0. $\Delta p = p_1 - p_0$ is determined by measuring the water pressure head, because we have

$$\Delta p = \rho g\,(\Delta H_1 + \Delta H_0) \quad \text{(see Figure 6)}. \tag{3.19}$$

Since $\Delta p = [p]$ holds in equilibrium we can measure the $([p], r)$ - curve, if for every head we determine the radius of the balloon.

3.5 A Suggestive Method of Determining Stability

Now suppose that we choose a bigger F_0, i.e. a smaller λ than in the previous section, so that not all values r correspond to stable equilibria. We proceed to describe a suggestive graphical method for the determination of the unstable region.

The key to this argument is the curve $([p], N_2)$ of Figure 3 and the observation that Δp in (3.19) is also a function of N_2. Indeed, (3.19) may be written as

$$\Delta p = \rho g\left(\frac{1}{F_1} + \frac{1}{F_0}\right)(V_1 - V_{10}) \tag{3.20}$$

and by the thermal equation of state of ideal gases we have

$$V_1 = \frac{N_1 kT}{p_1} = \frac{N_1 kT}{p_0\left(1+\frac{\Delta p}{p_0}\right)} \quad . \tag{3.21}$$

Since by Figure 2 the maximum possible Δp (or $[p]$) is equal to about 3% of the atmospheric pressure $p_0 = 1$ bar, we neglect the second term in the denominator of $(3.21)_2$ †) and obtain by combining (3.20) and (3.21)

---

†) We can also obtain a relation between Δp and N_2, if do not make that approximation, but it will not be nice and linear as it will be with the approximation (see (3.22) below).

$$\Delta p = \frac{1}{F_1^2} \quad \lambda \left[\frac{kT}{p_0} (N-N_2) - V_{10}\right] \tag{3.22}$$

where (18) was used to introduce λ and N_1 has been replaced by $N-N_2$. Thus Δp is a linear function of N_2 with a negative slope determined by λ and with an abscissa intercept determined by V_{10}. Figure 7. shows a graph of that function.

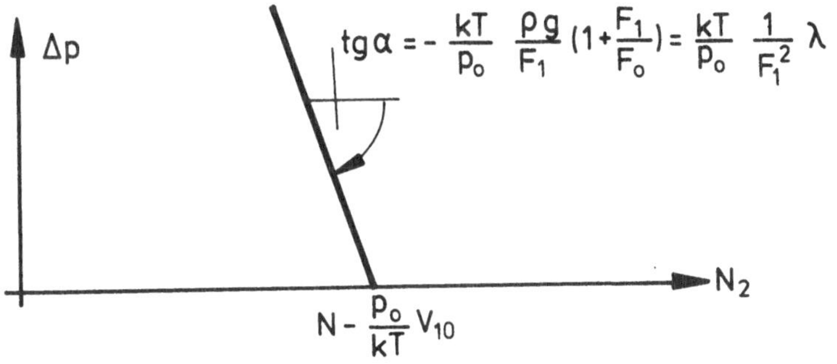

Figure 7: The pressure $\Delta p = p_1 - p_0$ as a function of N_2.

We shall now plot [p] and Δp as functions of N_2 on the same diagram, i.e., we shall superpose the diagrams of Figures 3 and 7 as shown in Figure 8. Since in equilibrium we must have Δp = [p], the abscissa of the point of intersection gives us the equilibrium value of N_2. As we keep filling water into tube 0 the volume V_{10} decreases and the abscissa intercept of the Δp - line moves to the right. Figure 9 shows three different positions of the line and the

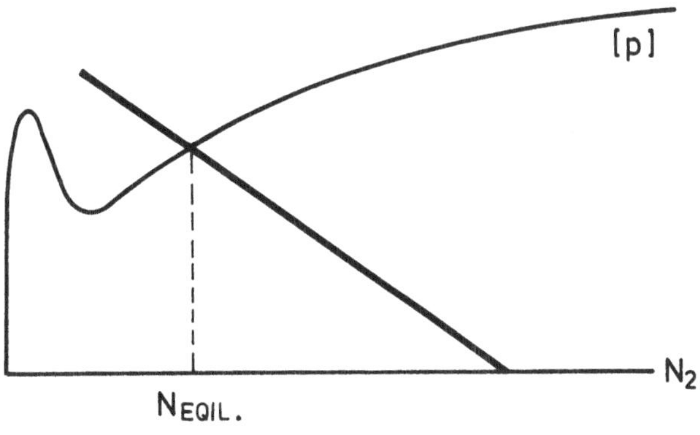

Figure 8. Δp and [p] as functions of N_2 and the equilibrium value of N_2.

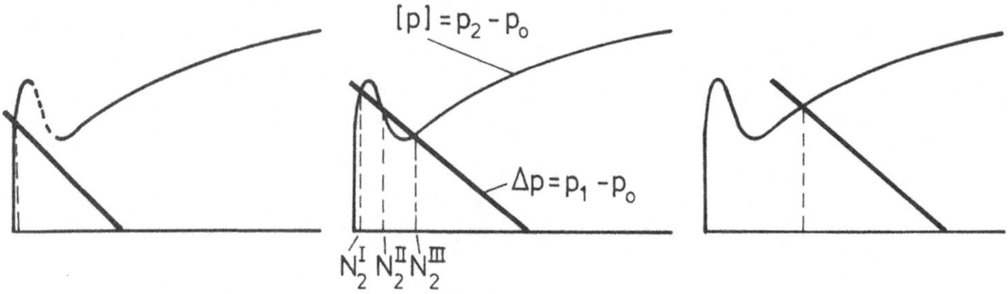

Figure 9. Three different equilibria appearing as water is added to tube 0.

corresponding equilibrium values of N_2. Note that there are three values of N_2 in the second diagram of Figure 9. The one in the middle corresponds to an unstable state. This can be seen as follows.

Suppose the equilbrium at N_2^{II} is established and now we change N_2 by δN_2 by squeezing air out of the balloon. Figure 9b shows that thereby p_1 and p_2 both increase, but p_2 increases <u>more</u> than p_1 so that more air will stream out of the balloon. Thus the equilibrium at N_2^{II} is unstable, a slight fluctuation making the balloon leave that state.

On the contrary, in Figure 9c we have a single stable equilibrium. Indeed, squeezing the balloon away from that state will increase p_1 and decrease p_2 so that the air will stream back.

In this manner each state can be investigated for stability. It turns out all states are unstable for which the slope of the ([p], N_2) - curve is smaller than the slope of the (Δp, N_2) - curve. The corresponding points are dotted in Figure 9a.

3.6 A Simple Device for Measuring [p].

Figure 10 shows photographs of the device that we have been discussing as it was actually built. The cross-section of the tube 0 on the left in that device is too big to stabilize all points of the balloon curve so that when the dotted part of Figure 9a is reached, corresponding to a certain water pressure head, the balloon is inflated with the pressure head dropping until it comes to rest at the beginning of the right solid branch. Figure 10a shows the device when the balloon is still close to its natural state and the other three photographs show it with the balloon in passage between the stable branches. The drop in the pressure head during the passage is clearly visible.

Actually the device as it was built has another attachment. There is a three-way valve that can be used to break the connection between tube 1 and tube 0 and connect tube 1 to a hose of small cross-section instead. Turning the valve does two things to the $(\Delta p, N_2)$ - curve of the Figures 9. It decreases the abscissa intercept, because V_{10} increases, and it decreases the slope (making it more negative), because F_0 is replaced by the cross-section of the hose. Therefore turning the valve from tube 0 to the hose can arrest the passage across the unstable branch.

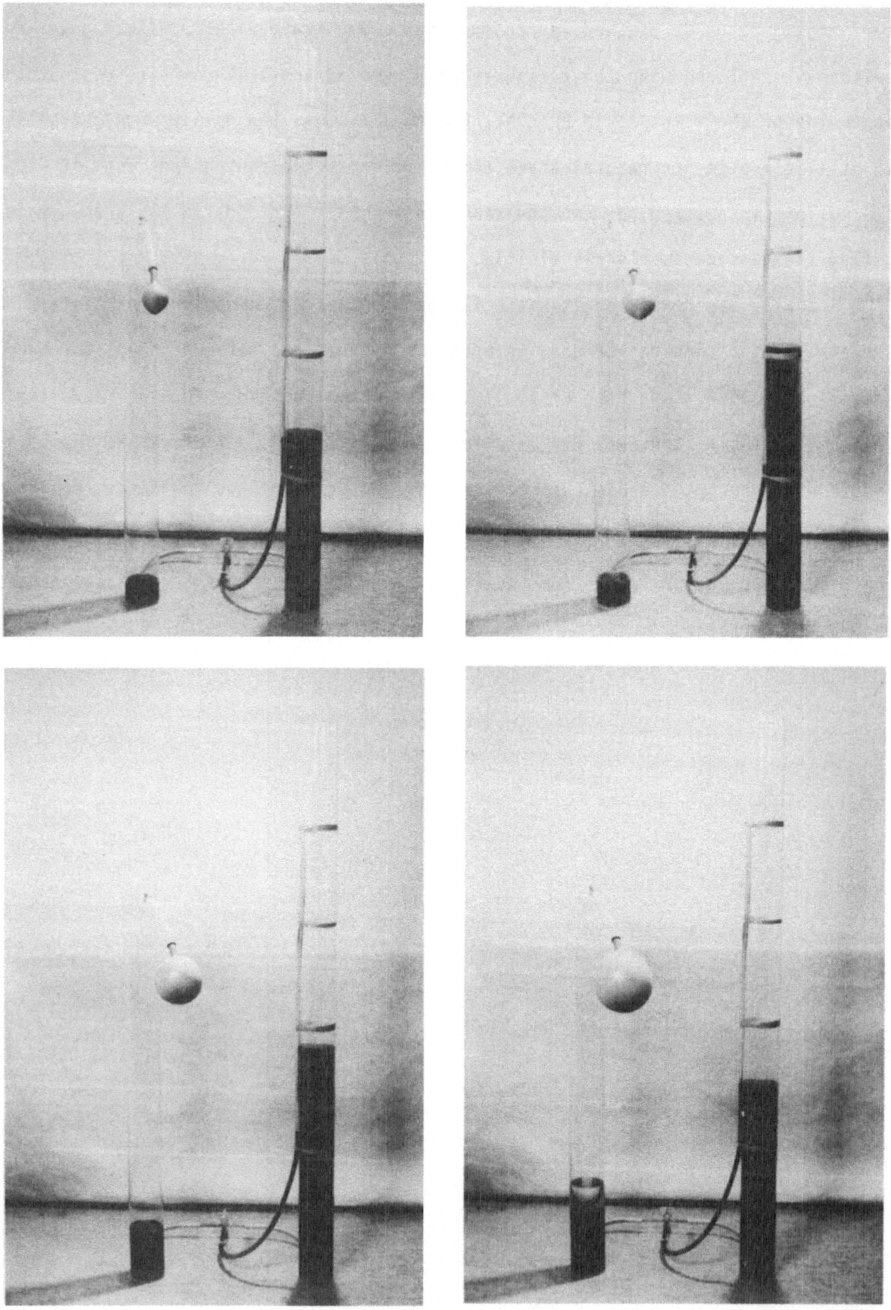

Figure 10: Measuring [p].

4. Two Balloons in Contact

4.7. Pressure versus Equivalent Strain.

When two balloons in contact are inflated together, as shown schematically in Figure 11, they will both have the same pressure of course, but they may have

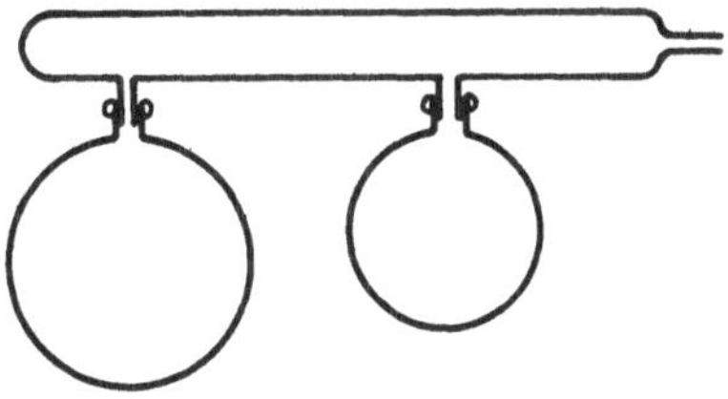

Figure 11: Two connected balloons.

different radii in the range of pressures where the ([p], r) - curve has three branches. Figure 12 shows in a schematic manner the six possibilities that can occur, when the two balloons either lie both on the same branch or on different ones. These six possibilities are identified by letters a through f.

Given a particular situation, say b, we can read off from Figure 2 the radii r_1 and r_2 of the balloons for a given pressure and plot [p] versus

$$\lambda_e = \sqrt[3]{\frac{r_1^3 + r_2^3}{2R^3}} \; , \tag{4.7}$$

which we call the equivalent strain. The resulting curves are shown in Figure 13

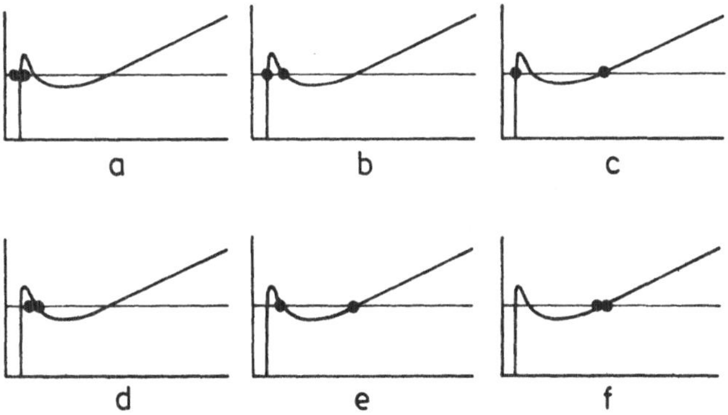

Figure 12: Possible states of two balloons under the same pressure.

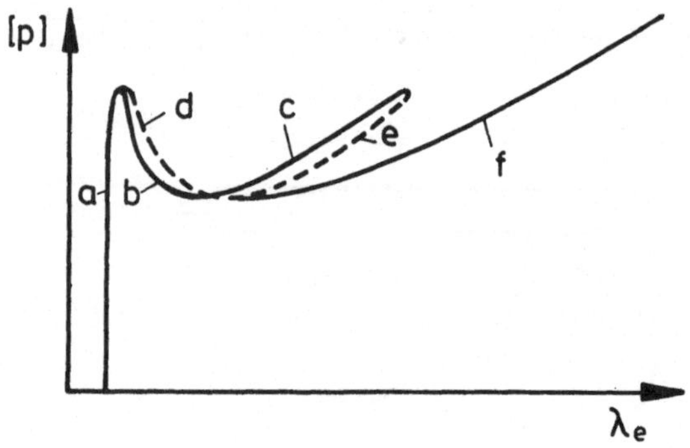

Figure 13: Pressure [p] as a function of equivalent strain. (Distance between curves c and e is exaggerated).

with letters attached to the different branches matching the letters of Figure 12. These curves must be constructed graphically or numerically, since we do not have the inversion of equation (2.5) to give us r as a function of [p]. The two curves c and e are virtually identical in their upper parts if drawn properly. For greater clarity the distance between these curves has been exaggerated arbitrarily.

4.2 Stability.

Not all situations in Figure 12 are stable, in fact d and e are not, which is why the corresponding curves in Figure 13 have been dotted. Unfortunately the stability analysis in this case does not lead to a condition as simple as (3.17) for the case of a balloon in contact with a cylinder. The condition we get in the present case has been written down in [5] and exploited there numerically. Here we may determine stability by a graphical method that is akin to the method described in Section 3.5. We proceed to discuss this.

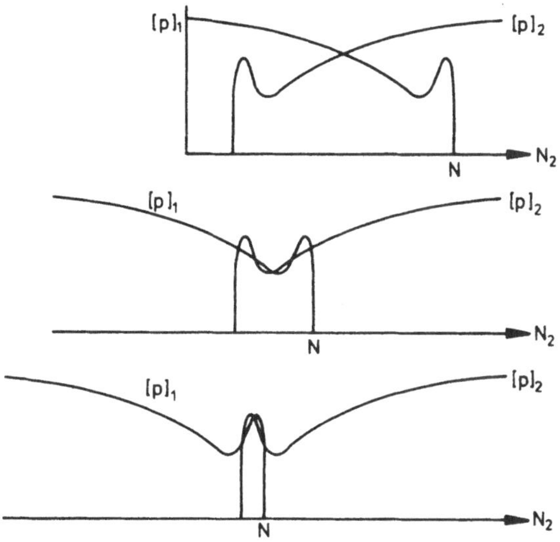

Figure 14. On the stability consideration

We must realize that the pressure jumps $[p]_1$ and $[p]_2$ across the two balloons are both functions of N_2, the particle number on balloon 2, since $N_1 = N - N_2$ holds. Figure 14 shows these curves for three different values of N. Possible equilibria occur where the two curves intersect and the equilibrium is stable, if at the intersection the $[p]_2$ - curve has a bigger slope than the $[p_1]$ - curve. This fact is established in the same manner as in Section 3.5 by considering a displacement δN_2 by squeezing balloon 2. Thus the one equilibrium of Figure 14a is a stable one of type f in the nomenclature of Figures 12 and 13. Of the five equilibria of Figure 14b the two of type c are stable and so is the one of type f but the two of type e are unstable. Figure 14c shows three equilibria, an unstable one of type d and two stable ones of type c.

Particularly surprising of course is the stability of situation b since here one balloon is placed on the descending part of the ([p], r) - curve.

4.3 Hysteresis.

Figure 15 is essentially the same as Figure 13 except that the unstable branches have been dropped. In addition the arrows in Figure 15 indicate what happens in an inflation - deflation cycle.

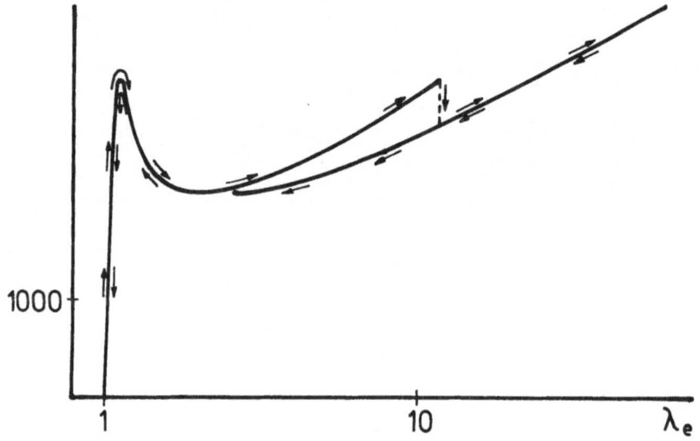

Figure 15: Hysteresis in an inflation - deflation cycle with two balloons.

As we start increasing N both balloons move up the initial ascending branch of their ([p],r) - curves until they reach the maximum. Further increase of N pushes one balloon (say the first one) onto the descending branch while the second one moves backwards onto the initial branch until the minimum pressure is reached. After that the first balloon moves upwards on the second ascending branch and the second balloon moves upwards on the initial branch again. As the maximum is reached again, the second balloon is filled at the expense of the first. The pressure drops as this happens and both balloons will assume the same radii on the second ascending branch. Further increase of N will let both balloons move upwards together.

Reduction of N will let both balloons stay on the right branch until the minimal pressure is reached. Once that has happened, one balloon (say again the first one) will deflate until it reaches the initial branch, thereby inflating the second one so that the pressure rises. Further reduction of N will see both balloons ascending to the maximum but from different directions. Eventually both assume equal size again and move down the initial curve together.

The result is the peculiar $([p],\lambda_e)$ - curve of Figure 15 which includes a hysteresis loop.

5. Further Development of the Hysteresis in Many Balloons

5.1 Inflation and Deflation Curves for Ten Balloons.

Figure 16 shows the system we consider next and at first we shall investigate the case $D = 10$, i.e. ten balloons. Having understood how the $([p],\lambda_e)$ - curves of Figure 13 for two balloons have been constructed, we find it conceptually easy to do the same for ten balloons. λ_e in the case of D balloons is defined as

$$\lambda_e = \sqrt[3]{\frac{1}{D}\sum_{i=1}^{D}\left(\frac{r_i}{R^3}\right)^3} . \tag{5.1}$$

Of course there are stable and unstable curves. Figure 17 shows only stable ones and of those only those parts that are seen in an inflation - deflation process. In Figure 20 for $D = 100$ we have shown all stable branches.

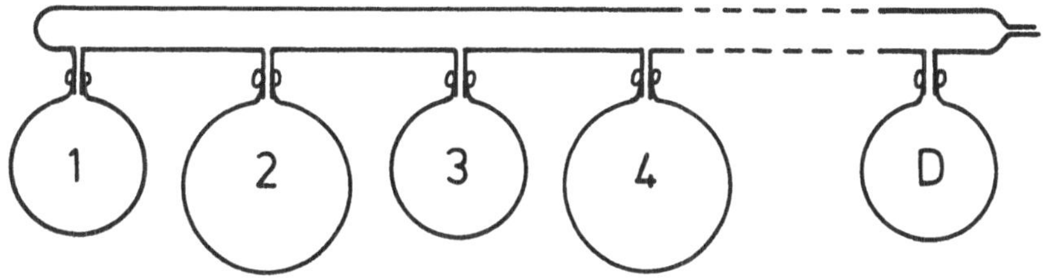

Figure 16. Many connected balloons

Figure 17 has to be compared with Figure 15 in that all non-stable curves have been eliminated. The upper curve is the one pertaining to inflation which proceeds until all balloons are together on their right branches. The tips occur wherever a balloon is surmounting the maximum of its ([p],r) - curve. In the deflation process there are also 10 tips, but the ones to the right are too small to be seen.

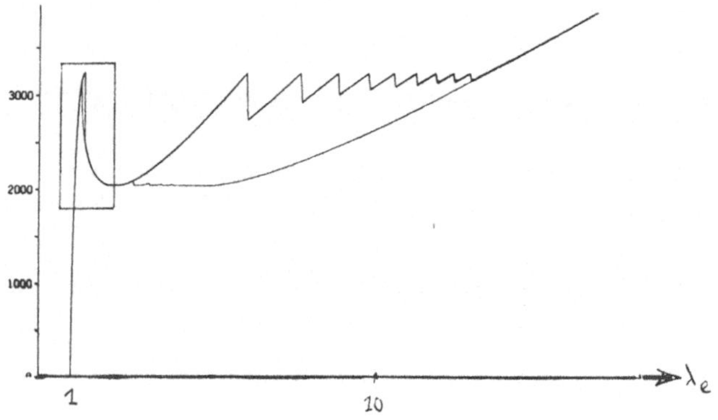

Figure 17: Hysteresis in an inflation - deflation cycle with 10 balloons. For blow-up of box see Figure 18a.

There is one striking difference between the curves of Figures 15 and 17 which goes beyond the obvious one that the curves refer to different numbers of balloons. Indeed, in Figure 17 the first balloon to surmount the maximum is obviously not let down on the other side smoothly as in Figure 15. Rather there are jumps both in the inflation and in the deflation curve. Also in the deflation process the last spike on the left does not quite reach the maximum height of the ([p],r) - curves. Instead there is a jump upward to a value less than the maximum as the last balloon is moving back to the initial branch.

To explain this behaviour properly, in Figure 18a we have drawn an enlargement of the curves [p] versus λ_e in the region that is indicated by the box in Figure 17. The λ_e - axis is stretched by a factor 6 and the [p] axis by a factor 2 and, in order to make things clearer, the unstable branch is included.

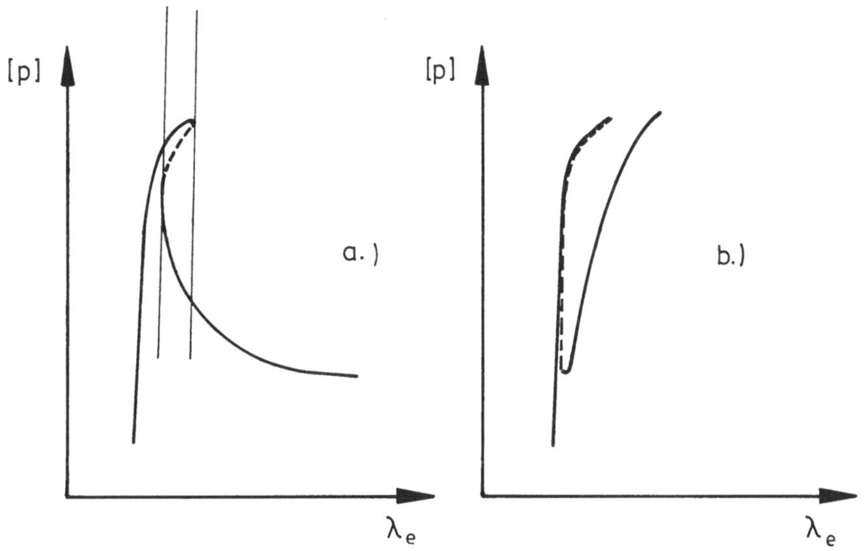

Figure 18: Blow-ups of the boxes in figures 17 and 21

The jumps occur on the vertical lines drawn in Figure 18a. It is clear what is happening here: The nine balloons that remain on the initial branch loose air which is used to inflate the one balloon that surmounts the maximum and is sent a good way along the descending branch before pressures become equal. During the deflation the opposite process is going on. The last balloon descending branch distributes its air to the nine balloons that have already arrived on the initial branch thereby raising the pressures in all of them.

This phenomenon becomes more pronounced as more balloons are considered. Indeed, Figure 18b refers to 1000 balloons and it shows that the 999 balloons which remain on the initial branch will have so much air to loose that they can send the one balloon that surmounts the maximum a long way onto the second ascending branch. As the following figures show this is still not so at D = 100 but after that the inflation curves no longer go through the minimum of the pressure curves.

5.2. Many Balloons.

Figure 19 shows the inflation - deflation process in a system of 100 balloons. No feature occurs here that has not yet been discussed. But figure 20 is interesting in that it shows, again for D = 100, what happens, if one stops the inflation process after any number of balloons have been inflated and then start to deflate. We thus observe that the hysteresis region is filled with stable curves of which we only see small parts when we go all the way in the inflation process or the deflation process. The vertical jumps that we need to get from one stable curve to the other and have been omitted in Figure 20 to make the picture clearer.

For D = 1000 and D = 4000 we show only the hysteresis loops in Figure 21. The small box in the diagram for D = 1000 is the one of which an enlargement has been shown in Figure 18, and was discussed in Section 5.1. The discussion revealed why the spikes in the curves become less pronounced as D grows.

Figure 22 shows how in a real inflation experiment the balloons neatly separate into small and big ones just as we expect it to happen according to the preceding theory.

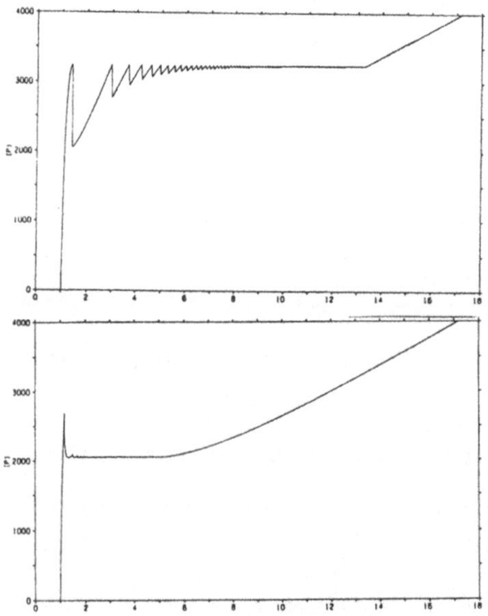

Figure 19: Inflation and deflation of 100 balloons.

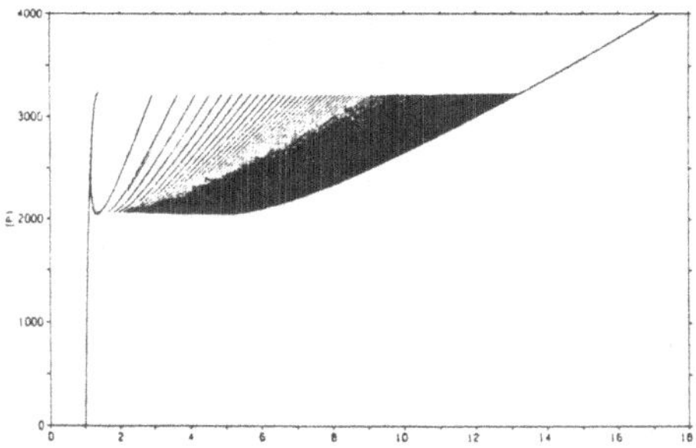

Figure 20: Inflation - deflation curves inside the hysteresis region of 100 balloons

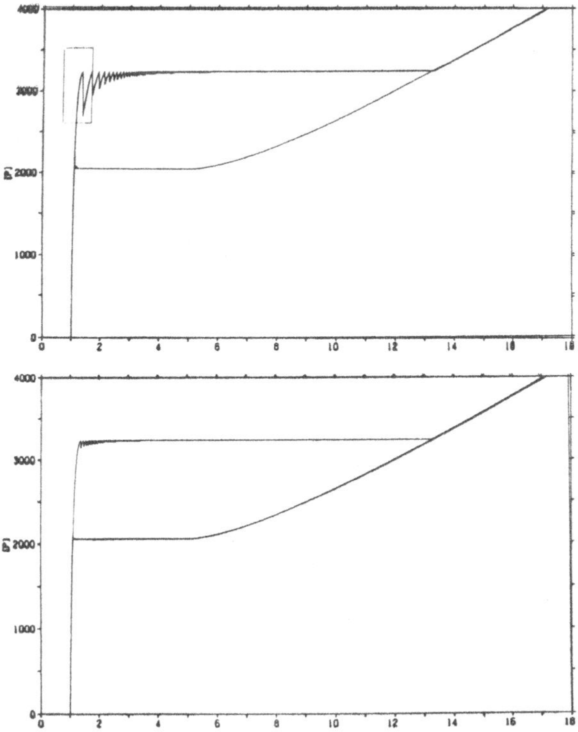

Figure 21: Hysteresis for D = 1000 and D = 4000. For blow-up of box see Figure 18b.

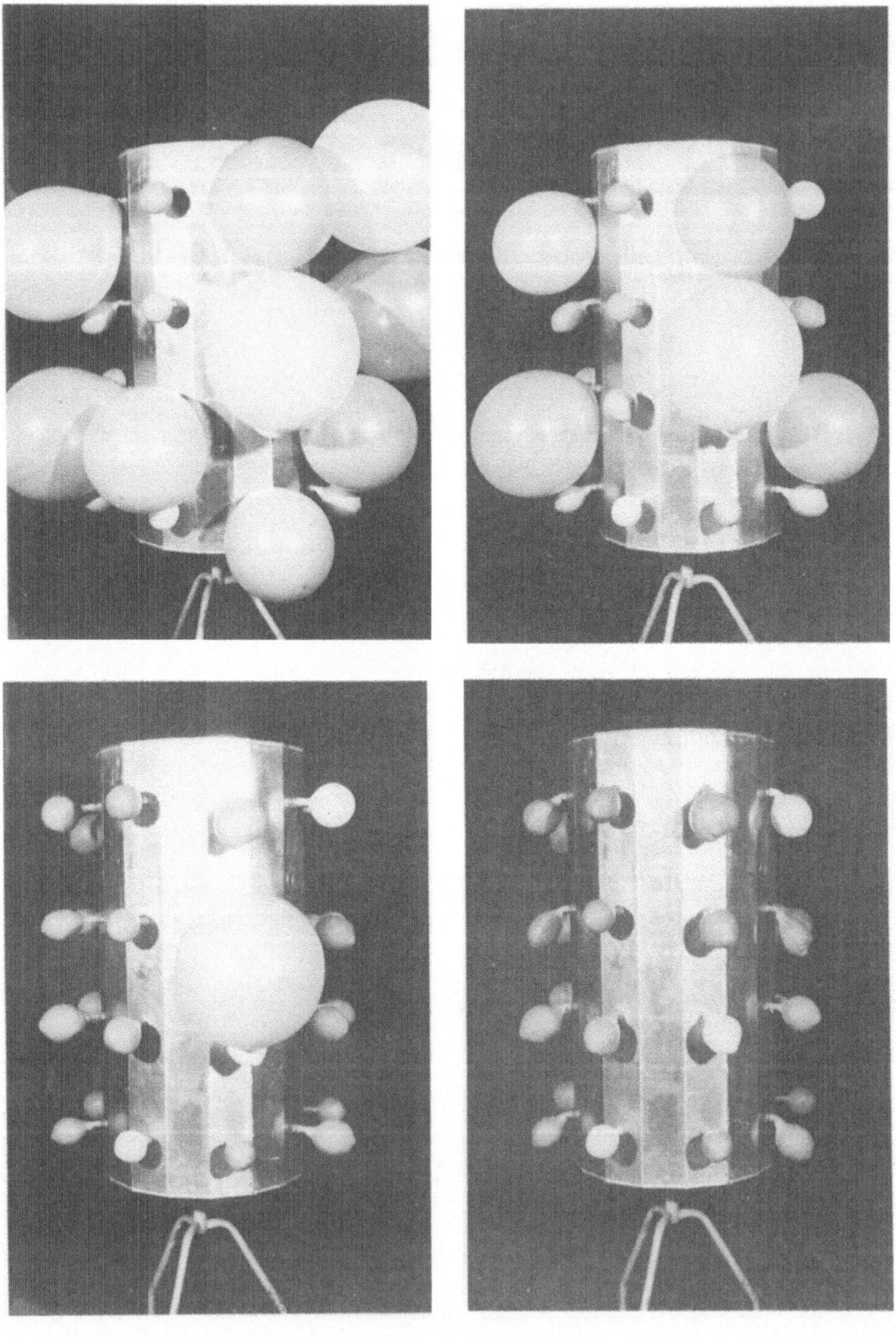

Figure 22: Separation of Balloons into small and big ones in a deflation experiment

5.3 Discussion

The hysteresis that has finally emerged for large numbers of balloons is reminiscent of the load-deformation curve of pseudo-elastic bodies such as shape memory alloys at higher temperatures, e.g. see [6]. In fact, it was our interest in phase transitions and hysteresis phenomena in memory alloys that has prompted this study.

However, having performed the work, we must admit that we are not quite certain what exactly we have learned about pseudo-elasticity. It seems conceivable that the slow emergence of the hysteresis - as illustrated by the progressive development in Figures 15, 17, 20 and 21 - may upon further reflection teach us something about the onset of a phase transition and the relaxation of the load this often entails. The serration of the "yield limit" and the "recovery limit" in Figure 19 and 21 certainly is reminiscent of a similar phenomenon in deformation controlled, hard-device experiments.

However, we do not want to over emphasize the possible relevance of the work. May it stand as an interesting non-trivial exercise on stability, possibly of interest to a student who studies non-monotone load-deformation curves and hystereses.

References

[1] Kitsche, W. Untersuchung der Stabilität von Gummimembranen. Studienarbit FB 12, TU Berlin (1984).

[2] Kitsche, W. Modellierung eines Phasenübergangs an einem System vieler Ballons. Diplomarbeit FB 12, TV Berlin. (1985)

[3] Adkins, J.E., Rivlin R.S. Large Elastic Deformations of Isotropic Materials IX. The Deformation of Thin Shells. Phil. Trans. A 244 (1951).

[4] Müller, I., Thermodynamics, Pitman Publ. (1985).

[5] Dreyer, W., Müller, I. Strehlow, P. A Study of Equilibria of Interconnected Balloons. Quarterly J. Mech. Appl. Math XXXV (1982).

[6] Muller, I. Pseudoelasticity in Shape Memory Alloys - An Extreme Case of Thermoelasticity. IMA Preprint No. 168. University of Minnesota 1985.

Acknowledgement. The author of this article (I.M.) gratefully acknowledges the support of the Institute for Mathematics and its Applications.

ASYMPTOTIC PROBLEMS IN DISTRIBUTED SYSTEMS

Riviere Memorial Lecture 1985

J.-L. Lions

Collége de France

Introduction

Distributed systems are systems governed by Partial Differential Equations; this terminology is classical in the framework of control theory; we use this terminology in order to emphasize that we are concerned, in this paper, with some asymptotic questions which arise in connection with the optimal control of distributed systems.

The main difficulty in dealing with problems of optimal control of distributed systems is the complexity; this complexity may be due to the complexity of the materials which constitute the system, or to the complexity of the model, or to the complexity of the geometry etc.

In general these questions are the same for the analysis of problems without control as for the control of the systems. Here we want to give some examples where the "control aspect" leads to some slightly unusual questions.

The examples are chosen among those leading to open questions.

The plan is as follows:

1. Composite materials and boundary control.
2. A thin domain.
3. Singular perturbations.
4. A fourth problem.

   Bibliography

1. Composite Materials and Boundary Control.

1.1 Statement of the problem.

Let $\Omega \subset \mathbb{R}^3$ (the dimension is taken equal to 3 to fix ideas) be a domain which consists of a composite material which can be modelled as follows.

Let $a_{ij}(y)$, $i,j = 1,2,3$, be a family of (smooth) functions which are Y-periodic ($Y =]0,1[^3$; a_{ij} admits period 1 in all variables) and which are such that

(1.1)
$$\left| \begin{array}{l} a_{ij} = a_{ji} \quad \forall_{i,j} , \\ a_{ij}\zeta_i\zeta_j \geq \alpha\, \zeta_i\zeta_i \quad \forall \zeta_i \in R,\ \alpha > 0 \end{array} \right. \qquad (^1) .$$

We consider the system with the state equation given by

(1.2)
$$\frac{\partial^2 y_\varepsilon}{\partial t^2} + A_\varepsilon y_\varepsilon = 0 \qquad \text{in } \Omega \times]0,\tau[,$$

where

(1.3)
$$A_\varepsilon = - \frac{\partial}{\partial x_i}\left(a_{ij}\ (x/\varepsilon)\ \frac{\partial}{\partial x_j}\right) .$$

The initial conditions are

(1.4)
$$y_\varepsilon(x,0) = \frac{\partial y_\varepsilon}{\partial t}(x,0) = 0 \qquad \text{in } \Omega$$

and one wants to control the system by a boundary control i.e.

(1.5)
$$y_\varepsilon = v \quad \text{on } \Sigma = \Gamma \times]0,T[,\ \Gamma = \partial\Omega .$$

Let us assume that the cost function is given by

(1.6)
$$J_\varepsilon(v) = \int_\Omega [y_\varepsilon(x,T;v) - z_d(x)]^2\, dx + N\int_\Sigma v^2\, d\Sigma$$

where

$$z_d \text{ is given in } L^2(\Omega) ,$$

$$N \text{ is given} > 0 .$$

The structure of (1.6) shows that we have to take

(1.7)
$$v \in L^2(\Sigma) .$$

---

(1) We use the summation convention of repeated indices.

Given v in $L^2(\Sigma)$, (1.2) (1.6) (1.5) admits a unique solution, which is denoted by $y_\varepsilon(x,t;v) = y_\varepsilon(v)$, and it is this function which is used in the first integral in (1.6). But one has to make this precise, cf. section 1.2 below. Assuming for the time being that the formulation (1.6) makes sense, the problem of optimal control is to find

(1.8) $$\inf_{v \in U_{ad}} J_\varepsilon(v) ,$$

where

(1.9) $$U_{ad} = \text{closed convex subset of } L^2(\Sigma) .$$

A few remarks are in order:

Remark 1.1

Given v , the computation of $y_\varepsilon(v)$ is - if we do not use asymptotic methods - very complicated due to the rapid oscillations of $a_{ij}(x/\varepsilon)$.

Remark 1.2

The goal of this section 1 is to seek an asymptotic expansion for

(1.10) $$\mathcal{Y}_\varepsilon = \inf_{v \in U_{ad}} J_\varepsilon(v)$$

as $\varepsilon \to 0$ and to seek the solution $v = u_\varepsilon$ of (1.8). As we shall see this question is essentially open!

1.2 Solution of the problem for ε fixed.

One can show that, given $v \in L^2(\Sigma)$, there exists a unique function $y_\varepsilon(v)$, a solution of (1.2) (1.6) (1.5), which satisfies

(1.11) $$y_\varepsilon(v) \text{ is continuous from } [0,T] \to L^2(\Omega) .$$

The proof (cf. J.-L. Lions [1]) is obtained by transposition of a result of regularity .

Let us consider ϕ_ε , the solution of

$$(1.12)\qquad \left|\begin{array}{l} \dfrac{\partial^2\phi_\varepsilon}{\partial t^2} + A_\varepsilon\, \phi_\varepsilon = f \quad \text{in} \quad \Omega\times]0,T[\ , \\[2ex] \phi_\varepsilon(x,0) = \dfrac{\partial\phi_\varepsilon}{\partial t}(x,0) = 0 \quad \text{in} \quad \Omega\ , \\[2ex] \phi_\varepsilon = 0 \quad \text{on} \quad \Sigma\ ; \end{array}\right.$$

let us assume that

$$(1.13)\qquad f \in L^1(0,T;\ L^2(\Omega))\ .$$

Then there exists a unique solution which satisfies

$$(1.14)\qquad \phi_\varepsilon \in C([0,T];\ H_0^1(\Omega))\ ,\ \frac{\partial\phi_\varepsilon}{\partial t} \in C([0,T];\ L^2(\Omega))$$ (1) ;

this is classical; but an interesting regularity result is that (2)

$$(1.15)\qquad \frac{\partial\phi_\varepsilon}{\partial\nu_{A_\varepsilon}} \in L^2(\Sigma)\ ,$$

where $\frac{\partial}{\partial\nu_{A_\varepsilon}}$ stands for the normal derivative associated with A_ε .

<u>Remark 1.3</u>

The main difficulty for what follows is that the estimate

$$(1.16)\qquad \left\| \frac{\partial\phi_\varepsilon}{\partial\nu_{A_\varepsilon}} \right\|_{L^2(\Sigma)} \leq C(\varepsilon)\, \|f\|_{L^1(0,T;L^2(\Omega))}$$

contains a constant $C(\varepsilon)$ which increases as $1/\varepsilon$ as $\varepsilon \to 0$

$$(C(\varepsilon) \text{ depends on } \frac{\partial}{\partial x_k}(a_{ij}(x/\varepsilon)))\ .$$

---

(1) $C([0,T];\ X)$ = continuous functions from $[0,T] \to X$.
$H_0^1(\Omega) = \{\phi \mid \phi, \frac{\partial\phi}{\partial x_i}\ \ L^2(\Omega),\ \phi=0 \text{ on } \Gamma\}$.

(2) cf. I. Lasiecka, J.-L. Lions, R. Triggiani [1] for other results along these lines

If we return to problem (1.8) we see that, by virtue of (1.11), $v \to J_\varepsilon(v)$ is continuous from $L^2(\Sigma) \to R$, so that there exists a unique element $u_\varepsilon \in U_{ad}$ such that

$$J_\varepsilon(u_\varepsilon) = \inf \ J_\varepsilon(v) \ , \ v \in U_{ad} \ . \tag{1.17}$$

This is the optimal control which is characterized by the optimality system which can be written as follows:

$$\left| \begin{aligned} &\frac{\partial^2 y_\varepsilon}{\partial t^2} + A_\varepsilon y_\varepsilon = 0 \ , \\ &\frac{\partial^2 p_\varepsilon}{\partial t^2} + A_\varepsilon p_\varepsilon = 0 \quad \text{in} \quad \Omega\times]0,T[\ , \\ &y_\varepsilon(x,0) = \frac{\partial y_\varepsilon}{\partial t}(x,0) = 0 \ , \\ &p_\varepsilon(x,T) = 0 \ , \ \frac{\partial p_\varepsilon}{\partial t}(x,T) = y_\varepsilon(x,T) - z_d(x) \quad \text{in} \quad \Omega \ , \\ &y_\varepsilon = u_\varepsilon \quad \text{on} \quad \Sigma \ , \ p_\varepsilon = 0 \quad \text{on} \quad \Sigma \end{aligned} \right. \tag{1.18}$$

and

$$\int_\Sigma \left(\frac{\partial p_\varepsilon}{\partial \nu_{A_\varepsilon}} + Nu\right)(v-u)\,d\Sigma \geq 0 \quad \forall\, v \in U_{ad} \ . \tag{1.19}$$

Remark 1.4

The "adjoint state" p_ε is given by the solution of the backward wave equation, with $\frac{\partial p_\varepsilon}{\partial t}(x,T) = y_\varepsilon(x,T) - z_d(x) \in L^2(\Omega)$; then $\frac{\partial p_\varepsilon}{\partial \nu_{A_\varepsilon}} \in L^2(\Sigma)$ so that the integrals in (1.19) make sense.

The main question is now: Is it possible to simplify the problem by using asymptotic expansions?

Remark 1.5

There are many situations where this is indeed possible, as shown in J.-L.

Lions [2], [3]. We have chosen here to present an (apparently) tricky situation.

1.3 Homogenization theory.

Let us return to problem (1.12), where f is given fixed. Then, as $\varepsilon \to 0$, one has:

(1.20)
$$\begin{cases} \phi_\varepsilon \to \phi \quad \text{in } L^\infty(0,T;H_0^1(\Omega)) \text{ weak star } (^1), \\ \dfrac{\partial \phi_\varepsilon}{\partial t} \to \dfrac{\partial \phi}{\partial t} \quad \text{in } L^\infty(0,T;L^2(\Omega)) \text{ weak star} , \end{cases}$$

where ϕ is the solution of

(1.21)
$$\begin{cases} \dfrac{\partial^2 \phi}{\partial t^2} + \mathcal{A}\phi = f \quad \text{in } \Omega\times]0,T[, \\ \phi(x,0) = \dfrac{\partial \phi}{\partial t}(x,0) = 0 \quad \text{in } \Omega , \\ \phi = 0 \quad \text{on } \Sigma ; \end{cases}$$

in (1.21) $\mathcal{A}$ is the homogenized operator, which is given by

(1.22)
$$\mathcal{A} = - q_{ij} \frac{\partial^2}{\partial x_i \partial x_j} ;$$

the q_{ij} 's are constants - the effective coefficients - which are given by explicit (constructive) formulas. We refer to E. Sanchez-Palencia [1], A. Bensoussan, J.-L. Lions and G. Papanicolaou [1] for the formulas.

Remark 1.6

The homogenized operator $\mathcal{A}$ is elliptic .

Remark 1.7

The coefficients q_{ij} do not depend on Ω but only on the material. There are codes to compute the q_{ij}'s .

---

(1) I.e. $\int_0^T(\phi_\varepsilon,\psi)_{H_0^1} g(t)dt \to \int_0^T(\phi,\psi)_{H_0^1} g(t)\,dt \quad \forall\psi \in H_0^1(\Omega), \forall g \in L^1(0,T)$.

1.4. A natural conjecture.

At this stage, it is very natural to introduce the "homogenized control problem", defined as follows. The state equation is given by

(1.23) $$\frac{\partial^2 y}{\partial t^2} + \mathcal{A}y = 0 \quad \text{in} \quad \Omega\times]0,T[\ ,$$

(1.24) $$y(x,0) = \frac{\partial y}{\partial t}(x,0) = 0 \quad \text{in} \quad \Omega \ ,$$

(1.25) $$y = v \quad \text{on} \quad \Sigma = \Gamma \times]0,T[\ .$$

This system admits a unique solution $y(v)$, which is continuous from $[0,T] \to L^2(\Omega)$; we consider the cost function

(1.26) $$\mathcal{J}(v) = \int_\Omega [y(x,T;v) - z_d(x)]^2 \, dx + N \int_\Sigma v^2 d\Sigma$$

and the problem

(1.27) $$\inf \mathcal{J}(v) \quad , \quad v \in U_{ad} \ .$$

It seems natural to conjecture that [1]

(1.28) $$\inf_v J_\varepsilon(v) \underset{\varepsilon\to 0}{\to} \inf \mathcal{J}(v) \ , \ v \in U_{ad}$$

and that, if u_ε denotes the unique solution of (1.27), then

(1.29) $$u_\varepsilon \to u \quad \text{in} \quad L^2(\Sigma) \quad \text{weakly} \ .$$

Remark 1.8.

Of course problem (1.27) is significantly simpler than the initial problem, since $\mathcal{A}$ is much simpler than A_ε. This is the interest of homogenization theory!

---

(1) Added in Proof. Very interesting and negative results along these lines have been obtained by M. Avellaneda and F.H. Lin. Several papers by these A. will appear soon. Conjecture as its stands is not correct, which makes things more interesting.

Remark 1.9

Of course one can raise similar problems in more general situations where we have nonperiodic structures cf. Homogenization and effective moduli of materials, IMA Volumes 1, J.L. Ericksen, D. Kinderlehrer, R. Kohn, J.-L. Lions, (eds), 1986.

Remark 1.10

The difficulty arises from Remark 1.3 . It is not known, whether or not, for fixed v, $y_\varepsilon(v) \to y(v)$ (in, say, $L^\infty(0,T; L^2(\Omega))$ weak star), where $y(v)$ is the solution of (1.23) (1.24)(1.25).

Remark 1.11

In this direction we also mention the following open question: Let ϕ_ε be the solution of the stationary problem

$$(1.30) \qquad \left| \begin{array}{l} A_\varepsilon \phi_\varepsilon = 0 \quad \text{in } \Omega, \\ \phi_\varepsilon = g \quad \text{on } \Gamma \end{array} \right.$$

where g is given in $L^2(\Gamma)$. Is it true that $\phi_\varepsilon \to \phi$ in $L^2(\Omega)$ weakly, where ϕ is the solution of

$$(1.31) \qquad \left| \begin{array}{l} \mathcal{A}\phi = 0 \quad \text{in } \Omega, \\ \phi = g \quad \text{on } \Gamma \end{array} \right.$$

Remark 1.12

Similar questions will arise for parabolic systems, with the boundary control in $L^2(\Sigma)$. □

2. A Thin Domain

2.1. A preliminary problem.

Let Ω be an open set in R^n with boundary

$$\partial\Omega = \Gamma_0 \cup \Gamma_1$$

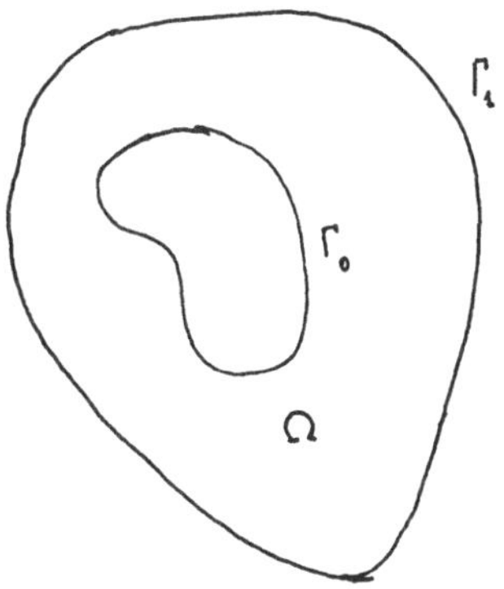

Fig. 1

as represented on Fig. 1.

We are interested in the problem

(2.1) $\Delta^2 u = f$ in Ω ,

(2.2) $\left| \begin{array}{l} u = \dfrac{\partial u}{\partial \nu} = 0 \text{ on } \Gamma_0 , \\ \Delta u = \dfrac{\partial \Delta u}{\partial \nu} = 0 \text{ on } \Gamma_1 . \end{array} \right.$

A natural approach is to introduce the space

(2.3) $$E_0 = \{\phi | \ \phi \in C^2(\overline{\Omega}),\ \phi = \frac{\partial \phi}{\partial \nu} = 0 \text{ on } \Gamma_0\} ;$$

we provide E_0 with

(2.4) $$\|\phi\|_E = \|\Delta\phi\|_{L^2(\Omega)} ;$$

this defines <u>a norm</u> on E_0 , since if $\Delta\phi = 0$ in Ω and $\phi = \frac{\partial\phi}{\partial\nu} = 0$ on Γ_0 , then $\phi = 0$ in Ω . We then introduce

(2.5) $$E = \text{completion of } E_0 \text{ for } \|\phi\|_E .$$

Then <u>if</u> $v \to (f,v)$ <u>defines a continuous linear form on</u> E , problem (2.1)(2.2) admits a unique solution in E , defined by

(2.6) $$(\Delta u, \Delta v) = (f,v) \quad \forall\, v \in E .$$

A question which does not seem to be settled is the following: What are the properties of u E near Γ_1 ? Another form of the same question is: What are the properties needed on f near Γ_1 in order for $v \to (f,v)$ to define a continuous linear form on E ?

We are now going to consider a problem of this type in a thin structure.

2.2 An asymptotic problem in a thin domain

Let $\mathcal{O} \subset \mathbb{R}^2$, $\partial\mathcal{O} = S$,

$\Omega_\varepsilon = \mathcal{O} \times]0,\varepsilon[\in \mathcal{O}$

$x' = \{x_1, x_2\} \in \mathcal{O}$,

$x_3 \in]0,\varepsilon[$,

$S_\varepsilon = S \times]0,\varepsilon[$.

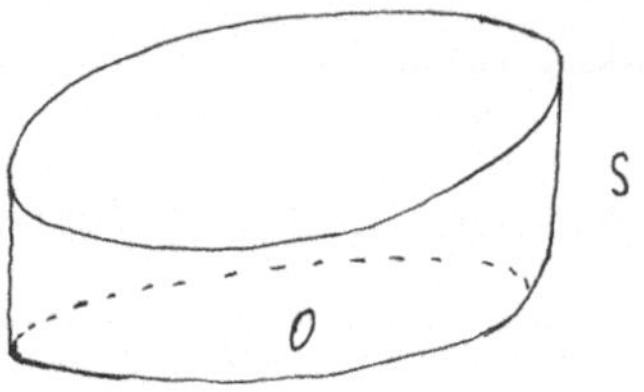

Fig. 2

In Ω_ε we consider the problem

(2.7) $$\Delta^2 u_\varepsilon = f \quad \text{in} \quad \Omega_\varepsilon ,$$

(2.8) $$u_\varepsilon = \frac{\partial u_\varepsilon}{\partial \nu} = 0 \quad \text{on} \quad S_\varepsilon ,$$

(2.9) $$\left| \begin{array}{l} \Delta u_\varepsilon = \frac{\partial}{\partial \nu} \Delta u_\varepsilon = 0 \quad \text{on} \quad \partial_\pm \Omega_\varepsilon , \quad \text{where} \\ \partial_+ \Omega_\varepsilon = \mathcal{O} \times \{\varepsilon\}, \quad \partial_- \Omega_\varepsilon = \mathcal{O} \times \{0\} . \end{array} \right.$$

Remark 2.1

We are going to work in a formal fashion; problem (2.7), (2.8), (2.9) is a variant of the problem considered in Section 2.1 and in order that this problem make sense it seems that f should satisfy "some conditions" near $x_3 = 0$ and near $x_3 = \varepsilon$.

Our goal is to have some kind of indication on the conditions by a (formal) asymptotic expansion.

2.3 An ansatz.

We introduce

(2.10) $$y = x_3/\varepsilon ;$$

in the new variables x', y , Ω_ε is replaced by

$$\Omega_1 = 0 \times]0,1[;$$

we look for u_ε in the form

(2.11) $u_\varepsilon = u_0 + \varepsilon^2 u_1 + \varepsilon^4 u_2 + \ldots$,

where

$$u_j = u_j(x',y) \text{ is defined in } \Omega_1$$

and where at the end of the computation we replace y by x_3/ε, and with $u_j = \frac{\partial u_j}{\partial \nu} = 0$ if $x' \in S = \partial O$.

We set

$$\Delta' = \frac{\partial^2}{\partial x_1^2} + \frac{\partial^2}{\partial x_2^2}, \; D = \partial/\partial y .$$

Then (2.7) becomes

$$(\varepsilon^{-2}D^2 + \Delta')^2(u_0 + \varepsilon^2 u_1 + \ldots) = f_\varepsilon = f(x', \varepsilon y) \tag{2.12}$$

and the boundary conditions (2.9) become

$$\begin{cases} (\varepsilon^{-2}D^2 + \Delta')(u_0 + \varepsilon^2 u_1 + \ldots) = 0 \text{ for } y = 0, 1, \\ (\varepsilon^{-2}D^3 + D\Delta')(u_0 + \varepsilon^2 u_1 + \ldots) = 0 \text{ for } y = 0, 1. \end{cases} \tag{2.13}$$

Identifying in (2.12) the powers of ε gives

$$\begin{cases} D^4 u_0 = 0 , \\ D^4 u_1 + 2D^2 \Delta' u_0 = 0 , \\ D^4 u_2 + 2D^2 \Delta' u_1 + \Delta'^2 u_0 = f_\varepsilon \end{cases} \tag{2.14}$$

and the boundary conditions

$$\begin{cases} D^2 u_0 = 0 , \; D^3 u_0 = 0 \quad \text{for } y=0 , 1 \\ D^2 u_1 + \Delta' u_0 = 0 , \; D^3 u_1 + D\Delta' u_0 = 0 \text{ for } y = 0 , 1 \\ D^2 u_2 + \Delta' u_1 = 0 , \; D^3 u_2 + D\Delta' u_1 = 0 \text{ for } y = 0, 1, \end{cases} \tag{2.15}$$

It follows from $(2.14)_1$, $(2.15)_1$ that

$$D^2 u_0 = 0 \quad \text{in } \mathcal{O} \times]0,1[, \tag{2.16}$$

i.e.,

(2.17) $$u_0 = u_0(x') + yw_0(x').$$

Then $(2.14)_2$, $(2.15)_2$ reduce to

$$D^4u_1 = 0 \ , \ D^2u_1 + \Delta'u_0 = 0 \ \text{ for } \ y = 0,1,$$

$$D^3u_1 + D\Delta'u_0 = 0 \ \text{ for } \ y = 0,1$$

i.e.

(2.18) $$D^2u_1 + \Delta'u_0 = 0 \ \text{ in } \ 0 \times]0,1[\ ,$$

i.e.

(2.19) $$u_1 + \Delta'(\frac{y^2}{2} v_0 + \frac{y^3}{6} w_0) = v_1(x') + yw_1(x').$$

Equations $(2.14)_3$ $(2.15)_3$ reduce to

(2.20) $$\left| \begin{array}{l} D^4u_2 - \Delta'^2 u_0 = f_\varepsilon(x',y) \ , \\ D^2u_2 + \Delta'u_1 = 0 \ , \ D^3u_2 + D\Delta'u_1 = 0 \ \text{ for } \ y=0,\ 1. \end{array} \right.$$

In order (2.20) to admit a solution u_2 one has to have <u>compatibility conditions</u> which are obtained by writing that

$$\int_0^1 (D^4u_2 - \Delta'^2u_0)dy = \int_0^1 f_\varepsilon(x',y)\ dy$$

and

$$\int_0^1 y(D^4u_2 - \Delta'^2 u_0)dy = \int_0^1 y\ f_\varepsilon(x',y)\ dy \ .$$

Using the boundary conditions, one verifies that these conditions reduce to

$$\int_0^1 f_\varepsilon(x',y)dy = 0 \ , \ \int_0^1 yf_\varepsilon(x',y)dy = 0$$

i.e.

(2.21) $$\int_0^\varepsilon f(x',x_3)dx_3 = 0 \ , \ \int_0^\varepsilon x_3f(x',x_3)dx_3 = 0 \ .$$

This leads to the following question: are they any connections between (2.21) and the condition that f belongs to the dual of the analog of the space E introduced in Section 2.1?

Remark 2.2

The control of thin structures, or of structures which contain some parts which are thin, is quite an important problem in the applications. □

3. Singular Perturbations.

3.1. Optimal control and regular approximation.

Let us consider the system whose state is given by

(3.1) $$\frac{\partial^2 y}{\partial t^2} - \Delta y = v \quad \text{in} \quad Q = \Omega \times]0,T[,$$

subject to

(3.2) $$y(x,0) = \frac{\partial y}{\partial t}(x,0) = 0 \quad \text{in} \quad \Omega ,$$

(3.3) $$y = 0 \quad \text{on} \quad \Sigma = \Gamma \times]0,T[, \quad \Gamma = \partial\Omega \ .$$

The cost function is given by

(3.4) $$J(v) = \int_\Sigma \left| \frac{\partial y}{\partial \nu}(v) - z_d \right|^2 d\Sigma + N \int_Q v^2 dxdt \ .$$

We remark that given v in $L^2(Q)$, the unique solution $y(v)$ of (3.1), (3.2), (3.3) satisfies $\frac{\partial y}{\partial \nu}(v) \in L^2(\Sigma) \in$ ([1]) so that (3.4) defines a continuous function on $L^2(Q)$.

The problem of optimal control

(3.5) $$\inf \ J(v), \ v \in U_{ad} = \text{closed convex subset of } L^2(Q)$$

([1]) Cf. (1.15). It would be sufficient to have $v \in L^1(0,T;L^2(\Omega))$ to obtain the same conclusion.

admits a unique solution, denoted by u.

We do not write here the *optimality system* which characterizes u. The question we want to raise is the following: in looking for *numerical* approximation schemes, we shall need approximations y_h of y *which give "good" approximations of* $\frac{\partial y_h}{\partial \nu}$ *under the hypothesis* that $v \in L^2(Q)$.

This type of question - which does not seem to have been considered in the literature - leads to the problem considered in the following section.

3.2. *A singular perturbation problem.*

Let us consider the equation

$$\frac{\partial^2 u_\varepsilon}{\partial t^2} + \varepsilon \Delta^2 u_\varepsilon - \Delta u_\varepsilon = f \quad \text{in} \quad Q = \Omega \times]0,T[, \quad \varepsilon > 0 \tag{3.6}$$

where u_ε is subject to

$$u_\varepsilon(x,0) = \frac{\partial u_\varepsilon}{\partial t}(x,0) = 0 \quad \text{in} \quad \Omega , \tag{3.7}$$

$$\left| \begin{array}{l} u_\varepsilon = 0 \quad \text{on} \quad \Sigma , \\ \Delta u_\varepsilon = 0 \quad \text{on} \quad \Sigma . \end{array} \right. \tag{3.8}$$

We assume that

$$f \in L^1(0,T;L^2(\Omega)). \tag{3.9}$$

This problem admits a unique solution for every $\varepsilon > 0$, which satisfies

$$\left| \begin{array}{l} u_\varepsilon \in C([0,T]; H^2(\Omega) \quad H_0^1(\Omega)) , \\ \frac{\partial u_\varepsilon}{\partial t} \in C([0,T]; L^2(\Omega)). \end{array} \right. \tag{3.10}$$

Moreover,

$$(3.11)\quad \left|\begin{array}{l} \|u_\varepsilon\|_{C([0,T];\, H_0^1(\Omega))} < C \quad \text{(independent of } \varepsilon) \\ \left\|\dfrac{\partial u_\varepsilon}{\partial t}\right\|_{C([0,T];\, L^2(\Omega))} < C , \end{array}\right.$$

and

$$(3.12)\qquad \sqrt{\varepsilon}\ \|u_\varepsilon\|_{C([0,T];H^2(\Omega))} < C .$$

One can easily show that, as $\varepsilon \to 0$,

$$(3.13)\quad \left|\begin{array}{l} u_\varepsilon \to u \ \text{ in } \ L^\infty(0,T;\, H_0^1(\Omega)) \ \text{ weak star,} \\ \dfrac{\partial u_\varepsilon}{\partial t} \to \dfrac{\partial u}{\partial t} \ \text{ in } \ L^\infty(0,T;\ L^2(\Omega)) \ \text{ weak star} \end{array}\right.$$

where u is the solution of

$$(3.14)\quad \left|\begin{array}{l} \dfrac{\partial^2 u}{\partial t^2} - \Delta u = f , \\ u(x,0) = \dfrac{\partial u}{\partial t}(x,0) = 0 \ \text{ in } \ \Omega , \\ u = 0 \ \text{ on } \ \Sigma . \end{array}\right.$$

□

A (seemingly) more difficult question is the following. It follows from $(3.10)_1$ that

$$(3.15)\qquad \frac{\partial u_\varepsilon}{\partial \nu} \in C([0,T];\, H^{1/2}(\Gamma))$$

so that <u>in particular</u>

$$(3.16)\qquad \frac{\partial u_\varepsilon}{\partial \nu} \in L^2(\Sigma) .$$

On the other hand we know that under the assumptions (3.19)

$$(3.17)\qquad \frac{\partial u}{\partial \nu} \in L^2(\Sigma)$$

Hence the natural question: Do we have

(3.18) $$\frac{\partial u_\varepsilon}{\partial \nu} \to \frac{\partial u}{\partial \nu} \quad \text{in } L^2(\Sigma) \ ?$$

Remark 3.1

If (3.18) were true, then we could use standard (finite element) approximations for $\frac{\partial^2 y}{\partial t^2} + \varepsilon \Delta^2 y - \Delta y = v$, $y = 0$ $\Delta y = 0$ on Σ , in numerical algorithms for computing inf $J(v)$ as given by (3.4). □

Remark 3.2.

Of course the question (3.18) amounts to finding an a priori estimate of the type

$$\left\| \frac{\partial u_\varepsilon}{\partial \nu} \right\|_{L^2(\Sigma)} \leq C ,$$

but this is not known.

A priori estimates of this sort - but for different boundary conditions - have been given in J.-L. Lions [4]. □

4. A Fourth Problem.

We end up by mentioning a problem which we have already raised several years ago and which may be of some relevance in vibrations problems.

We consider a perforated domain Ω_ε which consists of a domain Ω where we take out "holes" of size ε in a periodic manner, with period ε .

The boundary of Ω_ε consists of two parts:

$$\partial\Omega_\varepsilon = \Gamma_\varepsilon \cup S_\varepsilon$$

where

Γ_ε = what remains of $\Gamma = \partial\Omega$ after taking out the holes,

S_ε = union of the boundaries of the holes which intersect Ω .

Let us consider now the spectral problem

$$\Delta^2 u_\varepsilon = \lambda(\varepsilon) u_\varepsilon \quad \text{in} \quad \Omega_\varepsilon , \tag{4.1}$$

$$u_\varepsilon = \frac{\partial u_\varepsilon}{\partial \nu} = 0 \quad \text{on} \quad \Gamma_\varepsilon \cup S_\varepsilon . \tag{4.2}$$

What is the behaviour of - in particular - the discrete spectrum

$$0 < \lambda_1(\varepsilon) < \lambda_2(\varepsilon) < \dots < \lambda_m(\varepsilon) < \dots \tag{4.3}$$

of (4.1) (4.2) as $\varepsilon \to 0$?

Remark 4.1

The analogous problem for $-\Delta$:

$$-\Delta u_\varepsilon = \lambda(\varepsilon) u_\varepsilon \quad \text{in} \quad \Omega_\varepsilon , \tag{4.4}$$

$$u_\varepsilon = 0 \quad \text{on} \quad \partial\Omega_\varepsilon \tag{4.5}$$

has been solved by S. Vanninathan [1] and L. Tartar [1] (cf. a presentation of the results in J.-L. Lions [5]). The solution uses in a seemingly essential manner the fact that the first eigenfunction of $-\Delta$ is a positive function and does not seem to extend to problem (4.1), (4.2). □

Additional Remark

We have introduced in J.-L. Lions [1], Additional Bibliography, a general method to study exact controllability. It leads naturally to a systematic study of the "stability" of exact controllability under perturbations, which can be of a singular nature (cf. J.-L. Lions [2], Additional Bibliography) or of an "homogenization" nature (cf. M. Avellaneda 1, Additional Bibliography). Cf. a general review in J.-L. Lions, [3], Additional Bibliography.

Bibliography

A. Bensoussan, J.-L. Lions and G. Papanicolaou [1] Asymptotic Analysis for periodic structures - Studies in Math. and its Applications - 5 - (1978) - North Holland.

I. Lasiecka, J.-L. Lions, R. Triggiani [1] Boundary control of hyperbolic systems - J.M.P.A. 1985.

J.-L. Lions [1] Contrôle des systèmes distribués Singuliers. Paris, Gauthier Villars, 1983.

[2] Some methods in the analysis of systems and their control. Science Press, Beijing (1981). Gordon and Breach.

[3] Some asymptotic problems in the optimal control of distributed systems. NASA-IEEE conference in San Diego, 1984.

[4] Un résultat de régularité pour l'opérateur $\frac{\partial^2}{\partial t^2} + \Delta^2$. Dedicated to S. Mizohata for his 60th birthday. To appear - Kyoto Univ.

[5] Lectures at the course CEA-EDF-INRIA, Bréau Sans Nappes, Summer 1983. CEA Publication, 1985.

E. Sanchez-Palencia [1] Non homogeneous media and vibration theory. Lecture Notes in Physics 127 (1980), Springer Verlag.

L. Tartar [1] Unpublished.

S. Vanninathan [1] C.R. Acad. Sc. 1979. 1980.

Additional Bibliography

M. Avellaneda and F.-H. Lin [1] Counter examples related to high frequency oscillation of Poisson Kernels, Appl. Math. and Opt. (to appear).

M. Avellaneda and F.-H. Lin [1] Homogenization of elliptic problems with L^p boundary data. Appl. Math. and Opt. (to appear).

J.-L. Lions [1] Contrôlabilite exacte de systemes distribues. C.R.A.S., Paris, t. 302, 1986, pp. 471-475.

[2] Lecture at P. Lax Symposium. Berkeley, June 1986.

[3] J. Von Neumann lecture. SIAM, Boston, July 1986.

STABILITY OF NONLINEAR WAVES

Tai-Ping Liu

Department of Mathematics and Institute of
Physical Science and Technology
University of Maryland
College Park, MD 20742

Abstract

Various stability results for nonlinear hyperbolic waves are described. Basic elements of a general theory such as decomposition into normal modes, time-invariants, hyperbolic-parabolic methods and time-asymptotic equivalence of physical systems are explained for physical models.

1. Introduction

Substantial progress on the behavior of nonlinear waves has been made in recent years. The present article is concerned with hyperbolic waves, so called because of the important role played by the important effect of certain characteristic values on wave propagation. Nonlinear hyperbolic waves occur in the interiors of compressible media and on the surfaces of incompressible materials. Other types of nonlinear waves such as reaction-diffusion waves and solitons may be regarded as degenerate hyperbolic waves and will not be included in the present study. A rich variety of wave phenomena may be classified as hyperbolic waves. Consequently, it is neither impossible, nor desirable, to try to find a single analytical tool to study them. The purpose of the present article is to illustrate by examples a general approach in studying the qualitative behavior of hyperbolic waves. The main ingredients include the decomposition into normal modes, conservation laws, hyperbolic-parabolic method and the time-asymptotic equivalence of physical systems.

2. Hyperbolic Conservation Laws.

The simplest system which carries hyperbolic waves is the conservation law:

$$\frac{\partial u}{\partial t} + \frac{\partial f(u)}{\partial x} = 0, \quad t > 0, \; -\infty < x < \infty, \tag{2.1}$$

where u, the density and $f(u)$, the flux, are n-vectors. The system is assumed to be strictly hyperbolic, that is, $\partial f/\partial u$ has real and distinct eigenvalues $\lambda_1(u) < \lambda_2(u) < \dots < \lambda_n(u)$ for all u:

$$\frac{\partial f(u)}{\partial u} r_i(u) = \lambda_i(u) r_i(u), \quad i = 1,2,\dots,n. \tag{2.2}$$

In general, solutions of (2.1) with smooth initial data develop shock waves. An important class of nonlinear waves for (2.1) is the class of simple waves. An i-simple wave $\psi(x,t)$ takes values in the r_i direction:

$$\frac{\partial \psi}{\partial t} = \text{scalar } r_i(\psi); \; \frac{\partial \psi}{\partial x} = \text{scalar } r_i(\psi) .$$

When an i-characteristic field is genuinely nonlinear (Lax [4])

$$\nabla\lambda_i \cdot r_i \neq 0$$

and r_i is so normalized that $\nabla\lambda_i \cdot r_i \equiv 1$ then we have

$$\frac{\partial \psi}{\partial t} = \frac{\partial \lambda_i(\psi)}{\partial t} r_i(\psi), \quad \frac{\partial \psi}{\partial x} = \frac{\partial \lambda_i(\psi)}{\partial x} r_i(\psi).$$

This and (2.1), (2.2) yield

$$\frac{\partial \lambda_i(\psi)}{\partial t} + \lambda_i(\psi) \frac{\partial \lambda_i(\psi)}{\partial t} = 0 , \tag{2.3}$$

which is the inviscid Burgers equation. When an i-characteristic field is linearly degenerate:

$$\nabla\lambda_i(u) \cdot r_i(u) \equiv 0 ,$$

we may choose a nonsingular parameter τ_i in the r_i direction:

$$\nabla \tau_i(u) \cdot r_i(u) = 1$$

so that, in place of (2.3), we have

$$\frac{\partial \tau_i(\psi)}{\partial t} + \lambda_i \frac{\partial \tau_i(\psi)}{\partial x} = 0. \tag{2.4}$$

Here $\lambda_i = \lambda_i(\psi)$ is a constant and so (2.4) is the linear wave equation.

Equations (2.3) and (2.4) are derived for simple waves. System (2.1) is nonlinear. A general solution contains many shock waves and expansion waves. These waves interact nonlinearly. The result of the nonlinear interactions and the hyperbolicity of the system is that waves eventually decompose into n normal modes. Each normal mode is either a shock wave or a simple wave, time-asymptotically. The shock waves are governed by the Rankine-Hugoniot condition, and simple waves by (2.3) or (2.4). This is the idea of deriving simpler systems which are equivalent to the original system time-asymptotically. The idea is justified analytically in [5] using the principle of nonlinear superpositions, [2], [3], [6].

3. Viscous Conservation Laws.

Conservation laws with dissipative mechanisms, such as the compressible Navier-Stokes equations, often take the form

$$\frac{\partial u}{\partial t} + \frac{\partial f(u)}{\partial x} = \frac{\partial}{\partial x} \left(B(u) \frac{\partial u}{\partial x}\right) \tag{3.1}$$

where $B(u)$ is the viscosity matrix. System (3.1) carries viscous shock waves. The nonlinear stability of viscous shock waves has been shown only recently, [9],[10]. The basic idea is first to recognize that a perturbation of shock waves, in addition to translating the shock waves, also gives rise to diffusion waves. Diffusion waves satisfy, time-asymptotically either the Burgers equation or the linear heat equation:

$$\frac{\partial \lambda_i}{\partial t} + \lambda_u \frac{\partial \lambda_i}{\partial x} = \alpha_i \frac{\partial^2 \lambda_i}{\partial x^2} \quad \text{when} \quad \nabla\lambda_i \cdot r_i \neq 0, \tag{3.2}$$

$$\frac{\partial \lambda_i}{\partial t} + c\,\frac{\partial \lambda_i}{\partial x} = \alpha_i\,\frac{\partial^2 \lambda_i}{\partial x^2} \quad \text{when} \quad \nabla\lambda_i \cdot r_i \equiv 0 . \tag{3.3}$$

The derivation of (3.2) and (3.3) is somewhat involved. The reader is referred to [9]. A hyperbolic method is also introduced in [9] to study the stability of shock waves and diffusion waves. Though (3.1) is basically parabolic, but may not be uniformly parabolic, it carries hyperbolic waves. Thus it is no surprise that an effective technique in studying the qualitative behavior of nonlinear waves for (3.1) should contain a hyperbolic method such as the characteristic method. System (3.1) has n time invariants. These are used to obtain an a priori estimate of the location and strength of the normal modes in the asymptotic state of a general solution.

4. Conservation Laws with Damping

When a uniform damping is added to a conservation laws, such as

$$\frac{\partial u}{\partial t} + \frac{\partial f(u)}{\partial x} = -\,\alpha\, u,$$

for some positive constant α, then u tends to the zero state as time goes to infinity, [1]. Physical dampings are not uniform, however. Consider the following simple system, [11],

$$\begin{aligned} &\frac{\partial v}{\partial t} - \frac{\partial u}{\partial x} = 0 , \\ &\frac{\partial u}{\partial t} + \frac{\partial p(v)}{\partial x} = -\,\alpha\, u, \quad \alpha > 0, \ p'(v) < 0, \ u \in R^1, \ v \in R^1. \end{aligned} \tag{4.1}$$

The system is hyperbolic with characteristic speeds $\pm(-p'(v))^{1/2}$. We are interested in the qualitative behavior, in particular the large-time behavior of the solution. For simplicity we assume that the initial data $u(x,0)$ and $v(x,0)$ have compact support. Since the system is hyperbolic, the solution has finite speed of propagation and therefore is of compact support for all time. Thus we may integrate (4.1) with respect to x to get:

$$\int_{-\infty}^{\infty} v(x,t)\, dx = \int_{-\infty}^{\infty} v(x,0)dx, \tag{4.2}$$

(4.3) $$\int_{-\infty}^{\infty} u(x,t)dx = e^{-\alpha t}\int_{-\infty}^{\infty} u(x,0)dx.$$

Thus we have only one time-invariant (4.2). Since the support of $v(x,t)$ grows at most linearly in t, (4.2) implies that $v(x,t)$ decays pointwise at a rate no faster than t^{-1}. There is a strong coupling of u and v in (4.1) and so $u(x,t)$ decays at most like $O(1)t^{-1}$. On the other hand, (4.3) yields an exponential decay for the integral of $u(x,t)$. Consequently there is a strong possibility that $u(x,t)$ tends to an odd function in x.

To obtain more precise information on the qualitative behavior of $(u,v)(x,t)$, more analysis is necessary. Notice there is one time-invariant, (4.2), and the system (4.1) is symmetric in x with respect to $x = 0$. Thus we expect only one mode to survive eventually and the surviving mode propagates mostly along the t-axis. That is,

(4.4) $$\frac{\partial}{\partial t} << \frac{\partial}{\partial x} \quad \text{for} \quad t >> 1.$$

We want to derive a simplified system which governs the large-time behavior of solutions for (4.1). This is done as follows: Approximate u by ψ and v by ϕ, i.e.,

(4.5) $$|v-\phi| + |u - \psi| \to 0 \quad \text{as} \quad t \to \infty.$$

We want to respect the time-invariant (4.2), which comes from the first equation (1.1), by insisting that

(4.6) $$\frac{\partial\phi}{\partial t} - \frac{\partial\psi}{\partial t} = 0\ ,$$

(4.7) $$\int_{-\infty}^{\infty} \phi(x,t)dx = \int_{-\infty}^{\infty} \phi(x,0)dx = \int_{-\infty}^{\infty} v(x,0)dt.$$

We next find the time-asymptotic approximation of the second equation in (4.1). Note that from (4.4) and (4.6)

(4.8) $$\phi >> \psi \quad \text{for} \quad t >> 1.$$

Thus in

$$\frac{\partial\psi}{\partial t} + \frac{\partial p(\phi)}{\partial x} + \alpha\psi = 0$$

the first term is small, for $t >> 1$, compared with other terms. Consequently the second equation in (1.1) is simplified into

(4.9) $$\frac{\partial p(\phi)}{\partial t} \approx - \alpha\psi .$$

From (4.6) and (4.9) we have

(4.10) $$\frac{\partial\phi}{\partial t} \cong - \frac{1}{\alpha} \frac{\partial^2 p(\phi)}{\partial x^2} .$$

Since ϕ and ψ are expected to decay due to damping, the right-hand side of (4.10) can be approximated by

$$- \frac{1}{\alpha} \frac{\partial^2 P(\phi)}{\partial x^2} = - \frac{1}{\alpha} P'(\phi) \frac{\partial^2 f}{\partial x^2} - \frac{1}{\alpha} P''(\phi)\left(\frac{\partial\phi}{\partial x} \right)^2$$

$$\sim - \frac{1}{\alpha} P'(0) \frac{\partial^2\phi}{\partial x^2} .$$

Finally we define ϕ and ψ by (cf. (4.7), (4.9) and (4.10))

(4.11) $$\frac{\partial\phi}{\partial t} = \frac{|P'(0)|}{\alpha} \frac{\partial^2\phi}{\partial x^2} ,$$

(4.12) $$\int_{-\infty}^{\infty} \phi(x,t)dx = \int_{-\infty}^{\infty} v(x,0)dx,$$

(4.13) $$\psi = \frac{|p'(0)|}{\alpha} \frac{\partial\phi}{\partial x}$$

so that (4.6) and (4.7) hold and

(4.14) $$\frac{\partial\psi}{\partial t} + \frac{\partial p(\phi)}{\partial x} = - \alpha\psi + \text{error}$$

for some error term which turns out to tend to zero at a faster rate than the rest of the terms as $t \to \infty$. To see that, we first note that (4.11) and (4.12) do not define $\phi(x,t)$ uniquely. Nevertheless, since we are interested in the large-time behavior, it does not matter how $\phi(x,0)$ is distributed so long as the right hand side of (4.12) is known. Thus we may choose ϕ to be the heat kernel:

$$(4.15) \qquad \phi(x,t) = \left(\int_{-\infty}^{\infty} v(x,0)dx \right) \cdot \left(4\pi \frac{|p'(0)|}{\alpha} t\right)^{-1/2} \exp\left(-\frac{\alpha x^2}{4|p'(0)|t}\right).$$

With $\phi(x,t)$ given, $\psi(x,t)$ is calculated from (4.13) and the error term can be shown to tend to zero at faster rate than the rest of terms in (4.14).

Notice that while the original system (4.1) is strictly hyperbolic, the system (4.11) is parabolic. In an upcoming paper I will show rigorously that the two systems are time-asymptotically equivalent in the sense of (4.5). In the process of deriving (4.11) and (4.13) we have assumed (4.4) and (4.8) which are consistent with (4.15). The estimate (4.3) is also explained by the fact that $\psi(x,t)$ is an odd function in x as a consequence of (4.15) and (4.13).

5. Conservation Laws with Moving Sources

In the previous examples, the number of surviving modes equals the number of time invariants and all waves are nonlinearly stable. The following example is different in these aspects. Consider scalar conservation law with a moving source

$$(5.1) \qquad \frac{\partial u}{\partial t} + \frac{\partial f(u)}{\partial x} = g(x-ct,u), \quad -\infty < x < \infty,\ t > 0,\ u \in R^1.$$

The model was proposed in [8] to capture some of the essential qualitative features of gas flows through a nozzle, [7]. For simplicity we assume that

$$f(0) = f'(0) = 0,\ f''(u) \neq 0 \text{ for all } u.$$

The source term has fixed speed c and is assumed to be a finite source, for simplicity,

$$g(\xi,u) \equiv 0 \text{ for } \xi \notin (0,1),\ \xi \equiv x-ct.$$

The main assumption is that

$$g(\xi,u) \neq 0,\ \frac{\partial g(\xi u)}{\partial u} \neq 0 \quad \text{for} \quad \xi \in (0,1) \quad \text{and} \quad u \in (-\infty,\infty).$$

For $\xi \equiv x-ct \notin (0,1)$, u satisfies the conservation law

$$\frac{\partial u}{\partial t} + \frac{\partial f(u)}{\partial t} = 0. \tag{5.2}$$

For $\xi \in (0,1)$, time-asymptotically the source term $g(x-ct,u)$ eventually helps to generate a wave with speed c and is governed by

$$-\frac{cdu}{d\xi} + \frac{d(f(u))}{d\xi} = g(\xi,u), \tag{5.3}$$

$$u \equiv u(\xi) \ , \quad 0 < \xi \equiv x - ct < 1.$$

Thus, time-asymptotically the solution carries a traveling wave with speed c for $x - ct \in (0,1)$ and waves for the conservation law (5.2) for $x-ct \notin (0,1)$. As asymptotic states, these waves should not interact. Thus waves for $x - ct < 0$ (or $x-ct > 1$) have speed greater (or less than) the speed c of the moving source. Waves for (5.2) are governed by the characteristic speed $f'(u)$, which depends on u. When $f'(u)$ is around the speed c of the moving source, there may exist several waves in the asymptotic state. This is so even though the equation (5.1) has no time-invariant. Another new feature is that when waves for (5.2) enter the region $\xi \in (0,1)$ they may change their types and instability results. In particular a shock wave traveling with speed c in $\xi \in (0,1)$ is nonlinearly unstable if $\partial g/\partial u > 0$ there. The stability and instability of a given wave pattern can be predicted by the study of all possible time asymptotic states with given end states.

6. Concluding Remarks

There is a rich variety of wave phenomena concerning hyperbolic waves. The present article describes a general approach in obtaining the large-time behavior of a general solution and studying the nonlinear stability (or instability) of waves. The approach can also be applied to wave phenomena under other physical effects such as relaxation and external friction. Undoubtedly, much more remains to be done. For instance in a combustion process, convection, dissipation and reactions interact nonlinearly and give rise to interesing and complicated wave phenomena which remain to be analyzed.

References

1. Dafermos, C. and Hsiao, L. Hyperbolic systems of balance laws with inhomogeneity and dissipation, Indiana U. Math. J. 31 (1982), 471-491.

2. Glimm, J., Solutions in the large for nonlinear hyperbolic systems of conservation laws, Comm. Pure. Appl. Math. 18 (1965), 697-715.

3. Glimm, J. and Lax, P.D., Decay of solutions of nonlinear hyperbolic conservation laws, Memoirs, A.M.S., 101 (1970).

4. Lax, P.D., Hyperbolic system of conservation laws, II, Comm. Pure. Appl. Math. 10 (1957), 537-566.

5. Liu, T.-P., Linear and nonlinear large-time behavior of solutions of general systems of hyperbolic conservation laws, Comm. Pure Appl. Math. 30 (1977), 767-796.

6. _______, Deterministic version of Glimm scheme, Comm. Math. Phys. 57 (1977), 767-796.

7. _______, Nonlinear stability and instability of transonic gas flow through a nozzle, Comm. Math. Phys. 83 (1982), 243-260.

8. _______, Resonance for quasilinear hyerbolic equation, Bulletin, A.M.S., 6 (1982), 462-465.

9. _______, Nonlinear stability of shock waves for viscous conservation laws, Memoirs, A.M.S. No. 328 (1985).

10. _______, Shock waves for compressible Navier-Stokes equations are nonlinearly stable, Comm. Pure. Appl. Math. (to appear).

11. Nishida, T., Global smooth solutions for the second order quasilinear wave equations with first order dissipation, Publications Math. d'Orsay (1978).

THE NASH-MOSER TECHNIQUE FOR AN INVERSE PROBLEM IN POTENTIAL THEORY RELATED TO GEODESY

C. Maderna

Dipartimento di Matematica
Federigo Enriques
Via Saldini 50
20133 Milano Italy

and

C. Pagani, and S. Salsa

Dipartimento di Matematica
Politecnico di Milano
Via Bonardi, 9
20133 Milano Italy

1. Introduction

We shall consider the following inverse problem of potential theory: to determine the shape of a body from measurements of the Newtonian potential at its surface, given some information about its distribution. More precisely, we consider a class of bodies G that can be parametrized as follows. Let u be a smooth mapping of $S^2 \to R$ with $|u| <$ constant $< r_0$. (Here S^2 is the surface of the unit ball in R^3). For $w \in S^2$, define

$$\phi_u(w) = r_0 + u(w), \; \Gamma_u = \phi_u\,(S^2).$$

Then $G \equiv G_u$ is the bounded domain whose boundary is Γ_u. (Thus Γ_0 is reference sphere with center at the origin and radius r_0).

Let now $\delta: R^3 \to R$ be a given strictly positive function and V_u the potential created by a mass of density δ distributed over G_u; we can write

$$(1) \qquad V_u(x) = \int_{S^2} dw \int_0^{|\phi_u(w)|} t^2\; \delta(tw)\,|x-tw|^{-1}\, dt.$$

The measured datum (the potential on Γ_u) is then a function v defined on S^2.

We wanted to find a function u such that

$$(2) \qquad A(u) \equiv V_u \circ \phi_u = v \quad \text{on} \quad S^2$$

where $V_u \circ \phi_u$ denotes the composition of V_u with ϕ_u. We prove a local result by linearizing near suitable surfaces and using a version of the Nash-Moser inverse function theory.

The use of ordinary Banach implicit function theorem is prevented by a continuous loss of regularity in the iteration procedure involved in the solution of the equation. A point of interest is that the solvability of the linearized equation depends on the mass distribution; namely, for particular densities, δ, including $\delta \equiv$ constant, the presence of non trivial eigenspaces with dimension N, depending on the density itself, forces the introduction of projections onto those eigenspaces. This combination of the Nash-Moser technique and projections on finite-dimensional spaces leads us to solve a modified equation of the form

$$A(u) + w = v$$

where w is an eigenvector of the linearized operator. We shall describe the results in the particular case of small perturbations of an homogeneous sphere.

A problem similar to that described by equation (2) is the determination of the shape of the body where the potential is known on a surface surrounding (and far from) the body itself. A description of classes of stable solutions in both cases is proved in [P]. Two other problems showing some features in common with ours are described in the papers of [S] and [HO]; the first one arising in electrostatics and the second one in geodesy. In the latter, the problem is to find the shape of the earth from the knwoledge of the potential and the gravity vector on its surface, while no assumptions are made on its internal structure. Both these problems have been dealt with by versions of the same inverse function theorem.

2. The Linearized Equation

We consider now the linearized equation at a fixed u. Standard computations give

$$A'(u)\rho = V'_u \rho \circ \phi_u + \langle \mathrm{grad}_x V_u \circ \phi_u , \phi'_u \rho \rangle$$

where $\langle , \rangle$ denotes the scalar product on R^3 . We have

$$V'_u \rho(x) = \int_{S^2} \delta \circ \phi_u(w) \rho(w) \frac{|\phi_u(w)|^2}{|x - \phi_u(w)|} dw$$

and

$$f_u(w) = - \int_{S^2} dw' \int_0^{|\phi_u(w')|} t^2 \delta(tw) \frac{\langle \phi_u(w) - tw', w\rangle}{|\phi_u(w) - tw'|^3} dt.$$

With $M_u \rho = V'_u \rho \circ \phi_u$ and $f_u(w) = \langle \mathrm{grad}_x V_u \circ \phi_u, w\rangle$, the linearized equation at u is as follows:

$$(3) \qquad M_u \rho(w) + f_u(w)\rho(w) = h(w) \qquad (w \in S^2),$$

a Fredholm integral equation of the second kind.

To see how this equation behaves we first consider the particular case in which u is a constant and $\delta = \delta(|x|)$ is a radial function. In this case, with $r = r_0 + u$,

$$f_u(w) = - \frac{4\pi}{r^2} \int_0^r t^2 \delta(t)dt \equiv - \frac{m}{r^2}$$

and equation (3) becomes

$$(3') \qquad r\, \delta(r) \int_{S^2} \rho(w') |w - w'|^{-1} dw' - mr^{-2} \rho(w) = h(w) \qquad (w \in S^2).$$

Observe now that the eigenvalues of the integral operator

$$\rho(\cdot) \to \int_{S^2} \rho(w')|\cdot - w'|^{-1} dw'$$

are $4\pi/(2n+1)$, $n = 0,1,\ldots$. The n-th eigenvalue has a $(2n+1)$ - fold degeneracy and the corresponding eigenfunctions are the $(2n+1)$-surface spherical harmonics of degree n. Hence the discussion of (3) depends on δ, more precisely on the fact that the equation

$$(4) \qquad \int_0^r t^2 \delta(t)\, dt = r^3 \delta(r)/(2n+1)$$

is satisfied or not for some n. Therefore we distinguish two cases:

Case I. No integer satisfies equation (4). By expanding ρ and h in spherical harmonics:

$$\rho = \sum_0^\infty \rho_n , \quad h = \sum_0^\infty h_n ,$$

we obtain from (3') that

$$\{4\pi/(2n+1)r\delta(r) - mr^{-2}\}\rho_n = h_n \qquad (n = 0,1,2....)$$

and hence, for any h in the Sobolev space $H^s(S^2) = H^s$ $(s > 0)$ (3') has a unique solution ρ and the estimate

$$\|\rho\|_s < c\|h\|_s$$

holds for every $s > 0$, with c independent of s.

Case II. Equation (4) is satisfied for some $\bar{n}$. In this case (3') admits a solution if and only if $h_{\bar{n}} = 0$, i.e., if and only if h belongs to a space of codimension $2\bar{n}+1$. The map

$$h \to r^2 m^{-1} \sum_{n \neq \bar{n}} \frac{2n+1}{2(\bar{n}-n)} h_n$$

satisfies (3') modulo $\bar{n}$ - degree harmonics. A unique solution can be selected by imp 2$\bar{n}$+1 linearly independent conditions. Observe that when δ is a constant, $n = 1$ satisfies (a) for every value of r.

Our results on the linearized equation can be stated as follows:

THEOREM 1. Assume that Case I holds for some radial function $\bar{\delta}$ and for $r = r_0$. Then, for every $h \in L^2$, equation (3) has a unique solution $\rho \in L^2$, provided that $|u|_{C^1(S^2)}$ and $|\delta - \bar{\delta}|_{C^1(G_0)}$ are sufficiently small. Moreover the estimate

$$\|\rho\|_0 < c\,\|h\|_0$$

holds.

As an example of the analogue of Theorem 1 when - fu takes on values belonging to the spectrum of M_u, we consider a perturbation of an homogeneous sphere. Thus, we assume δ to be close to a constant $\bar{\delta}$ and consider equation (3) together with the conditions

$$(5) \qquad B_{u,\delta}\rho = \int_{S^2} |\phi_u(w)|^3 \, \delta\circ\phi_u(w)\rho(w)w_j dw = 0 \,, \quad j = 1,2,3.$$

The meaning of conditions (5) is that the center of mass of G_u is the origin.

We have the following result:

THEOREM 2. Let $\bar{\delta}$ be a constant. Then, for every $h \in L^2$ orthogonal to the first degree spherical harmonics, system (3), (5) has a unique solution, provided that $|u|_{C^1(S^2)}$ and $|\delta - \bar{\delta}|_{C^1(G_0)}$ are sufficiently small. Moreover the estimate

$$\|\rho\|_0 < C\cdot\|h\|_0$$

holds.

3. The Nonlinear Equation

Consider first the case in which the linearized equation is uniquely solvable. Thus, let u_0 and δ be functions satisfying the conditions in Theorem 1 and set $A(u_0) = V_0$.

Our purpose is to show that if V is sufficently close to V_0, then there exists a solution u of $A(u) = V$ close to u_0.

As we already said the use of the ordinary inverse function theorem is prevented by the phenomenon of a loss of regularity. In fact, the derivative $A'(u)$ turns out to be discontinuous on Sobolev spaces. An estimate of the type

$$\|(A'(u_1) - A'(u_2))\rho\|_s < C \, \|\rho\|_s$$

is possible only with constants depending on a number of derivatives of $u_1 - u_2$ greater than s.

The appropriate tool to handle equation (2) is a Nash-Moser type technique. The one we have used is presented in [H].

We briefly describe the procedure which gives the solution. We first introduce a class of smoothing operators S_t, $t > 0$, by taking a function $\psi \in C^\infty(R)$, $0 < \psi(t) < 1$, $\psi = 0$ for $t < 0$, $\psi(t) = 1$ for $t > 1$ and defining for $u \in L^2(S^2)$,

$$u = \sum_{n=0}^{\infty} \sum_{j=1}^{2n+1} c_{nj} Y_{nj} , \qquad S_t u = \sum_{n=0}^{\infty} \sum_{j=1}^{2n+1} d_{nj} Y_{nj}$$

where $d_{nj} \varepsilon c_{nj} \cdot \psi(e^t - n - 1)$.

We can easily prove that $S_t u = 0$ if $t < 0$ and that

(i) $\|S_t u\|_s < e^{t(s-r)} \|u\|_r \qquad (r < s)$,

(ii) $\|S_t u - u\|_s < e^{t(r-s)} \|u\|_r \qquad (r > s)$.

Consider now the iteration procedure given by the following differential equation for a path u_t :

$$(6) \qquad u_t' = \mu \Lambda (S_t u_t) S_t [v - A(u_t)] \qquad (u_t' = \frac{d}{dt} u_t)$$

with u_0 as a starting point, where μ is a positive number and Λ denotes the inverse of A. We want to show that a solution exists for all times $t > 0$ and converges to a solution of $A(u) = v$ as $t \to +\infty$. Theorem 15.6.3 of [H] assures existence and uniqueness for small times. To go further we need uniform estimates in a neighborhood of u_0 for the operators A, A', Λ and for increments of the first derivative. We have the following results:

THEOREM 3. Let u_0 and δ satisfy the conditions of Theorem 1. Furthermore let $\|u - u_0\|_4 < 1$. For $s > 0$ the following inequalities hold:

$\|A(u) - A(u_0)\|_s < C \|u - u_0\|_{s+3}$,

$\|A'(u)\rho\|_s < C \cdot (\|\rho\|_2 \cdot \|u\|_{s+3} + \|\rho\|_{s+2})$,

$\|\Lambda(u)h\|_s < C \cdot (\|h\|_2 \cdot \|u\|_{s+3} + \|h\|_{s+2})$ (provided $\|u-u_0\|_3$ is small, depending on s),

$\|A'(u+\sigma)\rho - A'(u)\rho\|_s < C(\|\rho\|_{s+2}\|\sigma\|_3 + \|\rho\|_2 \cdot \|\sigma\|_{s+3})$,

where $C = C(s)$.

THEOREM 4. Let u_0 and σ satisfy the conditions of Theorem 1. Let $\mu > q > 9$, $s>9$, $a>2$. If $\|v - A(u_0)\|_q < \eta$, where η is sufficiently small, then the path u_t defined by (6) exists for all times, converges in H_{s-a} norm as $t \to \infty$ to a solution u_∞ of $A(u) = v$ and satisfies the following estimates:

$$\|u_t - u_\infty\|_{s-a} < c \cdot e^{-t} \|v - A(u_0)\|_{s+1} ,$$

$$\|u_t - u_\infty\|_{s-a} < c \cdot \|v - A(u_0)\|_s ,$$

with c independent of t.

We observe that, in general, η depends on s. If the solution ρ of the linearized equation (3) at $u = u_0$ satisfies the inequality $\|\rho\|_s < c \cdot \|h\|_s$ with c independent of s (this is true for instance, when u_0 = constant), then it is possible to choose η independent of s.

4. The Perturbation of a Homegeneous Sphere

We consider now the case of a perturbation of an homogeneous sphere (of radius r_0) where $-f_u$ takes on values in the spectrum of M_u.

Let $\bar{\delta}$ be a positive constant and take u_0 and δ as in Theorem 2. Our purpose is to show that if v is sufficiently close to v_0, then there exists a u close to u_0 and there are small constants a^1, a^2, a^3 such that the equation

$$A(u) - \sum_{j=1}^{3} a^j y_{1j} = v \tag{7}$$

is satisfied together with the conditions

$$\int dw \int^{|\phi_0(w)|} t^3 w_j \, \delta(tw) dt = 0 \qquad (j=1,2,3). \tag{8}$$

Again the meaning of (8) is that the center of mass of G_u is placed at 0.

The Nash-Moser iteration is to be set up for the unknown $V = (a^1, a^2, a^3, u)$ and the operator

$$\bar{A}(U) = A(u) - \sum_{j=1}^{3} a^j y_{1j} .$$

The uniform estimates for $\bar{A}$, $\bar{A}^l$, $\bar{A}^{l-1} \equiv \tilde{\Lambda}$, and the variation of $\bar{A}'$ hold in the same form. In fact the linearized system is now

$$A'(u)\rho = h + \sum_{j=1}^{3} \alpha^j y_{1j} \quad \text{together with conditions (5)} .$$

Here α^j denotes the variation of a^j. The construction of the solution develops as before and the conclusion is the following result:

THEOREM 5. Let δ and u_0 satisfy the conditions of Theorem 2. Choose numbers $\mu > q > 9$, $s > 9$, $a > 2$. If $\|v - A(u_0)\|_q < \eta$, for η sufficently small, the path $U_t = (a_t^1, a_t^2, a_t^3, u_t)$ defined by

$$U_t' = \mu \ \tilde{\Lambda}(S_t U_t) S_t [v - \tilde{A}(U_t)] \qquad (S_t U = (a^1, a^2, a^3, S_t u))$$

with initial condition $U_0 = (0,0,0,u_0)$, converges as $t \to \infty$ in H_{s-a} to a solution U_∞ of equation (7). Moreover the following estimates hold:

$$\|u_t - u_\infty\|_{s-a} + \sum_{j=1}^{3} |a_\infty^j| \leq c \cdot \|v - A(u_0)\|_s ,$$

$$\|u_t - u_\infty\|_{s-a} + \sum_{j=1}^{3} |a_\infty^j - a_t^j| \leq c \cdot e^{-t} \|v - A(u_0)\|_s$$

with c independent of t.

5. A Uniqueness Result

We conclude with a uniqueness result for equation (2). An analogous result holds for equation (7).

THEOREM 6. Take u_0 and δ as in Theorem 1. If $\|u - u_0\|_4 \leq 1$, there exists a number ρ_0 such that, if $\|\rho\|_3 \leq \rho_0$, then $\|\rho\|_0 \leq c \ \|A(u+\rho) - A(u)\|_2$.

The detailed proofs of the results stated above can be found in [MPS].

References

[H] R.S. Hamilton, "The inverse function theorem of Nash and Moser", Bull Amer. Math. Soc. (1982).

[HÖ] L. Hörmander, "The boundary problems of physical geodesy", Arch. Rat. Mech. Anal. (1976).

[MPS] C. Maderna, C. Pagani and S. Salsa, "Nonlinear analysis PMA. Vol. 10, 1986.

[P] C.D. Pagani "Stability of a surface determined from measures of potential", to appear in SIAM.

[S] D.G. Schaeffer, "The capacitor problem", Indiana Univ. Math. J. (1975).

VARIATIONAL STABILITY AND RELAXED DIRICHLET PROBLEMS

by

Umberto Mosco

Dipartimento di Matematica
Universitá di Roma
00185 ROMA (Italy)

Several papers have recently been devoted to the study of Dirichlet problems in highly perturbed domains of R^n and to related convergence matters, e.g. see [17, 19, 21, 27, 22, 8, 5, 6, 15, 11, 26, 9, 7, 10, 3, 2, 4].

In particular, it has been recognized that the relevant asymptotic behavior of the solutions is of variational nature and cannot be exhaustively described in explicit purely geometric or analytical terms.

In this article we give a survey of the main results of some forthcoming joint papers with Gianni Dal Maso, [12], [13], [14]. In our papers we introduce a class of problems, called relaxed Dirichlet problems and we relate a convergence of variational type to them, called γ-convergence. We prove compactness and density results. We also study the pointwise behavior of solutions. We establish a variational generalization of the classical Wiener's criterion of potential theory and we extend Maz'ja's estimate to it. We also obtain decay estimates for the energy.

In order to better clarify the problems dealt with in [12] [13] [14], we give a discussion of some examples, taken from [12].

1. Some Examples

In all the examples below we denote by $B_\rho(x_0)$ the open disk in R^2, with center at the point $x_0 \in R^2$ and radius $\rho>0$, and we put $B_\rho = B_\rho(0)$.

Example 1 Let $\Omega = B_2$. For every $h \in N$ we consider the open domain $\Omega - E_h$, where

$$E_h = \bigcup_{j,l \in \mathbb{N}} E_h^{jl}$$

and E_h^{jl} is the closed disk centered at the point with polar coordinates $(\exp(-\frac{2\pi j}{h}), \frac{2\pi l}{h})$ with radius $r^j = \exp(-\frac{2\pi j}{h} - \frac{h^2}{2\pi c})$, $c>0$ being some fixed constant.

We then consider the sequence $\{u_h\}$ of the weak solutions $u_h \in H^1(\Omega-E_h)$ of the Dirichlet problems

(1) $$-\Delta u_h = 0 \text{ in } \Omega-E_h \text{ , } u_h = 0 \text{ on } \partial E_h$$

(2) $$u_h = g \quad \text{on} \quad \partial\Omega \text{ ,}$$

where g is some given function in $H^1(\Omega)$. We consider u_h as being defined on the whole of Ω, by putting $u_h = 0$ on E_h.

Two main questions may be raised at this point. The first one concerns the global behavior of the u_h in Ω as $h \to +\infty$: Does u_h converge to some limit function u in Ω for some reasonable topology (necessarily a weak one, due to the highly oscillatory behavior of u_h as $h \to +\infty$)? The second question refers to the pointwise asymptotic behavior of the u_h at a given point x_0 of Ω, in particular at the origin.

The answer to the first question is provided by the theory of Γ-convergence for quadratic functionals of Dirichlet type, see [15],[2].

This theory can be easily applied to domains of the form Ω-E, with E a closed set formed of periodically distributed "holes" in R^2. The domains $\Omega-E_h$ of our example can be reduced to a periodic geometry, by performing the change of variables $y_1 = \ln|x|$, $y_2 = \arg x$.

Without entering into details, for which we refer to [12], the conclusion is that our sequence $\{u_h\}$ converges strongly in $L^2(\Omega)$ and weakly in $H^1(\Omega)$ to the (unique) weak solution $u \in H^1(\Omega) \cap L^2(qdx)$ of the Schrödinger equation

(3) $$-\Delta u + q(x)\, u = 0 \quad \text{in } \Omega$$

satisfying the boundary condition

(4) $$u = g \quad \text{on} \quad \partial\Omega \text{ .}$$

The (non-negative) potential $q(x)$ that appears in (3) is the function

(5) $$q(x) = c|x|^{-2} \text{ if } |x| < 1, \quad q(x) = 0 \text{ if } |x|>1 \text{ ,}$$

c being the constant associated with the sets E_h.

Now let us come to the second question raised above. By classical regularity results, u_h and u are continuous in the region $\Omega-\{0\}$. Moreover, the u_h are harmonic in $\Omega-\overline{B}_1$; therefore, since they converge to u strongly in L^2, they also converge uniformly to u on every compact subset $K \subseteq \Omega - \overline{B}_1$. The asymptotic behavior of the u_h , however, differs drastically at every other point $x_0 \in \overline{B}_1 - \{0\}$. By assuming for simplicity that $g>0$ on $\partial\Omega$, by the strong maximum principle we have $u(x_0) > 0$. Since the u_h are continuous functions converging strongly in L^2 to the continuous function u in a neighborhood of x_0 and since the u_h vanish on the sets E_h that accumulate at x_0 as $h \to +\infty$, we conclude that for every value $t \in [0,u(x_0)]$ we can find a sequence $x_h \to x_0$ as $h \to +\infty$ such that $u_h(x_h) \to t$ as $t \to +\infty$. This shows that the u_h oscillate wildly in every neighborhood of the given $x_0 \in \overline{B}_1 - \{0\}$.

A finer analysis is required at the origin. In order to prove that u_h is continuous at 0 we have to rely on the classical Wiener's criterion for the boundary regularity of solutions of Dirichlet problems, [29]. We first remark that for every $0 < \rho < 1$ and every $j \in N$ we have

$$\rho_h^j\,(E_h \cap B_\rho) = E_h \cap B_{\rho\rho_h^j} \quad , \quad \rho_h^j = \exp\left(-\frac{2\pi j}{h}\right);$$

therefore

$$\mathrm{cap}\,(E_h \cap B_{\rho\rho_h^j}\,,\,B_{2\rho\rho_h^j}) = \mathrm{cap}(E_h \cap B_\rho, B_{2\rho}),$$

whence

$$\inf_{0<\rho<1} \mathrm{cap}(E_h \cap B_\rho, B_{2\rho}) > 0\ .$$

This implies that Wiener's condition

$$\int_0^{1/2} \frac{\mathrm{cap}(E \cap B_\rho, B_{2\rho})}{\mathrm{cap}(B_\rho, B_{2\rho})}\,\frac{d\rho}{\rho} = +\infty \tag{6}$$

is satisfied for $E = E_h$ at the boundary point 0 of $\Omega-E_h$; therefore u_h is continuous at 0 in Ω. To carry our analysis further on, and study the limit of $u_h(0)$ as $h \to +\infty$, we use Maz'ja estimate, [23] see also [12]. This is an estimate of the modulus of continuity of u_h at 0, in terms of the rate of divergence of the Wiener's integral (6). Namely, there exist two constants k and $\beta>0$ such that

(7) $$\sup_{B_r} |u_h|^2 \leq k \|u_h\|^2_{L^2(B_{1/2})} \omega_h(r)^\beta ,$$

for every $0 < r < \frac{1}{4}$ and every $h \in \mathbb{N}$, where

(8) $$\omega_h(r) = \exp(- \int_r^{1/2} \mathrm{cap}(E_n \cap B_\rho, B_{2\rho}) \frac{d\rho}{\rho}).$$

In order to go to the limit in (7) as $h \to +\infty$, we have to control the exponential factor (8).

In [12] we prove that a capacity can be associated with the function $q(x)$ appearing in (3), such that

(9) $$\lim_{h\to+\infty} \mathrm{cap}\,(E_h \cap B_\rho, B_{2\rho}) = \mathrm{cap}_q(B_\rho, B_{2\rho})$$

for every $0 < \rho < \frac{1}{2}$, except possibly for a countable set of ρ's. This q-capacity is defined variationally, by setting

$$\mathrm{cap}_q(B_\rho, B_{2\rho}) = \min \{ \int_{B_{2\rho}} |\nabla v|^2 dx + \int_{B_\rho} q(x)\, v^2 dx ; v - 1 \in H^1_0(B_{2\rho})\} .$$

Therefore

(10) $$\lim_{h\to+\infty} \omega_h(r) = \omega_q(r)$$

for every $0 < r < \frac{1}{2}$, where $\omega_q(r)$ is defined by

(11) $$\omega_q(r) = \exp(- \int_r^{1/2} \mathrm{cap}_q(B_\rho, B_{2\rho}) \frac{d\rho}{\rho}).$$

We then can take the limit as $h \to +\infty$ in (7) and obtain the estimate

(12) $$\sup_{B_r} u^2 \leq k \|u\|^2_{L^2(B_{1/2})} \omega_q(r)^\beta$$

for every $0 < r < 1/4$.

An explicit calculation carried out in [14], which exploits the spherical symmetry of the function $q(x)$, shows that

$$\operatorname{cap}_q (B_\rho, B_{2\rho}) = \frac{2\pi\sqrt{c}}{1+\sqrt{c}\ \ln 2} = \alpha ,$$

for every $0 < \rho < 1/2$, where c is the constant appearing in (5). Thus,

$$\omega_q(r) = (2r)^\alpha$$

and we finally get

$$\sup_{B_r} u^2 \leq k \|u\|^2_{L^2(B_{1/2})} (2r)^{\alpha\beta} .$$

Therefore u is continuous at 0, and $u(0) = 0$. Moreover, it also follows from our previous estimates that the solutions u_h keep a stable behavior at the origin, with $u_h(x_h) \to 0$ for every sequence $x_h \to 0$ as $h \to +\infty$.

Example 2 We take Ω as in Example 1 and for every $h \in N$ we consider the set

$$E_h = \bigcup_{\ell \in \mathbb{N}} E_h^\ell$$

where E^ℓ denotes the closed ball with center at the point with polar coordinates $(\exp(-\frac{\ell}{h}), 0)$ of R^2 and radius $r_h^\ell = \exp(-\frac{\ell}{h} - \frac{\pi h}{c})$.

The solutions u_h of (1)(2), where E_h is now taken as above, converge (strongly in $L^2(\Omega)$ and weakly in $H^1(\Omega)$) to the (weak) solution u of the equation

$$(13) \qquad -\Delta u = 0 \quad \text{in} \quad \Omega\text{-}S,$$

where S is the line segment $[0,1]$ of the x-axis of R^2, and u satisfies the following boundary condition on S

$$(14) \qquad \frac{\partial u}{\partial x_2}(x_1,0+) - \frac{\partial u}{\partial x_2}(x_1,0-) = \frac{c}{x_1} u(x_1,0) \quad \forall x_1 \in [0,1]$$

together with the boundary condition (4) on $\partial\Omega$. Equations (13),(14) can be given the form of a Schrödinger equation, formally written as

$$-\Delta u + \mu u = 0 \quad \text{in } \Omega ,$$

where μ is the measure in R^2 supported by S and with line density $\frac{c}{x_1}$ on S.

Example 3. We consider a sequence of Schrödinger equations

(15) $$-\Delta u_h + q_h(x) u_h = 0 \quad \text{in } \Omega ,$$

where Ω is as in Example 1 and for every $h \in \mathbb{N}$ the function $q_h(x)$ is given by

$$q_h(x) = q_h(x_1,x_2) = \sigma_h(x_1) h \psi(h x_2)$$

where $\sigma_h \in C_0^\infty(]0,1[)$, $\sigma_h(x_1) \uparrow +\infty$ as $h \to +\infty$ for every $x_1 \in]0,1[$, and $\psi \in C_0^\infty(R)$, $\psi > 0$, $\int_R \psi\, dt = 1$. The solution u_h of (15), satisfying the boundary condition (4) on $\partial\Omega$, converges to the solution of the Dirichlet problem in the domain Ω-S:

$$-\Delta u = 0 \quad \text{in } \Omega\text{-}S , \quad u = 0 \quad \text{on } \partial S$$

with boundary condition (4) on $\partial\Omega$. □

Let us summarize the main features emerging from our examples.

The solutions u_h of Dirichet problems in varying domains Ω-E_h, $u_h = 0$ on ∂E_h, may have a wild oscillatory behavior as $h \to +\infty$. Therefore, they can be expected to converge to some function u in Ω only with respect to some weak topology, like that of the L^2 norm, or the weak topology of the Sobolev space H^1. Moreover, the limit function u may fail to be a solution of a Dirichlet problem in a domain of the form Ω-E, for some (possibly empty) limit set E. In fact, as our first example shows, the equation satisfied by u in Ω may be a Schrödinger equation in which a (non-negative) potential q(x) appears that can be singular at some point of Ω. Even if some sort of limit set E of the E_h

exists, as the set $E = S$ of the example 2, u may fail to satisfy the boundary condition $u = 0$ on ∂S. Instead, again a "Schrödinger equation" is satisfied by u in Ω, however now the "potential" that appears in such an equation is a measure in Ω. Furthermore, as our last example shows, the roles of the Dirichlet problems with "holes" and that of the Schrödinger equations with non-negative potentials can be reversed.

Concerning the asymptotic behavior of the solutions u_h at a given point x_0 of the domain, our examples show that there exist privileged points x_0 at which, independently of the boundary data, the solutions u_h are forced to take the value 0 continuously, that is,

$$u_h(x_h) \to 0 \quad = u(x_0) \quad \leftarrow u(x)$$

for every sequence $x_h \to x_0$ as $h \to +\infty$ and every $x \to x_0$. On the other hand, the same sequence u_h may wildly oscillate at every other point $x \neq x_0$ in a neighborhood of x_0. It is also clear from the examples that the reason for this stable behavior is that Maz'ja's estimates (7) hold uniformly in h, because of (10), and in the limit as $h \to +\infty$ they give the estimate (12) for u.

2. The Relaxed Dirichlet Problems

In [12], and in a more general setting in [13], we introduce a class of equations of the form

$$(16) \qquad -\Delta u + \mu u = 0$$

in an open region Ω of R^n, $n > 2$, where μ is an arbitrary given non-negative Borel measure in R^n, possibly $+\infty$ on some subset of R^n, such that $\mu(B) = 0$ for every Borel set B of (Newtonian) capacity zero in R^n.

We say that a function u is a local weak solution of (16) in Ω if

$u \in H^1_{loc}(\Omega) \cap L^2_{loc}(\Omega,\mu)$ and

$$\int_\Omega \nabla u \, \nabla v \, dx + \int_\Omega uv d\mu = 0$$

for every $v \in H^1(\Omega) \cap L^2(\Omega,\mu)$ having a compact support in Ω. Let us recall that the functions of the Sobolev space $H^1_{loc}(\Omega)$ can be defined up to sets of capacity zero in Ω; therefore, by our assumption on μ, the μ-integral above is well defined.

If E is a given compact subset of R^n and μ is of the special form

$$\mu = \infty_E , \tag{17}$$

where $\infty_E(B) = +\infty$ if $\text{cap}(B \cap E) > 0$ and $\infty_E(B) = 0$ if $\text{cap}(B \cap E) = 0$, then (16) reduces to the <u>Dirichlet problem "with holes"</u>

$$-\Delta u = 0 \quad \text{in } \Omega\text{-}E,\ u = 0 \quad \text{on} \quad \partial E \cap \Omega \ .$$

If

$$\mu = q(x)\mathcal{L}_n \tag{18}$$

where $\mathcal{L}_n$ denotes the Lebesgue measure in R^n and $q(x) \geq 0$ is a Borel function on R^n (i.e., $d\mu = q(x)dx$), then (16) reduces to the <u>Schrödinger equation</u>

$$-\Delta u + q(x)u = 0 \qquad \text{in} \quad \Omega .$$

We denote the class of all such measures μ by $\mathcal{M}_0$ and for a given $\mu \in \mathcal{M}_0$ we call (16) a <u>relaxed Dirichlet problem</u>.

In order to study the "variational" properties of the equation (16) under perturbations of the measure μ, in [12] we introduce a convergence for sequences $\{\mu_h\}$ of $\mathcal{M}_0$, called <u>γ-convergence</u> and denoted by $\mu_h \to \mu$, $\mu \in \mathcal{M}_0$.

In [12], we prove that the class $\mathcal{M}_0$ is (sequentially) <u>compact</u> under the γ-convergence. Moreover, we prove that the class of the "Dirichlet measures" (17), as well as that of the "Schrödinger measures" (18), is <u>dense</u> in $\mathcal{M}_0$.

In order to better understand these results and their relation to the examples of the previous section, let us define the γ-convergence and let us mention from [12] some of the simplest convergence properties that a sequence of perturbed solutions u_h inherits from the γ-convergence of the corresponding measures μ_h.

We say that a sequence $\{\mu_h\}$ in $\mathcal{M}_0$ γ-converges to the measure $\mu \in \mathcal{M}_0$ if for every bounded open subset Ω of R^n the sequence of functionals

$$F_{\mu_h}(v,\Omega) = \begin{cases} \int_\Omega |\nabla v|^2\,dx + \int_\Omega v^2 d\mu_h & \text{if} \quad v \in H_0^1(\Omega) \\ +\infty & \text{if} \quad v \in L^2(\Omega),\ v \notin H_0^1(\Omega) \end{cases}$$

Γ-converges in the space $L^2(\Omega)$ to the functional

$$F_{\mu}(v,\Omega) = \begin{cases} \int_\Omega |\nabla v|^2\,dx + \int_\Omega v^2 d\mu & \text{if} \quad v \in H_0^1(\Omega) \\ +\infty & \text{if} \quad v \in L^2(\Omega),\ v \notin H_0^1(\Omega) \end{cases}$$

as $h \to +\infty$.

We recall that a sequence of functionals $F_h(\cdot)$, possibly $+\infty$ at some point, Γ-converges in $L^2(\Omega)$ to the functional $F(\cdot)$ as $h \to +\infty$, if the following two conditions are satisfied:

(i) For every sequence $v_h \to v$ in $L^2(\Omega)$, we have $\liminf F_h(v_h) \geq F(v)$ as $h \to +\infty$;

(ii) For every $v \in L^2(\Omega)$, there exists a sequence $v_h \to v$ in $L^2(\Omega)$ such that $\limsup F_h(v_h) \leq F(v)$ as $h \to +\infty$ (hence also, by (i), $\lim F_h(v_h) = F(v)$ as $h \to +\infty$).

We identify two measures μ_1 and μ_2 of $\mathcal{M}_0$ if they give rise to the same functional $F_{\mu_1}(\cdot,\Omega) = F_{\mu_2}(\cdot,\Omega)$ for every Ω (that is, if $\int_\Omega u^2 d\mu_1 = \int_\Omega u^2 d\mu_2$ for every Ω and every $u \in H_0^1(\Omega)$).

The γ - convergence in $\mathcal{M}_0$ can be characterized in variational terms. In fact, a sequence μ_h γ-converges to μ in $\mathcal{M}_0$ if and only if for every bounded open set Ω of R^n and for every $f \in L^2(\Omega)$ the

(19) $$\min_{v \in H_0^1(\Omega)} [\int_\Omega |\nabla v|^2 dx + \int_\Omega v^2 d\mu_h + \int_\Omega fv dx]$$

converges to the

(20) $$\min_{v \in H_0^1(\Omega)} [\int_\Omega |\nabla v|^2 dx + \int_\Omega v^2 d\mu + \int_\Omega fv dx]$$

as $h \to +\infty$. An example of a γ-converging sequence in $\mathcal{M}_0$ is a sequence of Radon measures $\mu_h \in H^{-1}(R^n)$ that converges to a Radon measure $\mu \in H^{-1}(R^n)$ (where $H^{-1}(R^n)$ is the dual space of $H_0^1(R^n)$). Another example is a sequence of the form $\mu_h = \infty_{E_h}$ and $\mu = \infty_E$, with E_h and E closed subsets of R^n such that $H^1(R^n - E_h)$ converges to $H_0^1(R^n - E)$ in the space $H_0^1(R^n)$ according to the convergence of convex sets in [23].

If μ_h γ-converges to μ in $\mathcal{M}_0$, then the minimizing u_h in (19), which is the weak solution of the relaxed Dirichlet problem

$$-\Delta u_h + \mu_h u_h = f \quad \text{in} \quad \Omega ,$$

$$u_h = 0 \quad \text{on} \quad \partial\Omega ,$$

converges strongly in $L^2(\Omega)$ and weakly in $H^1(\Omega)$ to the minimizing u in (20), and u is the weak solution of the problem

$$-\Delta u + \mu u = f \quad \text{in} \quad \Omega ,$$

$$u = 0 \quad \text{on} \quad \partial\Omega.$$

A deeper result is given in [12] for the case when a nonhomogeneous boundary condition is assigned on $\partial\Omega$. In particular, we prove that if μ_h γ-converges to μ in $\mathcal{M}_0$ and supp $\mu_h \subset \Omega' \subset \Omega$, then the (weak) solutions u_h of the relaxed Dirichlet problems

$$-\Delta u_h + \mu_h u_h = 0 \quad \text{in} \quad \Omega ,$$

satisfying the boundary condition

$$u_h = g \quad \text{on } \partial\Omega$$

for a given $g \in H^1(\Omega)$, converge strongly in $L^2(\Omega)$ and weakly in $H^1(\Omega)$ to the solution u of

$$-\Delta u + \mu u = 0 \quad \text{in} \quad \Omega ,$$

satisfying

$$u = g \quad \text{on} \quad \partial\Omega .$$

3. Variational Estimates and Wiener's Criterion

In order to study the behavior of an arbitrary local weak solution u of the relaxed Dirichlet problem (16) at a given point x_0 of the domain, in [12] we associate a function

$$\omega_\mu(x_0;r,R), \qquad 0 < r < R,$$

with the measure $\mu \in \mathcal{M}_0$ appearing in (16) and with the point $x_0 \in R^n$. We call this function the Wiener modulus of μ at x_0. The function $\omega_\mu(x_0;r,R)$ is defined by setting

(21) $$\omega_\mu(x_0;r,R) = \exp\left(-\int_r^R \frac{\operatorname{cap}_\mu(B_\rho(x_0), B_{2\rho}(x_0))}{\operatorname{cap}(B_\rho(x_0), B_{2\rho}(x_0))}\frac{d\rho}{\rho}\right)$$

where

(22) $$\operatorname{cap}_\mu(B_\rho, B_{2\rho}) = \min\{\int_{B_{2\rho}} |\nabla v|^2 dx + \int_{B_\rho} v^2 d\mu \,|\, v-1 \in H_0^1(B_{2\rho})\}$$

for every $\rho>0$, $B_\rho = B_\rho(x_0)$, $B_{2\rho} = B_{2\rho}(x_0)$, and $\operatorname{cap}(B_\rho, B_{2\rho})$ is the usual Newtonian capacity of B_ρ in $B_{2\rho}$, that is,

$$\operatorname{cap}(B_\rho, B_{2\rho}) = \min\{\int_{B_{2\rho}} |\nabla v|^2 dx \,|\, v \in H_0^1(B_{2\rho}),\ v \geq 1 \text{ on } B_\rho\}.$$

Let us remark that the functions $\omega_h(r)$ and $\omega_q(r)$ occurring in Example 1 (see (8),(11)), are special cases of (21), for $\mu = \infty_{E_h}$ and $\mu = q(x)\mathcal{L}_n$, respectively.

The basic property of the Wiener modulus is its <u>stability</u> under γ - convergence: If $\mu_h \to \mu$ in $\mathcal{M}_0$, then

(23) $$\omega_{\mu_h}(x_0;r,R) \to \omega_\mu(x_0;r,R)$$

as $h \to +\infty$, for every $x_0 \in R^n$ and every $0 < r < R$.

Property (23) is proved in [12] by showing that the μ - capacity (22) has the stability property: If $\mu_h \to \mu$ in $\mathcal{M}_0$, then

(24) $$\mathrm{cap}_{\mu_h}(B_\rho(x_0), B_{2\rho}(x_0)) \to \mathrm{cap}_\mu(B_\rho(x_0), B_{2\rho}(x_0))$$

as $h \to +\infty$, for every $x_0 \in R^n$ and every $\rho>0$, except possibly a countable sets of ρ's depending only on x_0.

Properties (9) and (10) seen in Example 1 are special cases of (24), (23) above.

Examples are given in [12], showing that there exist sequences $\mu_h \to \mu$ in $\mathcal{M}_0$ for which (24) is indeed violated for some special values of ρ.

In [12] and [13] we study the behavior of an arbitrary local weak solution u in Ω at a given point $x_0 \in \Omega$. We obtain the following generalized <u>Maz'ja estimate</u>

(24) $$\sup_{B_r} u^2 + \int_{B_r} |\nabla u|^2 G(|x-x_0|)dx + \int_{B_r} u^2 G(|x-x_0|d\mu \leq k \frac{G(R)}{R^2} \|u\|^2_{L^2(B_R)} \omega_\mu(x_0;r,R)^\beta$$

for every $0 < r \leq \frac{R}{2}$ (with $R \leq \frac{1}{2}$ if $n=2$), where $G(\rho) = \rho^{2-n}$ if $n>2$ and $G(\rho) = \ln \frac{1}{\rho}$ if $n=2$ for every $\rho>0$, k and $\beta>0$ being suitable constants depending only on n.

Estimates (7) and (12) of Section 1 are special cases of (24).

An interesting feature of (24) is its stability under γ-convergence of μ in $\mathcal{M}_0$, which is inherited from the analogous stability (23) of ω_μ provided u remains bounded in the L^2-norm (as in the case of the L^2-norm converging sequences of perturbed solutions described in Section 2).

The estimate (24) is obtained in [12] by relying on the density results mentioned in Section 2, which in view of the stability property (23) reduce the proof of (24) to the special Dirichlet case of a μ of the form $\mu = \infty_E$. In [13], (24) is proved for a more general class of equations (16) by direct estimation techniques of the weak solutions.

Let us now state the variational Wiener's criterion for problem (16), whose sufficiency is proved in [12], [13] as a consequence of the estimate (24) and whose necessity is proved in [13] by a direct argument.

Given $\mu \in \mathcal{M}_0$ and $x_0 \in R^n$, we say that x_0 is a regular Dirichlet point of μ if every local weak solution u of (16) in a neighborhood of x_0 is continuous at x_0 and $u(x_0) = 0$. We say that x_0 is a Wiener point of μ if $\omega_\mu(x_0;r,R) \to 0$ as $r \to 0^+$ for some (hence all) $R>0$.

We prove that x_0 is a regular Dirichlet point of μ if and only if x_0 is a Wiener point of μ.

In the special case $\mu = \infty_E$, $x_0 \in \partial E$, the Wiener point property $\omega_\mu(x_0;r,R) \to 0$ as $r \to 0^+$ reduces to (6), where now $B_\rho = B_\rho(x_0)$ $B_{2\rho} = B_{2\rho}(x_0)$ so that we obtain the classical criterion.

In [13] we also prove decay estimates for the "μ-energy"

$$\mathcal{E}_\mu(r)(x_0) = \int_{B_r(x_0)} |\nabla u|^2 dx + \int_{B_r(x_0)} u^2 \, d\mu$$

associated with an arbitrary local solution u of (16) in a neighborhood of x_0, which are of the type of those occurring in the Saint-Venant principle of linear elasticity, see also [25].

We prove that

$$\mathcal{E}_\mu(r)(x_0) \le k\mathcal{E}_\mu(R) \quad \frac{r^{n-2}}{\operatorname{cap}_\mu(B_{2R},B_{4R})} \quad \omega_\mu(x_0;r,R)^\beta \tag{25}$$

where k and $\beta>0$ are structural constants depending only on the dimension n of the space $0 < r \le R$.

In [13], estimates (24), (25) are actually obtained for more general equations of the form (16), in which the Laplace operator $-\Delta$ is replaced by an arbitrary operator L of the form

$$Lu = -\sum_{i,j=1}^{n} \frac{\partial}{\partial x_i}(a_{ij}(x)\frac{\partial u}{\partial x_j})$$

with measurable coefficients $a_{ij}(x)$ satisfying the uniform ellipticity condition

$$\sum_{i,j=1}^{n} a_{ij}(x)\ \xi_i \xi_j \geq \lambda|\xi|^2 \quad \text{for } \xi \in \mathbb{R}^n \text{ and } x \in \mathbb{R}^n \text{ a.e,}$$

$$|a_{ij}(x)| \leq \Lambda \qquad \text{for } x \in \mathbb{R}^n \text{ a.e. and } x,j = 1\ldots,n,$$

for suitable constants $0 < \lambda \leq \Lambda$. In this case, the constants k and β in the estimates (24),(25) depend only on n and on λ, Λ.

In [13] we consider the special case of a <u>rotationally invariant</u> Radon measure $\mu \in \mathcal{M}_0$, such that $\mu(\{0\}) = 0$, and we study the Wiener modulus of μ at 0 in some more detail.

In particular, we prove that the function

$$\delta(\rho) = \frac{\text{cap}_\mu(B_\rho, B_{2\rho})}{\text{cap}(B_\rho, B_{2\rho})}, \qquad \rho>0,$$

can now be characterized as the (unique) solution of an ordinary differential equation involving the radial component of μ, and this allows us to carry out explicit estimates of the Wiener modulus

(26) $$\omega(r,R) = \exp(-\int_r^R \delta(\rho)\frac{d\rho}{\rho})$$

of μ at 0.

In the special case of μ of the form (18) with

(27) $$q(x) = q(|x|) \geq 0,\ q \in L^1_{loc}(]0,+\infty[),$$

we prove that 0 is a Wiener point of μ if and only if

(28) $$\int_0^R q(\rho)\rho\, d\rho = +\infty \qquad (n \geq 3),$$

or

$$(29) \qquad \int_0^R q(\rho)\rho \, \ln\frac{1}{\rho} \frac{d\rho}{\rho} = +\infty \qquad (n = 2)$$

This implies, by the Wiener's criterion, that if (27) and (28) ((29)) are satisfied, then an arbitrary local weak solution of the Schrödinger equation (3) in a neighborhood of 0 is continuous at 0, with $u(0) = 0$. If we combine this result with the regularity results of Kato [19] and Aizenman - Simon [1], which give the continuity of u at 0 when (28) is violated ($n \geq 3$), we obtain that for an <u>arbitrary</u> $q(x)$ satisfying (27), every local weak solution u of the equation (3) in a neighborhood of 0 is continuous at 0. For a different non-variational approach to the equation (3) with q of the form (27) see [28].

Let us finally mention that in [14] explicit examples of potentials $q(x) = q(|x|)$ are given, for which the corresponding Wiener modulus (26) is of <u>logarithmic type</u>, for instance

$$\omega(r,R) = \left[\frac{\ln\frac{1}{R}}{\ln\frac{1}{r}} \right]^{\alpha}$$

or

$$\omega(r,R) = \left[\frac{\ln(\ln\frac{1}{R})}{\ln(\ln\frac{1}{r})} \right]^{\alpha}$$

for some $\alpha > 0$.

This shows that for general relaxed Dirichlet problems (16) the modulus of continuity of the solutions and the decay of the local energy, as estimated by (24) and (25), may be only of logarithmic type at some point of the domain.

We conclude by pointing out that, when $\mu = \infty_E$ and x_0 are irregular point of ∂E shows, a solution u of (16) may well be discontinuous at some point of the domain, even if it is certainly always locally bounded as a consequence of (24).

Acknowledgement

The author wishes to thank the Institute for Mathematics and its Applications for hospitality and support and Professors J.L. Ericksen and D. Kinderlehrer for their useful comments.

References

[1] M. Aizenman, B. Simon: Brownian motion and Harnack inequality for Schrödinger Operators Comm. Pure Appl. Math. 35 (1982), 209-273.

[2] H. Attouch: Variational convergence for functions and operators. Pitman, London, 1984.

[3] H. Attouch, C. Picard: Variational inequalitities with varying obstacles: the general form of the limit problem. J. Funct. Anal. 50 (1983), 329-389.

[4] J.R. Baxter, R.V. Chacon and N.C. Jain: Weak limits of stopped diffusions, University of Minnesota, Mathematics Report 84-165 (1985), 1-44.

[5] L. Carbone, F. Colombini: On convergence of functionals with unilateral constraints. J. Math. Pures Appl. (9) 59 (1980), 465-500.

[6] D. Cioranescu: Calcul des variations sur des sous-espaces variables. C.R. Acad. Sci. Paris Série A 291 (1980), 19-22, 87-90.

[7] D. Cioranescu, F. Murat. Un terme étrange venu d'ailleurs, I and II. In "Nonlinear partial differential equations and their applications. Collége de France Seminar. Volume II and III. Ed. by H. Brezis and J.L. Lions. Research Notes in Mathematics 60 (1982), 98-138 and 70 (1983), 154-178.

[8] D. Cioranescu, J. Saint Jean Paulin. Homogenization in open sets with holes. J. Math. Pures Appl. 71 (1979), 590-607.

[9] G. Dal Maso: Asymptotic behaviour of minimum problems with bilateral obstacles. Amm. Mat. Pura Appl. (4) 129 (1981), 327-366.

[10] G. Dal Maso: Limiti di problemi di minimo con ostacoli. In Atti del Convegno su "Studio dei problemi-limiti in Analisi Funzionale", Bressanone, 7-9 Settembre 1981, Pitagora, Bologna, 1982, 79-100.

[11] G. Dal Maso, P. Longo: Γ-limits of obstacles. Ann. Mat. Pura Appl. (4) 128 (1980), 1-50

[12] G. Dal Maso, U. Mosco, Wiener's criterion and Γ-convergence. IMA Preprint Series #173.

[13] G. Dal Maso, U. Mosco: Wiener criteria and energy decay for relaxed Dirichlet problems. IMA Preprint Series #197.

[14] G. Dal Maso, U. Mosco:, The Wiener modulus of a radial measure. IMA Preprint Series #194.

[15] E. De Giorgi, G. Dal Maso, P. Longo: Γ-limiti di ostacoli. Atti Accad. Naz. Lincei, Rend. Cl. Sci. Fis. Mat. Natur., (8) 68 (1980), 481-487.

[16] E. De Giorgi, T. Franzoni: Su un tipo di convergenza variazionale. Atti. Accad. Naz. Lincei, Rend., C. Sci. Fis. Mat. Natur. (8) 58 (1975), 842-850 and Rend. Sem Mat. Brescia 3 (1979), 63-101.

[17] E. Ya. Hruslov. The method of orthogonal projections and the Dirichlet problem in domains with a fine grained boundary. Math. USSR Sb. 17 (1972), 37-59.

[18] E. Ya. Hruslov. The first boundary value problem in domains with a complicated boundary for higher order equations. Math. USSR Sb. 32 (1977), 535-549.

[19] M. Kac: Probabilistic methods in some problems of scattering theory. Rocky Mountain J. Math. 4 (1974), 511-538.

[20] T. Kato: Schrödinger operators with singular potentials. Israel J. Math. 13 (1973), 135-148.

[21] A.V. Marchenko, E. Ya. Hruslov: Boundary value problems in domains with close-grained boundaries (in Russian). Naukova Dumka, Kiev, 1974.

[22] A.V. Marchenko, E. Ya. Hruslov: New results in the theory of boundary value problems for regions with close-grained boundaries. Uspehi Mat. Nauk 33 (1978) 127.

[23] V.G. Maz'ja: On the continuity at a boundary point of solutions of quasilinear elliptic equations. Vestnik Leningrad Univ. Math. 3 (1976), 225-242.

[24] U. Mosco: Convergence of convex sets and of solutions of variational inequalities. Adv. in Math. 3 (1969), 510-585.

[25] U. Mosco: Wiener criterion and potential estimates for the obstacle problems. IMA Preprint Series # 135 1985, 1-56.

[26] G.C. Papanicolaou, S.R.S. Varadhan. Diffusion in regions with many small holes. In "Stochastic Differential Systems, Filtering and Control", Proceedings of the IFIP-WG 7/1 Working Conference, Vilnius, Lithuania, USSR, Aug. 27-Sept. 2, 1978. Ed. by B. Grigelionis, Lecture Notes in Control and Information Sciences 25, Springer, 1980, 190-206.

[27] J. Rauch, M. Taylor: Potential and scattering theory on mildly perturbed domains J. Funct. Anal., 18, (1975), 27-59.

[28] J.L. Vazquez, C. Yarur: Singularités isolées et comportement á l'infini des solutions de l'équation Schrödinger stationnaire. C.R. Acad. Sci. Paris, de Série I t.300 (1985), 105-108.

[29] N. Wiener: The Dirichlet problem. J. Math. and Phys. 3 (1924), 127-146).

A CONTRIBUTION TO THE DESCRIPTION OF NATURAL STATES FOR ELASTIC CRYSTALLINE SOLIDS

Mario Pitteri

Seminario Matematico
Via G.B. Belzoni, 7
35131 Padova ITALY

1. Introduction.

Recently ERICKSEN [2] proposed an approach to material symmetry which is, in a sense, more local than the generally accepted one of COLEMAN & NOLL [1]. That approach aims at a theory which is flexible enough to account for phase transitions, twinning, and other phenomena which occur in crystalline solids within the range of large or plastic deformations. In his approach to material symmetry, ERICKSEN [3] analyzes molecular models of crystals and their geometric symmetries, and proposes a suitable infinite, discrete and non compact group $\mathcal{G}$ as a reasonable candidate for the material symmetry group of crystalline solids which, from the molecular point of view, are represented by Bravais lattices. PITTERI [12] shows that the invariance of the constitutive equations associated with $\mathcal{G}$ reduces to the invariance proposed by COLEMAN & NOLL [1] if one restricts the domain of those equations to a suitable neighborhood N of a given assigned configuration C. On the other hand two configurations $\mathbf{C}_1$ and $\mathbf{C}_2$ may have different, non-isomorphic invariance groups for the restrictions of the constitutive equations to the analogues N_1 and N_2 of the neighborhood N , and ERICKSEN [4] studies the existence of continuous paths in the space of configurations, along which the aforementioned group of local invariance changes.

Following ERICKSEN [7], we henceforth denote by $\mathbf{C}$ the right Cauchy-Green deformation tensor; we denote by $\mathbf{H}^T$ the transpose of the tensor H; we consider a stored energy function $W = \hat{W}(C)$ which is invariant under $\mathcal{G}$:

$$(1.1) \qquad \hat{W}(H^T C\, H) = W(C) \qquad \forall\, H \in \mathcal{G},$$

and we assume $\hat{W}$ to have an isolated minimum at $\overline{C}$. Then any homogeneous configuration whose Cauchy-Green tensor is $\overline{\mathbf{C}}$ is a natural state, that is, a stable unstressed equilibrium configuration. The same can be said of any homogeneous configuration whose Cauchy-Green tensor is $\tilde{C} = H^T \bar{C} H$, for any element H of $\mathcal{G}$. If any two such natural states co-exist coherently in one body, then we have a pairwise homogeneous natural state. Here, coherence means that the displacement is continuous over the body, whereas the displacement gradient may suffer a jump discontinuity on a regular surface which, in this case, turns out to be planar. According to ERICKSEN [7], the deformation gradients F_1 and F_2 of the two co-existent, coherent natural states satisfy the conditions

$$(1.2) \qquad F_2 = (1 + a \otimes n)\, F_1 \quad \text{and} \quad F_2 = QF_1H \ ,$$

for a suitable choice of the vectors $\mathbf{a}$ and $\mathbf{n}$, and of the tensors Q and H such that

$$(1.3) \qquad a \cdot n = 0 \quad , \quad Q\,Q^T = 1 = Q^TQ \quad , \text{ and } \quad H \in \mathcal{G} .$$

Therefore, a pairwise homogeneous natural state as above exists if and only if F_1 satisfies the condition

$$(1.4) \qquad S\,F_1 = Q\,F_1\,H \ , \qquad S := 1 + a \otimes n \ ,$$

for some choice of $\mathbf{a}$, $\mathbf{n}$, Q and H as above.

Conditions equivalent to (1.4) are considered in various papers which deal with coherently co-existent phases of a solid and deal with twinning. Among these papers are those of ERICKSEN [5], GURTIN [9], JAMES [10], [11], and PITTERI [13], [14]. I shall adopt here the kinematic description of crystals in terms of lattice vectors instead of deformation gradients, following ERICKSEN [3] and PITTERI [13], [14]. In this case:

(i) The present configuration is described in terms of 3 linearly independent <u>lattice vectors</u> $\mathbf{e}_a$, $a = 1,2,3$.

(ii) The basic invariance group is the group $G := GL(3,Z)$ of unimodular 3 x 3 matrices with integral entries, and is $\mathcal{G}$ conjugate to G.

(iii) Let $\mathbf{C}$ be the 3 x 3 matrix whose entries are $C_{ab} = \mathbf{e}_a \cdot \mathbf{e}_b$, and notice that $\mathbf{C} = \mathbf{C}^T > 0$, that is, $\mathbf{C}$ is symmetric and positive definite.

The invariance of the stored energy $W = \hat{W} = \hat{W}(e_a) = W(C)$ is given by the following analogue of (1.1):

$$W\,(m^T C\, m) = W\,(C) \quad \forall\, m \in G \text{ , and} \tag{1.5}$$

(iv) the analogue of condition (1.4) is

$$S\, e_a = m^b{}_a\, Q\, e_b \quad , \quad m \in G \text{ ,} \tag{1.6}$$

where S and Q are as above.

Two classes of solutions of $(1.4)_1$ or, equivalently, of (1.6), are by now rather well understood. They correspond to what crystallographers and metallurgists call Type 1 and Type 2 twins. In the case of a pairwise homogeneous deformation constituting a Type 1 twin, PITTERI [13] shows that for any choice of lattice vectors e_a generating [deformation gradient F in] one of the two homogeneously deformed regions of the crystal, there is one element $m \in G$ [$H \in \mathcal{H}$] such that $m^2 = 1$ [$H^2 = 1$] and (1.6) [and (1.4)] holds when

$$Q = D(2n \otimes n - 1) \quad , \text{ and } D = \det Q = \det m = \pm 1 \text{ .} \tag{1.7}$$

Conversely, to within inessential limitations, <u>for any</u> $m \in G$ <u>such that</u> $m^2 = 1$ [<u>for any</u> $H \in \mathcal{H}$ such that $H^2 = 1$), <u>and any choice of lattice vectors</u> e_a [of F], we can uniquely determine $a \neq 0$ and a unit vector n such that (1.6) [(1.4)] and (1.7) hold. PITTERI [14] proves that the same results hold when we replace Q in (1.7) by

$$\tilde{Q} = D(2\|a\|^{-2}\, a \otimes a \; - \; 1\,) \text{ .} \tag{1.8}$$

ERICKSEN [5] shows that the choices (1.7) and (1.8) are the only ones compatible with (1.4) for a <u>special choice</u> of H orthogonal with $H^2 = 1$. GURTIN [9] proves an analogous Compatibility Theorem, by means of which he shows that the equality $(1.4)_1$ <u>for</u> H <u>orthogonal</u> and Q a rotation of π radians requires Q to be given by either (1.7) or (1.8). PITTERI [14] shows that (1.7) and (1.8) are the <u>only possible choices</u> of Q in (1.1) <u>if</u> m, or H <u>has order 2</u>. Conversely, if Q in (1.6) <u>has order</u> 2 and m, or H, <u>has finite order, then necessarily</u> $m^2 = 1$ <u>and</u>

Q is given by either (1.7) or (1.8). The hypothesis that m has finite order is necessary to prove the last statement: an example is proposed in [14] where $Q^2 = 1$ and (1.6) holds for some $m \in G$ of infinite order. In addition, according to ERICKSEN [8], the last underlined statement becomes false if Q and m are not of order 2 . For instance, there is a solution of (1.6) where m has order 4 and Q has order 3. Moreover the examples in [8] suggest that, for m [for H] of order greater than 2, (1.6) [(1.4)] need not have a solution for any choice of e_a [of F], and that, by suitably choosing e_a [F] we can solve (1.4) [(1.6)] for almost any rotation Q . Summarizing, we can say that the extension of GURTIN's Classification Theorem to the case that $H \in \mathcal{U}$ is valid if and only if either $H^2 = 1$ or $Q^2 = 1$ and H has finite order. In either case, for any choice of F there are exactly two solutions of (1.4) .

ERICKSEN [6] produces another class of elements m of G, besides those of order 2, for which (1.6) has a solution for any choice of e_a. These are the lattice-invariant shears, which can shown to be conjugate to

$$(1.9) \qquad \bar{m} = \begin{vmatrix} 1 & 0 & 0 \\ n & 1 & 0 \\ 0 & 0 & 1 \end{vmatrix} , \quad n \in Z .$$

For any such choice of m, (1.6) holds for Q = 1 .

2. Two Theorems.

In the preceding section we have examined two classes of elements m of G for which (1.6) holds for any choice of e_a , namely the lattice-invariant shears and the m's of order 2 . These two classes seem to be privileged also because for any element of them the solutions of (1.6) can be easily and explicitly produced. We have no clear understanding of the solutions of (1.6) when m does not belong to either class. In this case the examples mentioned in Section 1 seem to indicate that (1.6) will not hold for any choice of e_a ,if it holds at all, and that no easy classification is at hand. In this section we first prove that (1.6) has at least two solutions for any choice of $m \in G$. Secondly, we prove that the two aforementioned classes are indeed special: (1.6) holds for a given $m \in G$ and for any choice of e_a if and only if either $m^2 = 1$ or m is a lattice-invariant shear. Therefore, if m [H] neither has order 2 nor is a lattice-invariant shear, we should expect (1.6) [(1.4)] to hold only for a restricted, non-empty class of vectors $\mathbf{e}_a$ [of deformation gradients F]. Unfortunately we are not able to characterize this class, since the proof of neither of the underlined statements above is constructive.

The following theorems rest upon the definition of the matrix $\mathbf{C}$ that we introduced in Section 1, and upon the following easy modification of a theorem of ERICKSEN [7, (3.12)].

Theorem 1: Equality (1.6) holds if and only if

$$\det (m^T \mathbf{C}\, m - \mathbf{C}) = 0 . \tag{2.1}$$

Moreover, if m and e_a satisfy (2.1), then there are two choices of $\mathbf{a}$, $\mathbf{n}$ and Q for which (1.6) holds.

As ERICKSEN [7] points out, and we already mentioned, condition (2.1) holds for any symmetric, positive definite C if m either has order 2 or is a lattice-invariant shear.

Theorem 2: For any $m \in G$ there is $C = C^T > 0$ such that (1.6) holds; that is, if $\mathbf{e}_a$ is any set of lattice vectors such that $\mathbf{e}_a \cdot \mathbf{e}_b = C_{ab}$, then there are vectors $\mathbf{a}$ and $\mathbf{n}$, and an orthogonal tensor Q such that

$$m^b{}_a\, e_b = Q\, S\, \mathbf{e}_a\ , \quad S = \mathbf{1} + \mathbf{a} \otimes \mathbf{n}\ , \quad \mathbf{a} \cdot \mathbf{n} = 0 . \tag{2.2}$$

Proof: Let $B = m^T m - \mathbf{1}$. If $\det B = 0$, there is nothing to prove, so assume that $\det B \neq 0$, and look for

$$\mathbf{C} = \lambda\, \mathbf{a} \otimes \mathbf{a} + \mathbf{1}\ , \quad \lambda > -1, \ \ \mathbf{C} = \mathbf{C}^T > 0 . \tag{2.3}$$

We shall compute

(2.4) $$\Delta := \det (m^T C m - C)$$

for the choice (2.3) of C. To do this computation, call $b_i = (b_{i1}, b_{i2}, b_{i3})$ the rows of B. Then it is easy to see that

(2.5) $$\Delta = \det (B + \lambda(\alpha \otimes \alpha - a \otimes a)) \quad , \quad \alpha := m^T a .$$

We can write

$$\Delta = [b_1 + \lambda(\alpha_1\alpha - a_1 a)] \cdot [b_2 + \lambda(\alpha_2 a - a_2 a)] \times$$

$$[b_3 + \lambda(\alpha_3\alpha - a_3 a)] .$$

This equality can be written as follows:

(2.7) $$\Delta = \det B - \lambda K_1 + \lambda^2 K_2 - \lambda^3 K_3 \text{ , where}$$

(2.8) $$K_3 = (a_1 a - \alpha_1\alpha) \cdot (a_2\alpha_3\ \alpha \times a + \alpha_2\alpha_3\ a \times \alpha) = 0 ,$$

(2.9) $$K_2 = b_1 \cdot (a_2\, a - \alpha_2\, \alpha) \times (a_3\alpha - \alpha_3\alpha) + (a_1 a - \alpha_1 - \alpha).$$

$$(a_2\, a - \alpha_2\alpha) \times b_3 + (a_1 a - \alpha_1\alpha) \cdot b_2 \times (a_3\alpha - \alpha_3\alpha) ,$$

and

(2.10) $$K_1 = b_1 \cdot b_2 \times (a_3\, a - \alpha_3\, \alpha) + b_1 \cdot (a_2\, a - \alpha_2\, \alpha) \times b_3 + $$
$$+ (a_1\, a - \alpha_1) \cdot b_2 \times b_3 .$$

Using Cartesian components, and remembering that b_i are the rows of B, and the fact that, for instance [(1)]

(2.11) $$b_1 \times b_2 = (\text{adj } B)_3 = (B^{-1})_3 \det B ,$$

we can check that

(2.12) $$K_2 = -(a \times \alpha) \cdot B\, (a \times \alpha) \text{ and}$$

(2.13) $$K_1 = a \cdot \text{adj } B\, a - \alpha \cdot \text{adj } B\, \alpha = \det B\, (a \cdot B^{-1} a - \alpha \cdot B^{-1}\, \alpha) .$$

Since $\det B \neq 0$ and $\det m^T m = 1$, at least one proper value of B is positive and at least one is negative. We look for a such that

(2.14) $\mathbf{a} \times \alpha = \mathbf{b}$, $\|\mathbf{b}-\mathbf{e}\|$ small, $\mathbf{Be} = ((\lambda_1)^2 - 1)\mathbf{e}$, sgn $((\lambda_1)^2 - 1)$ = sgn det B.

Since $\alpha = \mathbf{m}^T\mathbf{a}$, $(2.14)_1$ is equivalent to

$$(2.15) \qquad \mathbf{b} \cdot \mathbf{a} = 0 = \mathbf{mb} \cdot \mathbf{a} \quad ,$$

so, for any $\mathbf{b}$, $(2.14)_1$ has one solution, to within the sign, if $\mathbf{b} \times \mathbf{mb} \neq 0$:

$$(2.16) \qquad \mathbf{a} = \pm \mu\, \mathbf{b} \times \mathbf{mb} \text{ for } \mu \text{ a real suitable number.}$$

In this case, by continuity, we can choose $\|\mathbf{b}-\mathbf{e}\|$ so small that

$$(2.17) \qquad \text{sgn } K_2 = - \text{ sgn det } B \ ;$$

hence in (2.7) we can make Δ vanish if we suitably choose λ, because Δ is a continuous function of λ which has the sign of det B for $\lambda = 0$ and is negative for sufficiently large λ. Of course the easiest choice is to take $\mathbf{b} \equiv \mathbf{e}$.

The same conclusion as above holds if $\mathbf{b} \times \mathbf{mb} = \mathbf{0}$, that is, if $\mathbf{b}$ is a proper vector of $\mathbf{m}$, provided $\mathbf{a} \times \mathbf{m}^T\mathbf{a} \neq \mathbf{0}$ for some $\bar{\mathbf{a}}$ orthogonal to $\mathbf{b}$. If so, take $a = \mu\, \bar{\mathbf{a}}$, and adjust μ to satisfy $(2.14)_1$. Again, we can assume that $\mathbf{b} \equiv \mathbf{e}$, in which case $\mathbf{e}$ is a proper vector of both $\mathbf{m}$ and $\mathbf{m}^T$:

$$(2.18) \qquad \mathbf{m}\,\mathbf{e} = \mathbf{m}^T\mathbf{e} = \lambda_1\mathbf{e} \quad .$$

This relation is also consistent with the notation in $(2.14)_3$.

The reasoning above does not work if (2.18) holds and

$$(2.19) \qquad \mathbf{a} \times \mathbf{m}^T\mathbf{a} = 0 \qquad \forall\, \mathbf{a} \text{ such that } \mathbf{a} \cdot \mathbf{e} = 0 \ .$$

But these conditions imply that

$$(2.20) \qquad \mathbf{m} = \mathbf{m}^T = \lambda_1\, \mathbf{e} \otimes \mathbf{e} + \lambda_2\, (\mathbf{f} \otimes \mathbf{f} + \mathbf{g} \otimes \mathbf{g}) \ ,$$

and $\mathbf{e}$, $\mathbf{f}$, $\mathbf{g}$ form an orthonormal basis. Any $\mathbf{a}$ can be written as

$$(2.21) \qquad \mathbf{a} = k\,\mathbf{e} + \mathbf{v} \quad , \quad \mathbf{v} \cdot \mathbf{e} = 0 \ .$$

Then, taking into account that $\mathbf{Bv} = ((\lambda_2)^2 - 1)\,\mathbf{v}$ for any $\mathbf{v}$ satisfying $(2.21)_2$, we obtain

(2.22) $$m^T a = K\lambda_1 e + \lambda_2 v \quad , \; a \times m^T a = K(\lambda_2 - \lambda_1)\, e \times v \quad , \text{ and}$$

(2.23) $$(a \times m^T a) \cdot B (a \times m^T a) = k^2(\lambda_2 - \lambda_1)^2 \, \| e \times v \|^2 \, ((\lambda_2)^2 - 1) \, ,$$

where $(2.21)_1$ implies that $\| e \times v \|^2 = \| v \|^2$. But by (2.20)

(2.24) $$\operatorname{sgn} \det B = \operatorname{sgn} ((\lambda_1)^2 - 1) \quad = \; - \operatorname{sgn} ((\lambda_2)^2 - 1) \, .$$

Therefore, K_2 and $\det B$ have the same sign, and we cannot make Δ vanish by choosing λ sufficiently large. On the other hand, taking (2.21) and $(2.22)_1$ into account, we find

$$K_1 = \det B \, [(k\, e + v) \cdot \frac{k\, e}{\lambda_1^2 - 1} + \frac{v}{\lambda_2^2 - 1} - (k\lambda_1 \, e + \lambda_2 v) \cdot \frac{k\, \lambda_1 e}{\lambda_1^2 - 1} + \frac{\lambda_2 v}{\lambda_2^2 - 1}$$

(2.25)
$$= -\det B \, [\, k^2 + v^2 \,] \, ,$$

where $v := \| v \|^2$. Hence

(2.26) $$\Delta = \det B + \det B \, (k^2 + v^2) \, \lambda + A \, (\operatorname{sgn} \det B) \, \lambda^2 \, ,$$
$$\text{where } A := k^2 \, (\lambda_2 - \lambda_1)^2 \, v^2 \, |(\lambda_2)^2 - 1| > 0 \, .$$

We look for solutions λ of $\Delta = 0$ which satisfy the condition that $\lambda > -1$. If

(2.27) $$\delta > 0 \, , \text{ where } \delta : (k^2 + v^2)^2 \, (\det B)^2 - 4A \, | \det B | \, ,$$

then $\Delta = 0$ has the solutions

(2.28) $$\lambda_{\pm} = \frac{-(k^2 + v^2) \, |\det B| \pm \sqrt{(K^2 + v^2)^2 (\det B)^2 - 4A |\det B|}}{2\, A} \, .$$

If we fix v, for instance v = 1, and we choose k sufficiently large, we deduce from $(2.26)_2$ and (2.28) that α is positive and $|\lambda_+|$ is sufficiently small, so the condition that $\lambda_+ > -1$ is satisfied. Such a choice of k and v allows us to prove the theorem when (2.18) and (2.19) hold.

<u>Theorem 3</u> : Given $m \in G$, condition (2.1) holds <u>for all $C=C^T > 0$</u> if and only if either $m^2 = 1$ or m is a lattice invariant shear. More explicitly, for a suitable choice of $\mathbf{v}_1$, $\mathbf{v}^1$, $\mathbf{a}$ and $\mathbf{n}$ in Q^3, either

$$m = -1 + 2\,\mathbf{v}_1 \otimes \mathbf{v}^1 , \quad \mathbf{v}_1 . \mathbf{v}^1 = 1 , \tag{2.29}$$

or

$$m = 1 + \mathbf{a} \otimes \mathbf{n} , \quad \mathbf{a} . \mathbf{n} = 0 . \tag{2.30}$$

<u>Proof.</u> Let us consider matrices **C** of the form (2.3). Then (2.7) to (2.10), $(2.11)_1$ and $(2.12)_1$ hold. If Δ has to vanish for all $\mathbf{C} = \mathbf{C}^T > 0$, then necessarily

$$\det B = 0 = K_1 \quad \text{and} \quad K_2 = 0 . \tag{2.31}$$

The first condition implies that for a suitable choice of the orthonormal left-handed basis **e**, **f**, **g**

$$\mathbf{m}^T \mathbf{m} = \mathbf{e} \otimes \mathbf{e} + \mu\, \mathbf{f} \otimes \mathbf{f} + \mu^{-1}\, \mathbf{g} \otimes \mathbf{g} , \quad \det \mathbf{m}^T\mathbf{m} = 1 . \tag{2.32}$$

There are two possibilities: <u>either</u> $\mathbf{m}^T\mathbf{m}$ is the identity, <u>or</u> B has a 1-dimensional null space, according to whether $\mu=1$ or $\mu\neq 1$.

We leave the case $\mu=1$, $\mathbf{m}^T = \mathbf{m}^{-1}$ to be analyzed later. Moreover, unless the contrary is explicitly stated, we understand that "components" of vectors and tensors shall be components with respect to the basis **e**, **f**, **g**. With this understanding

$$B = \begin{vmatrix} 0 & 0 & 0 \\ 0 & \mu_1 & 0 \\ 0 & 0 & \mu_2 \end{vmatrix} , \text{ where } \begin{cases} \mu_1 = \mu-1 \text{ and } \mu_2 = -\mu_1\, \mu^{-1} , \\ \\ \text{and } \mu_1 \neq 0 \neq \mu_2 , \ \mu_1 \neq \mu_2 . \end{cases} \tag{2.33}$$

It is easy to see that

$$\operatorname{adj} B = (\mu_1)^2 \begin{vmatrix} -\mu^{-1} & 0 & 0 \\ 0 & 0 & 0 \\ 0 & 0 & 0 \end{vmatrix} = -(\mu_1)^2\, \mu^{-1}\, \mathbf{e} \otimes \mathbf{e} . \tag{2.34}$$

This relation and (2.13) imply that

$$K_1 \equiv 0 \iff (\,\beta\cdot\mathbf{e}\,)^2 = (\mathbf{m}^T \beta\cdot\mathbf{e}\,)^2 \ \forall\, \beta \in R^3 . \tag{2.35}$$

Hence

$$(2.36) \qquad \beta \cdot e = 0 \quad \Longrightarrow \quad \beta \cdot m e = 0 \quad .$$

Consequently, for some real number $K \neq 0$

$$(2.37) \qquad m e = k e , \quad \text{hence } m^T e = k^{-1} m^T m e = k^{-1} e \quad .$$

On the other hand

$$(2.38) \qquad k = e \cdot m e = m^T e \cdot e = k^{-1} \quad , \text{ hence } k = \pm 1 \quad ,$$

Since

$$(2.39) \qquad e \cdot m f = 0 = e \cdot m g \quad ,$$

we see that [2]

$$(2.40) \qquad m = \begin{vmatrix} \pm 1 & 0 & 0 \\ 0 & m_{22} & m_{23} \\ 0 & m_{32} & m_{33} \end{vmatrix} \quad \text{and} \quad m^{-1} = \begin{vmatrix} \pm 1 & 0 & 0 \\ 0 & \pm m_{33} & \mp m_{23} \\ 0 & \pm m_{32} & \pm m_{22} \end{vmatrix} ,$$

where, for instance, $m_{22} = f \cdot m f$ and $m_{23} = f \cdot m g$.

We want to show that, if in (2.40) $m_{11} = 1$, then (2.30) holds, whereas if $m_{11} = -1$, then (2.29) holds. Equivalently, we should prove that, in the first case, m has an additional proper vector v, $v \cdot e = 0$, corresponding to the proper value 1 . In the second case, we should prove that m has a proper vector v ($\equiv v_1$) , $v \cdot e = 0$, corresponding to the proper value 1 . In either case, if $v = (0, v_2, v_3)$, then the system

$$(2.41) \qquad \begin{aligned} (m_{22} - 1) v_2 + m_{23} v_3 &= 0 \\ m_{32} v_2 + (m_{33} - 1) v_3 &= 0 \end{aligned}$$

must have a non-zero solution, hence a vanishing determinant D:

$$(2.42) \qquad 0 = D = 1 - m_{22} - m_{33} + m_{22} m_{33} - m_{23} m_{32} \cdot$$

Since det $m = 1$, it is easy to see that (2.42) is equivalent to

(2.43) $\quad m_{22} + m_{33} = 0 \,, \iff \quad \mathrm{tr}\, m = -1 \,, \quad \text{if } m_{11} = -1 \,,$

$\quad\quad\quad m_{22} + m_{33} = 2 \,, \iff \quad \mathrm{tr}\, m = 3 \,, \quad \text{if } m_{11} = 1 \,.$

We are going to show that (2.43) is a consequence of $K_2 \equiv 0$, that is,

$$(\beta \times m^T\beta) \cdot B(\beta \times m^T\beta) = 0 \quad \forall \beta \in R^3 \quad . \tag{2.44}$$

The equation $\mathbf{w} \cdot B\mathbf{w} = 0$ obviously has the solution $\mathbf{w} = \mathbf{e}$. It also has two linearly independent solutions $\mathbf{w}_1$ and $\mathbf{w}_2$ orthogonal to $\mathbf{e}$, since it can be written as follows:

$$(\mu - 1)\,[(\mathbf{w} \cdot \mathbf{f})^2 - \mu^{-1}(\mathbf{w} \cdot \mathbf{g})^2] = 0 \quad . \tag{2.45}$$

Therefore two linearly independent solutions orthogonal to $\mathbf{e}$ are

$$\mathbf{w}_1 = \mathbf{f} + \sqrt{\mu}\,\mathbf{g} \quad \text{and} \quad \mathbf{w}_2 = \mathbf{f} - \sqrt{\mu}\,\mathbf{g} \quad . \tag{2.46}$$

We deduce that (2.44) is equivalent to either

$$\beta \times m^T\beta \cdot \mathbf{u}_1 = 0 \quad \text{or} \quad \beta \times m^T\beta \cdot \mathbf{u}_2 = 0 \,, \quad \text{where} \tag{2.47}$$

$$\mathbf{u}_1 = \sqrt{\mu}\,\mathbf{f} - \mathbf{g} \quad \text{and } \mathbf{u}_2 = \sqrt{\mu}\,\mathbf{f} + \mathbf{g} \quad . \tag{2.48}$$

Condition $(2.47)_1$ for $m_{11} = \pm 1$, respectively, is equivalent to

$$0 = \det \begin{vmatrix} \beta_1 & \beta_2 & \beta_3 \\ \pm\beta_1 & (m^T\beta)_2 & (m^T\beta)_3 \\ 0 & \sqrt{\mu} & -1 \end{vmatrix} = -\beta_1[(m^T\beta)_2 + \sqrt{\mu}(m^T\beta)_3 \mp (\beta_2 + \sqrt{\mu}\beta_3)]. \tag{2.49}$$

By the arbitrariness of β, this condition requires that

$$m_{22} + \sqrt{\mu}\; m_{23} = \pm 1 \tag{2.50}$$

$$m_{32} + \sqrt{\mu}\; m_{33} = \pm\sqrt{\mu} \quad ,$$

hence

$$m_{22} = \pm 1 - \sqrt{\mu}\, m_{23} \quad \text{and} \quad m_{33} = \pm 1 - \mu^{-1/2} m_{32} \quad . \tag{2.51}$$

On the other hand, the condition that $\det \mathbf{m} = 1$ implies that

$$(2.52) \qquad \pm 1 = (\pm 1 - \sqrt{\mu}\, m_{23})(\pm 1 - \mu^{-1/2}\, m_{32}) - m_{23}m_{32} = 1 \mp \sqrt{\mu}\, m_{23} \mp \mu^{-1/2}\, m_{32} \;.$$

Therefore

$$(2.53) \qquad \begin{aligned} m_{32} &= -\mu\, m_{23} && \text{if } m_{11} = 1 \,, \\ m_{32} &= -\mu\, m_{23} - 2\sqrt{\mu} && \text{if } m_{11} = -1 \,. \end{aligned}$$

Hence, respectively,

$$(2.54) \qquad \begin{aligned} \operatorname{tr} m &= 1 + 1 - \sqrt{\mu}\, m_{23} + 1 - \mu^{-1/2}\, m_{32} = 3 \quad \text{and} \\ \operatorname{tr} m &= -1 - 1 - \sqrt{\mu}\, m_{23} - 1 - \mu^{-1/2}\, m_{32} = -1 \;, \end{aligned}$$

which is what we had to prove in the case of u_1 . Similarly, condition

$(2.47)_2$ is equivalent to

$$(2.55) \qquad 0 = \beta_1 \left[(m^T\beta)_2 - \sqrt{\mu}\, (m^T\beta)_3 \mp (\beta_2 - \sqrt{\mu}\, \beta_3) \right] \;,$$

which requires that

$$(2.56) \qquad m_{22} = \pm 1 + \sqrt{\mu}\, m_{23} \quad \text{and} \quad m_{33} = \pm 1 + \mu^{-1/2}\, m_{32} \;.$$

Now the condition that det m = 1 implies that

$$(2.57) \qquad \pm 1 = 1 \pm \sqrt{\mu}\, m_{23} \pm \mu^{-1/2}\, m_{32} \;;$$

hence

$$(2.58) \qquad \begin{aligned} m_{32} &= -\mu\, m_{23} && \text{if } m_{11} = 1 \quad \text{and} \\ m_{32} &= -\mu\, m_{23} + 2\sqrt{\mu} && \text{if } m_{11} = -1 \,. \end{aligned}$$

Therefore, respectively,

$$(2.59) \qquad \begin{aligned} \operatorname{tr} m &= 1 + 1 + \sqrt{\mu}\, m_{23} + 1 + \mu^{-1/2}\, m_{32} = 3 \quad \text{and} \\ \operatorname{tr} m &= -1 - 1 + \sqrt{\mu}\, m_{23} - 1 + \mu^{-1/2}\, m_{32} = -1 \;, \end{aligned}$$

which proves the theorem when $\mu \neq 1$.

When $\mu=1$ the argument above collapses because $B = 0 = \text{adj } B$, whence Δ in (2.4) vanishes for any choice of $C = C^T > 0$ of the form (2.3). Let us try to make Δ vanish identically in the class of matrices $C = C^T > 0$ given by

$$C = 1 + \lambda \, \mathbf{a} \otimes \mathbf{a} + \nu \, \mathbf{b} \otimes \mathbf{b} \ , \quad \lambda > -1 \ , \quad \nu > -1 \ , \tag{2.60}$$

where $\mathbf{a}$ and $\mathbf{b}$ are arbitrary vectors in R^3. If we set $\gamma_i = \alpha_i \, \alpha - a_i \, \mathbf{a}$, $\beta = m^T b$ and $\delta_i = \beta_i \, \beta - b_i \, b$, we can use a calculation similar to the one in Theorem 2 to show that, since $B = 0$, Δ is a cubic polynomial in λ and ν :

$$\Delta = -K_3 \, \lambda^3 + K \, \lambda^2 \mu + L \, \lambda \, \mu^2 - L_3 \, \nu^3 \ , \tag{2.61}$$

where K_3 is given by (2.8) and vanishes as well as its analogue L_3 , which is obtained from K_3 by replacing $\mathbf{a}$ with $\mathbf{b}$ everywhere in it. Moreover,

$$K = \gamma_1 \cdot \gamma_2 \times \delta_3 + \gamma_1 \cdot \delta_2 \times \gamma_3 + \delta_1 \cdot \gamma_2 \times \gamma_3$$

$$= \delta_1 \cdot \gamma_2 \times \gamma_3 + \delta_2 \cdot \gamma_3 \times \gamma_1 + \delta_3 \cdot \gamma_1 \times \gamma_2 \tag{2.62}$$

Since an easy computation shows that

$$- \delta_1 \cdot \gamma_2 \times \gamma_3 = \beta_1 \, (\mathbf{a} \times \alpha)_1 \, \beta \cdot \mathbf{a} \times \alpha - b_1 \, (\mathbf{a} \times \alpha)_1 \, b \cdot \mathbf{a} \times \alpha \ , \tag{2.63}$$

it is not difficult to verify that

$$K = -(\beta \cdot \alpha \times \mathbf{a})^2 + (b \cdot \alpha \times \mathbf{a})^2 \ . \tag{2.64}$$

Analogously

$$L = -(\alpha \cdot \beta \times b)^2 + (\mathbf{a} \cdot \beta \times b)^2 \ . \tag{2.65}$$

If Δ has to vanish for any C of the form (2.60), then both K and L have to vanish. Using the fact that $m^T m = 1$, whence $(m^T)^{-T} = m^T$, we see that $K = 0$ is equivalent to

$$(m^T a \cdot a \times b)^2 = (b \cdot m^T a \times a)^2 = (\beta \cdot \alpha \times a)^2 = (a \cdot \alpha \times \beta)^2$$

$$= (a \cdot m^T a \times m^T b)^2 = (a \cdot m^T (a \times b))^2 \tag{2.66}$$

$$= (m\, a \cdot a \times b)^2 ,$$

which has to hold for any vectors a and b. For any given a , a x b is an arbitrary vector perpendicular to a and b. Therefore (2.66) implies that

$$m^T a = \pm m a \qquad \forall a \in R^3 , \tag{2.67}$$

whence $m^T = m$, since $\det m^T = \det m$. We conclude that the theorem holds also when m is orthogonal, since in this case it is necessarily true that $m^2=1$.

Acknowledgments

This work was written while the author was on leave at the Institute for Mathematics and its Applications, University of Minnesota, partly supported by a fellowship of the Italian C.N.R.

Footnotes

([1]) adj B is the adjugate of B , that is, the matrix of co-factors of B .

([2]) Henceforth we choose either the upper or the lower signs everywhere, except in (2.67) .

References

[1] B.D. COLEMAN & W. NOLL, "Material symmetry and thermostatic inequalities in finite elastic deformations", Archive for Rational Mechanics and Analysis 15 (1964),87-111.

[2] J.L. ERICKSEN, "On the symmetry and stability of thermoelastic solids", Journal of Applied Mechanics 45 (1978), 740-743.

[3] J.L. ERICKSEN, "On the Symmetry of Deformable Crystals", Archive for Rational Mechanics and Analysis **72** (1979), 1-13.

[4] J.L. ERICKSEN, "Some phase transitions in crystals", Archive for Rational Mechanical and Analysis **73** (1980), 99-124.

[5] J.L. ERICKSEN, "Continuous martensitic transitions in thermoelastic solids", Journal of Thermal Stress **4** (1981), 107-119.

[6] J.L. ERICKSEN, "The Cauchy and Born hypotheses for crystals', in Phase Transformations and Material Instabilities in Solids, M.E. Gurtin ed., Academic Press, New York, 1984.

[7] J.L. ERICKSEN, "Some surface defects in unstressed thermoelastic solids", Archive for Rational Mechanics and Analysis **88** (1985), 337-345.

[8] J.L. ERICKSEN , "Stable equilibrium configurations of elastic crystals", to appear in the Archive for Rational Mechanics and Analysis.

[9] M.E. GURTIN , "Two-phase deformations of elastic solids", Archive for Rational Mechanics and Analysis **84** (1983), 1-29.

[10] R.D. JAMES, "Finite deformation by mechanical twinning". Archive for Rational Mechanics and Analysis **77** (1981), 143-176.

[11] R.D. JAMES, "Mechanics of coherent phase transformations in solids", MRL Report, Brown University, Division of Engineering, October 1982.

[12] M. PITTERI, "Reconciliation of Local and Global Symmetries of Crystals", Journal of Elasticity, **14** (1984), 175-190.

[13] M. PITTERI, "On the kinematics of mechanical twinning in crystals", Archive for Rational Mechanics and Analysis **88** (1985), 25-57.

[14] M. PITTERI, "On Type 2 twins", to appear in the International Journal of Plasticity.

Nonlocal Problems in Electromagnetism

Robert C. Rogers[1]
Mathematics Research Center and
Department of Mathematics
University of Wisconsin, Madison

1 Introduction

In this paper we consider two problems in classical electromagnetism that model nonlocal effects. The first is the problem of a conducting body. The Biot-Savart law implies that an electric current causes a magnetic field at distant points in a body. If the current is allowed to depend on this "self-field" through a generalized Ohm's law, then the magnetic field must be given as a function of the *global* values of the electric and magnetic potentials. This problem is examined in ROGERS & ANTMAN (1986); we summarize the results here. The second problem is that of a ferromagnetic material. The spontaneous alignment of atomic spins in such material is usually explained in terms of 'exchange forces' between neighboring particles. We propose a nonlocal constitutive equation to model the interaction between particles, and prove a general existence theorem for the resulting differential equations. In doing so we refine the theory of nonlocal constitutive equations for electromagnetic media as developed by ERINGEN (1973).

In section 2 we present the basic equations of steady-state electromagnetism in conducting bodies. In section 3 we show how nonlinear coupling between the electromagnetic fields leads to a system of functional-differential equations for a conducting body. We state a general existence theorem for such systems. In section 4 we give a phenomenological description of ferromagnetic materials, emphasizing the nonlocal nature of their behavior. We also give a brief outline of the theory of 'micromagnetics' which attempts to model such behavior. Finally, in section 5 we propose a nonlocal constitutive model for ferromagnetic materials. We state an existence theory for such materials and comment on the possibility of showing the existence of multiple solutions and bifurcations.

[1]Partially supported by the United States Army under Contract No. DAAG29-80-C-0041, and partially based on work supported by the National Science Foundation under Grant No. DMS-8210950, Mod 1.

2 Electromagnetostatics

The steady-state electromagnetic behavior of a rigid material body is governed by the folllowing field equations

$$div D = \sigma, \tag{1}$$

$$div B = 0, \tag{2}$$

$$curl E = 0, \tag{3}$$

$$curl H = J. \tag{4}$$

Here D is the *dielectric displacement*, B is the *magnetic induction*, E is the *electric field*, H is the *magnetic field*, J is the *current density*, and σ is the *free charge density*.

The classical boundary-value problems of electromagnetism are obtained by supplementing (1)-(4) with appropriate boundary conditions and constitutive equations. In the case of dielectric and paramagnetic materials we assume constitutive relations of the form:

$$D(x) = \epsilon_0 E(x) + \hat{P}(E(x), H(x), x) \equiv \hat{D}(E(x), H(x), x), \tag{5}$$

$$B(x) = \mu_0 H(x) + \hat{M}(E(x), H(x), x) \equiv \hat{B}(E(x), H(x), x). \tag{6}$$

Here ϵ_0 and μ_0 are universal constants and P and M are the *polarization* and *magnetization vectors* respectively.

Equations (3) and (4) imply that there exist scalar fields ϕ and ψ, called the *electric* and *magnetic scalar potentials* respectively, such that

$$E(x) = \nabla\phi(x), \tag{7}$$

and

$$H(x) = \nabla\psi(x) + \int_\Omega \frac{J(y) \times (y - x)}{| y - x |^3} dv_y. \tag{8}$$

Here Ω is the conducting body which we assume to be rigid. Of course rigidity eliminates the important 'self effects' caused when electromagnetic forces affect the deformation of the body and thus change the domain of integration in (8). However, our assumption is made only to simplify our exposition; the more general problem can be treated by using the methods of section 3 (*cf.* ROGERS & ANTMAN (1986) .

3 Nonlocal Problems in Conducting Bodies

If a body is nonconducting ($J \equiv 0$), we can combine (1)-(8) to get the following second-order quasilinear system of partial differential equations in divergence form:

$$div\hat{D}(\nabla\phi(x), \nabla\psi(x), x) = \sigma(x), \tag{9}$$

$$div\hat{B}(\nabla\phi(x), \nabla\psi(x), x) = 0. \tag{10}$$

We consider the following boundary conditions: At each $x \in \partial\Omega$ we prescribe

$$\text{either } \phi(x) = \tilde{\phi}(x) \quad \text{or } D(x)\cdot\eta(x) = \tilde{\delta}(x), \tag{11}$$

$$\text{either } \psi(x) = \tilde{\psi}(x) \quad \text{or } B(x)\cdot\eta(x) = \tilde{\beta}(x), \tag{12}$$

where $\tilde{\phi}, \tilde{\psi}, \tilde{\delta}$, and $\tilde{\beta}$ are given functions.

We say that the pair (ϕ, ψ) is a *weak solution* of (9)-(12) if it satisfies

$$\int_\Omega [\hat{D}\cdot\nabla\phi^\sharp + \hat{B}\cdot\nabla\psi^\sharp]dv + \int_\Omega \sigma\phi^\sharp dv = \int_{\partial\Omega}[\tilde{\delta}\phi^\sharp + \tilde{\beta}\psi^\sharp]da \tag{13}$$

for all reasonably nice *virtual fields* $\phi^\sharp, \psi^\sharp$ that respectively vanish in the sense of trace at all points $x \in \partial\Omega$ at which Dirichlet conditions are specified in (11) and (12). (The term 'reasonably nice' is made more specific below.)

The existence and regularity theory for such systems is well known, if not fully developed (*cf.* GIAQUINTA (1983)). Crucial to the development of the theory is some sort of ellipticity condition. For example one might make the following constitutive assumption

$$\sum_{i,j=1}^{3}(\frac{\partial\hat{D}_i}{\partial E_j}U_iU_j + \frac{\partial\hat{D}_i}{\partial H_j}U_iV_j + \frac{\partial\hat{B}_i}{\partial E_j}V_iU_j + \frac{\partial\hat{B}_i}{\partial H_j}V_iV_j) > 0 \ \forall\, U, V \in \Re^3/\{0\}. \tag{14}$$

Here D_i are the components of the vector D with respect to some fixed orthonormal basis for $\Re^3$, *etc.*

In a conducting body the problem is more complicated. We adopt as a constitutive equation for J a nonlinear version of Ohm's law:

$$J(x) = \hat{J}(E(x), H(x), x) \tag{15}$$

We substitute (15) into (8) to obtain

$$\begin{aligned} H(x) - \nabla\psi(x) &= \int_\Omega \frac{\hat{J}(\nabla\phi(y), H(y), y) \times (y-x)}{\mid y - x\mid^3} dv_y \\ &\equiv K(\nabla\phi(\cdot), H(\cdot), y). \end{aligned} \tag{16}$$

If $\hat{J}$ is independent of H, as it is in the classical form of Ohm's law, then (16) gives an explicit representation for H in terms of $\nabla\psi$ and $\nabla\phi$. When $\hat{J}$ depends on H, (16) represents a fixed-point problem. A typical representation result is the following:

Theorem 1 *Let $\alpha > 1$ and let Ω lie in a ball B_γ of radius γ and center 0. Let $\nabla\psi$ and $\nabla\phi$ be fixed in $L_\alpha(\Omega)$. Suppose that there are positive numbers μ, θ, ς with $3\varsigma > \alpha$ such that*

$$| \hat{J}(\nabla\phi, \nabla\psi, x) | \leq \mu(1 + | \nabla\phi |^{1+\varsigma} + | \nabla\psi |^{1+\varsigma}), \tag{17}$$

$$| \partial\hat{J}(\nabla\phi, \nabla\psi, x)/\partial H | \leq \theta(1 + | \nabla\phi |^{1+\varsigma} + | \nabla\psi |^{1+\varsigma}). \tag{18}$$

If γ and θ are small enough, then (16) has a unique solution of the form

$$H(x) = \nabla\psi(x) + \hat{K}(\nabla\phi(\cdot), \nabla\psi(\cdot))(x) \tag{19}$$

where $L_\alpha(\Omega) \ni (\nabla\phi, \nabla\psi) \mapsto \hat{K}(\nabla\phi, \nabla\psi)(\cdot) \in L_\alpha(\Omega)$ is continuous and compact.

The proof of the theorem follows immediately from ROGERS & ANTMAN (1986). A number of related results, including some for unbounded domains, can be based on techniques presented by SOBOLEV (1950, §6.9), STEIN (1970, Ch. V), and KANTOROVICH & AKILOV (1977, Ch. XI).

We now combine (1), (2), (5), (6), (7), and (19) to get the following second-order system of partial *functional* differential equations:

$$div\hat{D}(\nabla\phi(x), \nabla\psi(x) + \hat{K}(\nabla\phi(\cdot), \nabla\psi(\cdot))(x), x) = \sigma, \tag{20}$$

$$div\hat{B}(\nabla\phi(x), \nabla\psi(x) + \hat{K}(\nabla\phi(\cdot), \nabla\psi(\cdot))(x), x) = 0. \tag{21}$$

We now outline an existence theory that can be applied directly to this system. We present the theory in abstract form so that it can be applied directly to our model for ferromagnetic materials described below.

Let Ω, as before, be the closure of a domain in $\Re^3$. We assume that $\partial\Omega$ has a locally Lipschitz continuous graph. A typical point in Ω is denoted x. Let $\bar{U}(x) = (u_1(x), \ldots, u_m(x))$. For $p \in (1, \infty)$, let the operator

$$[L_p(\Omega)]^m \times [L_p(\Omega)]^{3m} \ni (\bar{U}, \nabla\bar{V}) \mapsto \hat{k}(\bar{U}, \nabla\bar{V})(\cdot) \in [L_p(\Omega)]^r \tag{22}$$

take bounded sets into bounded sets. Let

$$\Re^m \times \Re^{3m} \times \Re^r \times \Omega \ni (\xi, \eta, \varsigma, x) \mapsto \left\{ \begin{array}{l} A^i(\xi, \eta, \varsigma, x) \in \Re^3 \\ \beta^i(\xi, \eta, \varsigma, x) \in \Re \end{array} \right\}, i = 1, \ldots, m, \tag{23}$$

$$\Re^m \times \partial\Omega \ni (\xi, x) \mapsto \gamma^i(\xi, x) \in \Re, i = 1, \ldots, m \tag{24}$$

satisfy

(i) For almost all x in Ω, the functions $A^i(\cdot,\cdot,\cdot,x), \beta^i(\cdot,\cdot,\cdot,x)$ are continuous, and for all ξ,η,ς, the functions $A^i(\xi,\eta,\varsigma,\cdot), \beta^i(\xi,\eta,\varsigma,\cdot)$ are measurable. (These are the Carathéodory conditions.)

(ii) For almost all $x \in \partial\Omega$, the functions $\gamma^i(\cdot,x)$ are continuous, and for all ξ, the functions $\gamma^i(\xi,\cdot)$ are measurable on $\partial\Omega$ (with respect to two-dimensional Lebesgue measure).

(iii) There exist a constant $c_1 > 0$ and a function $k_1 \in L_q(\Omega)$ (with $q = p/(p-1)$) such that

$$|A^i(\xi,\eta,\varsigma,x)|,\ |\beta^i(\xi,\eta,\varsigma,x)| \le c_1[|\xi|^{p-1} + |\eta|^{p-1} + |\varsigma|^{p-1} + k_1(x)]$$

for $i = 1,\ldots,m$.

The Hölder inequality then implies that the functions

$$A^i(\bar{U}(\cdot),\nabla\bar{V}(\cdot),\hat{k}(\bar{U},\nabla\bar{V})(\cdot),\cdot),\quad \beta^i(\bar{U}(\cdot),\nabla\bar{V}(\cdot),\hat{k}(\bar{U},\nabla\bar{V})(\cdot),\cdot)$$

are in $L_q(\Omega)$ for all $\bar{U},\bar{V} \in [W_p^1(\Omega)]^m$. It follows that the functional

$$\begin{aligned} a(\bar{U},\bar{W}) \equiv & \int_\Omega [\sum_{i=1}^m A^i(\bar{U}(x),\nabla\bar{U}(x),\hat{k}(\bar{U},\nabla\bar{U})(x),x)\cdot\nabla w_i(x) \\ & + \beta^i(\bar{U}(x),\nabla\bar{U}(x),\hat{k}(\bar{U},\nabla\bar{U})(x),x)w_i(x)]dv(x) \\ & + \int_{\partial\Omega}\sum_{i=1}^m \gamma^i(\bar{U}(x))w_i(x)da(x) \end{aligned} \quad (25)$$

is well defined for all $\bar{U},\bar{W} \in [W_p^1(\Omega)]^m$.

We shall prescribe $u_1,\ldots,u_m$ respectively on subsets $S_1,\ldots,S_m$ of $\partial\Omega$. We assume that these subsets are measurable. Let $\mathcal{V}$ be the closed subspace of $[W_p^1(\Omega)]^m$ containing $[W_p^{1,0}(\Omega)]^m$ that consists of the functions $(w_1,\ldots,w_m)$ for which $w_1 = 0$ on $S_1,\cdots,w_m = 0$ on S_m in the sense of trace. Let $\bar{U}^*$ be a given element of $[W_p^1(\Omega)]^m$. We require that u_1 agree with u_1^* on S_1, etc. in the sense of trace by seeking solutions $\bar{U}$ of our equations in $[W_p^1(\Omega)]^m$ for which $\bar{U} - \bar{U}^* \in \mathcal{V}$. (This prescription of boundary conditions enables us to avoid the very delicate questions of whether functions defined on $S_1,\ldots,S_m$ can be extended to functions in $W_p^1(\Omega)$.)

We now define a map G from $\mathcal{V}$ to $\mathcal{V}^*$ by

$$< G(\bar{U}),\bar{W} > \equiv a(\bar{U},\bar{W}) \quad \forall \bar{W} \in \mathcal{V}. \quad (26)$$

Here $< \bar{V},\bar{W} >$ indicates the action of $\bar{V} \in \mathcal{V}^*$ on $\bar{W} \in \mathcal{V}$

Note that if the A^i are continuously differentiable and if $\bar{U}$ is twice continuously differentiable on Ω and vanishes on $\partial\Omega$, then we have

$$< G(\bar{U}), \bar{W} > \; = \int_\Omega \sum_{i=1}^{m} [-div A^i(\bar{U}, \nabla\bar{U}, \hat{k}(\bar{U}, \nabla\bar{U}), x) + \beta^i(Y, \nabla\bar{U}, \hat{k}(\bar{U}, \nabla\bar{U}), x)] w_i(x) dv(x). \tag{27}$$

Let us set $\Re^{3m} \ni \eta = (\vec{\eta}_1, \ldots, \vec{\eta}_m), \vec{\eta}_i \in \Re^3$. Our basic abstract result is the following:

Theorem 2 *Let $\partial\Omega$ have a locally Lipschitz graph. Let $p \in (0, \infty)$. Let (i),(ii), and (iii) hold. Suppose that*

$$\frac{a(\bar{V}, \bar{V})}{\|V, \mathcal{V}\|} \to \infty \; as \; \|\bar{V}, \mathcal{V}\| \to \infty \; for \; \bar{V} \in \mathcal{V}, \tag{28}$$

$$\sum_{i=1}^{m} [A^i(\xi, \eta + \rho, \varsigma, x) - A^i(\xi, \eta, \varsigma, x)] \cdot \vec{\rho}_i > 0 \; \forall \rho \equiv (\vec{\rho}_1, \ldots, \vec{\rho}_m) \neq 0. \tag{29}$$

If Ω is bounded, let

$$\sum_{i=1}^{m} \frac{A^i(\xi, \eta, \varsigma, x) \cdot \vec{\eta}_i}{[|\eta| + |\eta|^{p-1}]} \to \infty \quad as \; |\eta| \to \infty \tag{30}$$

for almost all x in Ω and for bounded ξ, η. If Ω is unbounded, let the following stronger relation hold: There is a number $c_2 > 0$ and a function $k_2 \in L_1(\Omega)$ such that

$$\sum_{i=1}^{m} A^i(\xi, \eta, \varsigma, x) \cdot \vec{\eta}_i \leq c_2 |\eta|_p - k_2(x). \tag{31}$$

Define $\tilde{k}$ by

$$[W_p^1(\Omega)]^m \ni \bar{U} \mapsto \tilde{k}(U)(\cdot) \equiv \hat{k}(\bar{U}, \nabla\bar{U})(\cdot) \tag{32}$$

where $\hat{k}$ is defined in (22). Let χ_C be the characteristic function of a set C in Ω. For every subdomain C of Ω with compact closure in Ω let

$$[W_p^1(\Omega)]^m \ni \bar{U} \mapsto \chi_C(\cdot)\tilde{k}(\bar{U})(\cdot) \in [L_p(C)]^r \tag{33}$$

be compact. Then for every $\bar{F} \in \mathcal{V}^*$ and for every $\bar{U}^* \in [W_p^1(\Omega)]^m$ there exists a $\bar{U} \in [W_p^1(\Omega)]^m$ with $\bar{U} - \bar{U}^* \in \mathcal{V}$ such that

$$< G(\bar{U}), \bar{V} > = < \bar{F}, \bar{V} > \quad \forall \bar{V} \in \mathcal{V}. \tag{34}$$

The proof of this theorem is obtained by making minor adjustments to those of BREZIS (1968) (*cf.* LIONS (1969, p. 297)) and BROWDER (1977). The key to the proof is the use of monotonicity in the local values of the highest order derivatives (*cf.* (29)) and compactness in the global values of the highest order derivatives (through $\tilde{k}$).

We now identify the variables appearing in theorem 2 with those used in the problem at the begining if the section. In particular we set

$$\bar{U} = (\phi, \psi), \tag{35}$$

$$\tilde{k}(\bar{U})(\cdot) = \hat{K}(\nabla\phi, \nabla\psi)(\cdot), \tag{36}$$

where $\hat{K}$ is defined in (19). We identify the variables appearing in (25) as follows:

$$\begin{aligned} A^1(\bar{U}(x), \nabla\bar{U}(x), \tilde{k}(\bar{U})(x), x) &= \hat{D}(\nabla\phi(x), \nabla\psi(x) + \hat{K}(\nabla\phi, \nabla\psi)(x), x), \\ A^2(\bar{U}(x), \nabla\bar{U}(x), \tilde{k}(\bar{U})(x), x) &= \hat{B}(\nabla\phi(x), \nabla\psi(x) + \hat{K}(\nabla\phi, \nabla\psi)(x), x), \end{aligned} \tag{37}$$

$$\begin{aligned} \beta^1(\bar{U}(x), \nabla\bar{U}(x), \tilde{k}(\bar{U})(x), x) &= \sigma(x), \\ \beta^2(\bar{U}(x), \nabla\bar{U}(x), \tilde{k}(\bar{U})(x), x) &= 0, \end{aligned} \tag{38}$$

$$\begin{aligned} \gamma^1(\bar{U}, x) &= -\tilde{\delta}(x), \\ \gamma^2(\bar{U}, x) &= -\tilde{\beta}(x). \end{aligned} \tag{39}$$

We identify $\bar{W}$ with $(\phi^\sharp, \psi^\sharp)$. Note that hypothesis (29) is ensured by the ellipticity condition (14). We then have

Theorem 3 *Let $\hat{D}, \hat{B}, \tilde{\delta}, \tilde{\beta}$ satisfy the hypotheses of Theorem 2 with the identifications (35), (36), (37), (38), and (39). Then (13) is satisfied for all $(\phi^\sharp, \psi^\sharp)$ in $\mathcal{V}$.*

The question of regularity of solutions for these nonlocal problems remains open. Since it is our intention to use this type of system to model ferromagnetic materials for which we expect highly discontinuous fields (see the comments on magnetic domains below) we do not find this disturbing.

4 Ferromagnetic Materials

The most remarkable and familiar characteristic of ferromagnetic materials is their ability to 'magnetize': They can become a source of magnetic fields even when no external fields are applied. This is illustrated by a hysteresis diagram

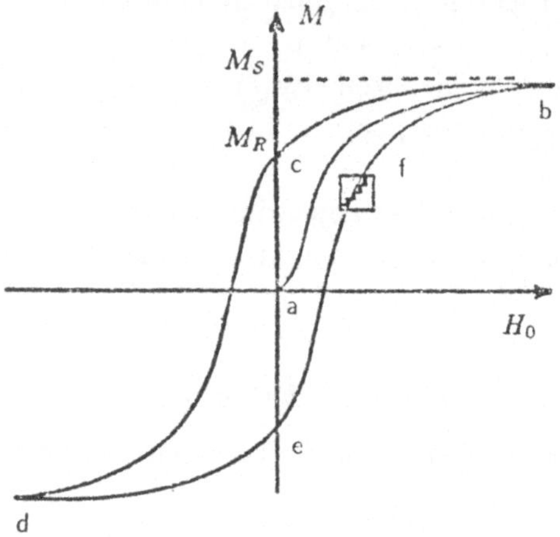

Figure 1: Main Hysteresis Loop

describing the results of a simple experiment: A specimen of ferromagnetic material is taken in an unmagnetized state; it produces no magnetic field, just as it would if its magnetization M were identically zero. An uniform field H_0 is applied, and the resultant field (the sum of H_0 and H_{sp}, the field produced by the specimen) is measured. From H_{sp} we compute the average magnetization M, the uniform magnetization of the specimen needed to produce the field H_{sp}. As H_0 is increased, M follows the curve a-b in Fig. 1. At some point a saturation magnetization M_S is reached. When H_0 is reduced, M remains high, following curve b-c and remaining at a residual magnetization M_R when H_0 is reduced to 0. At this point the body is said to be 'magnetized'. If we now apply H_0 in the opposite direction, M follows the path c-d, eventually saturating in this direction. Reversing the process maps out the curve d-e-f.

Hysteresis is a temperature-dependent phenomenon. At high temperatures θ ferromagnetic materials act in much the same way as paramagnetic materials; the magnetization M is a single-valued monotone function of the applied field H. But below a critical temperature, called the *Curie point* θ_C, quantum effects (called *exchange forces*) that cause atomic spins to align become more prominent than thermal effects, and hysteresis takes place. The hysteresis loop becomes larger as the temperature drops and the spins can become more closely aligned (*cf.* figure 2).

In the steep part of the curve (e-f in figure 1) the magnetization changes discontinuously in small steps as shown. This is know as the *Barkhausen effect.*

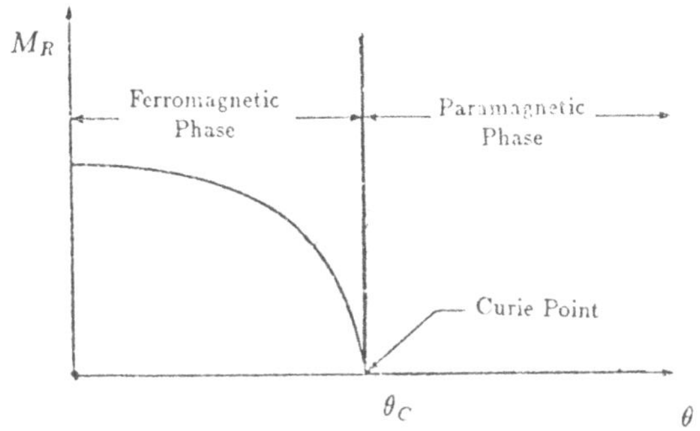

Figure 2: Onset of Hysteresis at the Curie Point

The small jumps are irreversible: If one starts to reduce the applied field short of the saturation point the original hysteresis loop is not retraced. Instead, one follows one of the smaller subloops seen in figure 3. These effects are usually explained phenomenologically by 'domain theory[2] ' (*cf.* TEBBLE (1969)), but we are concerned with developing a mathematical theory.

We now present a brief outline of the most widely accepted mathematical model of ferromagnetism (often called the theory of 'micromagnetics'). We consider a nonconducting body at a constant, uniform temperature. We assume that the free energy of a magnetized body is given by the functional:

$$\mathcal{E}(\psi, M) = \int_\Omega \{\mu_0(\frac{1}{2}|\nabla\psi|^2 + M \cdot \nabla\psi) + \mathcal{W}(M) + \frac{\lambda}{2}|\nabla M|^2\} dv. \tag{40}$$

The first term in the integrand is usually refered to as the *field energy*, the second as the *interaction energy*, the third as the *magnetization energy*, and the fourth as the *exchange energy*. The exchange energy is the term that causes the material to magnetize with no applied field, giving the material its ferromagnetic character. We discuss the derivation of this term below. The Euler-Lagrange equations for this functional are

$$div\ \ \mu_0(\nabla\psi(x) + M(x)) = 0, \tag{41}$$

$$\mu_0 \nabla\psi(x) + \frac{\partial \mathcal{W}}{\partial M}(M(x)) + \lambda \Delta M(x) = 0. \tag{42}$$

[2]According to this theory a specimen of ferromagnetic material is divided up into numerous tiny 'domains' (the scale of these regions is much larger than the atomic yet smaller than the macroscopic) which are uniformly magnetized to saturation. However, the direction of magnetization changes discontinuously from region to region, causing the overall magnetization of the specimen to depend on the arrangment of the domains. Changes in overall magnetization are caused by two processes: movement of the domain walls (which is reversible), and 'jumps' of domain walls across imperfections in the material (which are irreversible).

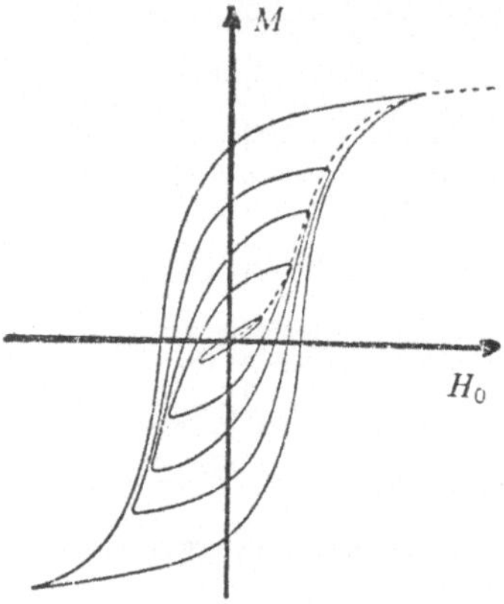

Figure 3: Subloops

To facilitate comparison to our constitutive model below we have misrepresented the theory somewhat. One usually assumes that the magnitude of M is constrained to be constant, and allows the magnetization and exchange energies to account for anisotropies in the material. We feel that constraint on M is unnecessarily restrictive. We shall take anisotropy into account in our constitutive model without direct reference to the micromagnetic theory.

5 A nonlocal constitutive theory of ferromagnetic materials

The theory of micromagnetics seeks to model exchange forces by relating the local values of H to higher-order derivatives of M. While progress has been made in obtaining global solutions to these equations (*cf.* BROWN (1963) and SHTRIKMAN & TREVES (1963)) these higher-order problems have proven quite difficult. (The same type of difficulties are encountered in the van der Waals-Cahn-Hilliard theory of phase transitions in fluids (*cf.* GURTIN (1985)).) We propose to replace the differential equations of (42) with a nonlocal constitutive equation as was first suggested by ERINGEN (1973). We suggest a constitutive functional of the following form:

$$M(x) = (1-\alpha(\theta))\check{M}(E(x), H(x), x) + \alpha(\theta)\int_{\Omega} \frac{e^{-\eta|x-y|}\hat{\mathcal{M}}(E(x), H(y), y)}{|x-y|} dv_y. \quad (43)$$

Here the temperature θ acts through the function α to scale the local and nonlocal contributions to the magnetization.

We can justify our choice through a physical rationale similar to that usually invoked when the energy relation (41) is derived [3]. The exchange energy term in (41) is derived from a discrete model by introducing a term of the form

$$\mathcal{E} = -\gamma \sum_{i,j} |\mathbf{m}_i - \mathbf{m}_{j(i)}|^2 + \cdots .$$

Here $\mathbf{m}_i = \mathbf{m}(p_i)$ is the magnetic moment of the ith lattice particle p_i and $\{j(i)\}$ is the set of indicies of nearest neighbors to the ith particle. To justify (43) in a similar fashion we could assume a relation of the form

$$\mathbf{m}_i = (1-\alpha)\tilde{\mathbf{m}}_i + \frac{\alpha}{n}\sum_{j(i)} \mathbf{m}_{j(i)} \equiv (1-\alpha)\tilde{\mathbf{m}}_i + \alpha \; avg \; \mathbf{m}_{j(i)},$$

asserting that the magnetization at a lattice point is a combination of a local term $\tilde{\mathbf{m}}$ and the average magnetization of the nearest neighbor particles. Here n is the number of nearest neighbor particles. We now rearrange and add a term $(1-\alpha)[\tilde{\mathbf{m}}_i - avg\; \tilde{\mathbf{m}}_{j(i)}]$ which is $o(h^2)$ where h is the lattice spacing. We obtain (after dropping the indicies and the correction term of order $o(h^2)$)

$$[(avg\; \mathbf{m} - \mathbf{m}) - \frac{1-\alpha}{\alpha}\mathbf{m}] - (1-\alpha)[(avg\; \tilde{\mathbf{m}} - \tilde{\mathbf{m}}) - \frac{1-\alpha}{\alpha}\tilde{\mathbf{m}}] = -(1-\alpha)\tilde{\mathbf{m}}.$$

Multiplying this by n/h^2 we get an approximation of the differential equation

$$\left[\Delta - \left(\frac{\beta}{h}\right)^2\right](\mathbf{m} - (1-\alpha)\tilde{\mathbf{m}}) = -\alpha\left(\frac{\beta}{h}\right)^2 \tilde{\mathbf{m}},$$

where $\beta = \sqrt{n(1-\alpha)/\alpha}$. This has the solution

$$\mathbf{m} - (1-\alpha)\tilde{\mathbf{m}} = \frac{\alpha}{4\pi}\int_\Omega \frac{e^{-\frac{\beta}{h}|x-y|}\beta^2\tilde{\mathbf{m}}(y)}{h^2|x-y|}dy, \tag{44}$$

which we identify with (43). We can thus argue that (43) quantifies the notion that ferromagnetism is a combination of local effects with effects of nearest neighbor particles.

The mathematical motivation for (43) is somewhat stronger. Theorem 2 can be applied directly to our new constitutive functional. For clarity we assume that the material is nonconducting ($J \equiv 0$). However, our existence theorem does

[3] While I find the justification of a constitutive law for a continuum through the use of a discrete model less compelling than the types of mathematical motivations given below, I recognize the appeal of such arguments to other readers (particularly those with a physics or engineering backround) and present the following for their benefit.

apply to the combination of these two nonlocal effects. We assure monotonicity in the local values of $\nabla\phi$ and $\nabla\psi$ by assuming that (14) holds with $\hat{B}$ replaced by

$$\check{B}(x) = \mu_0 H(x) + (1 - \alpha(\theta))\check{M}(E(x), H(x), x).$$

Then if $\mathcal{M}(\cdot,\cdot)$ is continuous, the compactness of our nonlocal term is assured by the compactness of $L_p \ni U \mapsto \int_\Omega [e^{-\eta|x-y|}U(y)/|x-y|]dv \in L_p$. Thus, if we make the identifications (35), (38), and (39) along with

$$\tilde{k}(\bar{U})(x) = \check{K}(\nabla\phi, \nabla\psi)(x) \equiv \alpha(\theta) \int_\Omega \frac{e^{-\eta|x-y|}\hat{\mathcal{M}}(\nabla\phi(y), \nabla\psi(y), y)}{|x-y|} dv_y, \tag{45}$$

$$\begin{aligned} A^1(\bar{U}(x), \nabla\bar{U}(x), \tilde{k}(\bar{U})(x), x) &= \hat{D}(\nabla\phi(x), \nabla\psi(x), x), \\ A^2(\bar{U}(x), \nabla\bar{U}(x), \tilde{k}(\bar{U})(x), x) &= \check{B}(\nabla\phi(x), \nabla\psi(x), x) + \check{K}(\nabla\phi, \nabla\psi)(x), \end{aligned} \tag{46}$$

we then have

Theorem 4 *Let $\hat{D}, \check{B}, \tilde{\delta}, \tilde{\beta}$ satisfy the hypotheses of theorem 2 with the identifications (35), (38), (39), (45), and (46). Then (13) is satisfied for all $(\phi^\sharp, \psi^\sharp)$ in $\mathcal{V}$.*

More importantly, (43) was designed to fit into the structure of modern bifurcation theory. Our goal is to be able to produce figures 1-3 as bifurcation diagrams for the differential equations analyzed in Theorem 4. In the case where the applied field is zero we would like to be able to show a bifurcation at the Curie point as indicated in figure 2. At lower temperatures we expect several solutions as indicated in figure 3. We conjecture that as the applied field H_0 is varied not only do the subloops of figure 3 branch out from these solutions, but that the Barkhausen jumps can be interpreted in terms of loss of stability of these branches. Of course, these equations are difficult to analyze. In particular, because of the nonlocal nature of the constitutive equation, we cannot reduce the partial differential equations to ordinary differential equations even for simple geometries and boundary conditions. Nevertheless, at least the problem of zero applied field seems tractable in its present form, and I am hopeful that the entire problem is approachable if only in some simplified (perhaps one-dimensional) form.

References

[1] Brezis, H. (1968), Équations et inéquations non linéaires dans les éspaces vectoriels en dualité, *Ann. Inst. Fourier*, **18**, 115-175.

[2] Browder, F.E. (1977), Pseudo-monotone operators and nonlinear elliptic boundary value problems on unbounded domains, *Proc. Natl. Acad. Sci.*, **74**, 2659-2661.

[3] Brown, W.F. (1963), *Micromagnetics*, John Wiley & Sons (Interscience).

[4] Eringen, A.C. (1973), Theory of Nonlocal Electromagnetic Solids, *J. Math. Phys.*, **14**,733-740.

[5] Giaquinta, M. (1983), *Multiple Integrals in the Calculus of Variations and Nonlinear Elliptic Systems*, Princeton Univ. Press.

[6] Gurtin, M.E. (1985), Some results and conjectures in the gradient theory of phase transitions, Inst. Math. Appl. Preprint No.156, Univ. Minnesota.

[7] Kantorovich L.V. & Akilov, G.P. (1977) *Functional Analysis*, 2nd ed., (in Russian), Nauka, Moscow; Engl. Transl. (1982), Pergamon.

[8] Lions, J.L. (1969), *Quelques Méthods de Résolution des Problèmes aux Limites non Linéaires*, Dunod, Gauthier-Villars.

[9] Rogers, R.C. & Antman, S.S. (1986), Steady-State Problems of Nonlinear Electro-Magneto-Thermo-Elasticity (to appear in Arch. Rat. Mech. Anal.).

[10] Shtrikman, S. & Treves, D. (1963), *Micromagnetics*, in *Magnetism, a treatise on modern theory and materials* (G.T. Rado & H. Suhl, editors), vol. III, Academic Press.

[11] Sobolev, S.L. (1950), *Applications of Functional Analysis in Mathematical Physics*, Leningrad State Univ. Press, English transl. 1963, Amer. Math. Soc.

[12] Stein, E.M. (1970), *Singular Integrals and Differentiability Properties of Functions*, Princeton Univ. Press.

HYPERBOLIC ASPECTS IN THE THEORY OF THE POROUS MEDIUM EQUATION

By

Juan Luis Vazquez[(1)]

Division de Matematicas
Universidad Autonoma
28049 Madrid, Spain

1. Introduction

The porous medium equation

$$u_t = \Delta(u^m),\ m>1\ , \tag{PME}$$

is one of the simplest models of nonlinear diffusion equations. It arises naturally in the study of a number of problems describing the evolution of a continuous quantity subject to a nonlinear diffusion mechanism, which we can for instance explain as caused by a diffusion coefficient of the form

$$c(u) = mu^{m-1} \tag{1.1}$$

if we write the PME as $u_t = \mathrm{div}(c(u)\nabla u)$. Among the applications of the PME we have

(i) Percolation of gas through porous media, where $m \geq 2$ [M],

(ii) Radiative heat transfer in ionized plasmas, where $m \approx 6$ [ZR],

(iii) Thin liquid films spreading under gravity, where $m = 4$ [Bu],

(iv) Crowd-avoiding population spreading, where $m > 1$ [GM].

In all of them nonnegative solutions of the PME are considered and m is a constant larger than 1. For motivation and terminology we shall stick to the first application. Thus we shall think of a gas with <u>density</u> $u = u(x,t)$ flowing through a porous medium that fills a region $\Omega \subset \mathbb{R}^n$, $n \geq 1$, for times $t \geq 0$. The <u>pressure</u> of the gas is then given by

---

(1) Partly supported by USA-Spain Cooperation Agreement under Joint Research Grant CCB-8402023. The paper was written while the author was a member of the Institute for Mathematics and its Applications, University of Minnesota, 1985.

(1.2) $$v = \frac{m}{m-1} u^{m-1}$$

and the <u>local velocity</u> by

(1.3) $$w = - \nabla v = - mu^{m-2} \nabla u$$

according to Darcy's law. The mathematical study of the PME seems to have begun in the late forties in Moscow, cf. [ZR]. In particular, Barenblatt [B] exhibited in 1952 a set of explicit self-similar solutions corresponding to source-type initial data (i.e. $u(x,0)$ is a Dirac mass) for $n=1,2,3$ and $m>1$, which play in some sense the role of the fundamental solution for the heat equation $u_t = \Delta u$.

The PME shares with the heat equation (i.e. $m = 1$) many important properties. In particular since $c(u) > 0$ at all points where $u > 0$ the PME is of parabolic type in the region occupied by the gas and consequently the solutions of (1) should be C^∞ functions there. However, at the points where $u = 0$ the equation <u>degenerates</u>. It is precisely the nature of this degeneracy that has been the object of much recent reseach and will be the main subject of this paper.

In fact if we look at the Barenblatt solutions, which can be most conveniently expressed in terms of the pressure as

(1.4) $$\bar{v}(x,t) = \frac{(r(t)^2 - x^2)_+}{2(m+1)t},$$

where $(\cdot)_+ = \max(\cdot,0)$ and

(1.5) $$r(t) = Ct^{1/(m+1)}, \quad C>0 \text{ arbitrary},$$

we observe that the positivity region $P = \{(x,t) \in \mathbb{R} \times (0,\infty) : \bar{v}(x,t) > 0\}$ is bounded by two curves $x_1 = -r(t)$, $x_2 = r(t)$, that we shall call <u>interfaces</u> or <u>free boundaries</u>. We also see that $\bar{v} \in C^\infty(P)$. Nevertheless $\bar{v}$ is only Lipschitz-continuous across these interfaces.

The appearance of interfaces causes the qualitative properties to differ markedly from those of the linear heat equation: It is well-known that any nonnegative solution of the heat equation in a cylindrical domain $Q = \Omega \times (t_1, t_2)$ is in fact positive in Q (a version of the strong maximum principle). This property goes against the physical evidence in most of the applications. In fact it is the existence of a finite propagation speed and the corresponding existence of interfaces which makes the study of the PME most interesting. As we shall see, these interfaces are responsible for the lack of regularity of the solutions to the PME and for the need to introduce a concept of generalized or weak solutions. Oleinik, Kalashnikov and Czhou [OKC] proved the existence and uniqueness of a generalized solution for the Cauchy problem

(1.6.a) $$u_t = \phi(x,u)_{xx} \quad \text{for } x \in \mathbb{R},\ t>0,$$

(1.6.b) $$u(x,0) = u_0(x) \quad \text{for } x \in \mathbb{R},$$

where ϕ is a function of $x \in \mathbb{R}$ and $u \geq 0$ that has a continuous derivative ϕ_u' and such that $\phi(x,0) = \phi_u'(x,0) = 0$, $\phi_u'(x,u) > 0$ if $u > 0$ and u_0 is a continuous, nonnegative and bounded function on $\mathbb{R}$. u is continuous, nonnegative, bounded in $\mathbb{R} \times (0,\infty)$ and satisfies an integral equality obtained from (1.6) by integration by parts. They also prove that if ϕ is a smooth function for $u > 0$, then $u(x,t)$ is also smooth whenever it is positive. However, since the Barenblatt solutions $\overline{v}(x,t+1)$ are generalized solutions, it is clear from the uniqueness statement that no classical solutions exist in general. The construction was generalized to several space variables by Sabinina [S].

It was also proved in [OKC] that, under conditions on ϕ that cover the cases $\phi(x,u) = u^m$, $m>1$, if the initial datum $u_0(x)$ has compact support, then for every $t>0$, $u(\cdot,t)$ also has compact support. Therefore there exist two finite interfaces

(1.7.a) $$x = s_1(t) = \inf\{x \in \mathbb{R}: u(x,t) > 0\}$$

and

(1.7.b) $$x = s_2(t) = \sup\{x \in \mathbb{R}: u(x,t) > 0\}$$

that bound the positivity region. [OKC] called this property the finite speed of propagation of disturbances.

The study of the interfaces has revealed that the PME shares a number of properties with Hamilton-Jacobi equations and nonlinear conservation laws. We will devote most of the rest of the paper to explain these connections in the case of the Cauchy problem in one space variable, n=1, which has been studied recently with great detail and is now well understood. A comment on the state of the art in several space dimensions will be made in the final Section 6.

2. Basic Equations and Estimates

We consider the problem

$$u_t = (u^m)_{xx} \quad \text{in} \quad Q = \mathbb{R}\times(0,\infty), \tag{2.1}$$

$$u(x,0) = u_0(x) \quad \text{for} \quad x \in \mathbb{R}, \tag{2.2}$$

where $m>1$ and $u_0 \in L^1(\mathbb{R})$ and is nonnegative. In terms of the pressure $v = (m/(m-1))u^{m-1}$, (2.1) becomes

$$v_t = (m-1)\, vv_{xx} + v_x^2 \,. \tag{2.3}$$

Differentiating (2.3) we obtain the equation for the velocity $w = -v_x$

$$w_t + (w^2)_x = (m-1)(vw_x)_x \,. \tag{2.4}$$

We want to view (2.3) as a perturbation of the Hamilton-Jacobi equation

$$v_t = v_x^2 \,, \tag{HJ}$$

sometimes called nonstationary eikonal equation. The perturbation $(m-1)vv_{xx}$ can be considered a nonlinear viscosity term. It degenerates for $v = 0$. In the same way the velocity equation (2.4) can be viewed as a viscous perturbation of the first order conservation law, which is the Burgers equation,

$$w_t + (w^2)_x = 0. \tag{CL}$$

It is our purpose to show how the effect of the degeneracy of (PME) at $u = 0$ makes the pressure of the solutions of (2.1), (2.2) resemble the solutions of (HJ) and how the corresponding velocity w behaves like the solutions to (CL). There are three instances at which this similarity appears:

i) at the interfaces.

ii) as $m \downarrow 1$.

(In both cases the perturbation term in (2.3) or (2.4) formally vanishes.)

iii) as $t \to \infty$.

Before we proceed with the description of those cases let us comment on another striking similarity. The nonnegative solutions to problem (2.1), (2.2) enjoy an a priori estimate of the form

$$(2.5.a) \qquad v_{xx} \geq -\frac{k}{t} \quad \text{with} \quad k = \frac{1}{m+1},$$

cf. [AB]. The estimate is even true for the Cauchy problem in n space dimensions in the form

$$(2.5.b) \qquad \Delta v \geq -\frac{k}{t} \quad , \quad k = (m-1+\frac{2}{n})^{-1} ,$$

and has played an important role in the theory of the PME. Observe that (2.5) is valid for <u>all</u> nonnegative solutions of (2.1)(2.2).

Similar conditions play an important role in the theory of Hamilton-Jacobi equations of the form

$$(2.6) \qquad v_t = H(Dv)$$

where $H: \mathbb{R}^n \to \mathbb{R}$. Since the weak solutions $v \in C(\overline{Q}) \cap W^{1,\infty}(Q)$ of the Cauchy problem associated with (2.6) are not unique, it is necessary to somehow select a class of <u>good</u> solutions in which the problem has a unique solution and that are physically relevant. Such a concept has been developed recently by Crandall and Lions [CL], cf. also [CEL], for general Hamiltonians H under the name of <u>viscosity solutions</u>. In case H is convex, it has been known for a time that the characterization of the good unique solution can be achieved by imposing an additional semiconvexity condition, which in n=1 takes the form (2.5.a) for some $k>0$, cf. [Li, Theorem 10.2].

A similar problem of nonuniqueness occurs for conservation laws like (CL). There the right additional condition can take the form

$$w_x < \frac{c}{t} \tag{2.7}$$

and is called an entropy condition, cf. [O] (for a general characterization in several space dimensions cf. [Kr]).

Using (2.5) we can obtain from (2.4) the inequality

$$v_t + \frac{m-1}{m+1} \cdot \frac{v}{t} > v_x^2 \ . \tag{2.8}$$

Let us now consider the transformation

$$\begin{aligned} x &= x \ , \\ \tau &= \frac{m+1}{2} t^{\frac{2}{m+1}} \ , \\ V(x,\tau) &= v(x,t) t^{\frac{m-1}{m+1}} . \end{aligned} \tag{2.9}$$

With this change of variables (2.8) and (2.5) become respectively

$$V_\tau \geq V_x^2 \quad \text{and} \quad V_{xx} > -\frac{1}{2\tau} \ . \tag{2.10}$$

This means that (2.8) transforms any nonnegative solution of (2.1)(2.2) into a viscosity supersolution of the equation (HJ). But the correspondence is more striking when applied to the Barenblatt solutions (1.5). In this case the estimate (2.5) is exact in the support; therefore (2.8) and (2.10) are also exact and $\bar{v}$ transforms into

$$\bar{V}(x,\tau) = \left(K - \frac{x^2}{4\tau}\right)_+ \ , \ K = \frac{C^2}{2(m+1)} \ . \tag{2.11}$$

$\bar{V}$ is a self-similar solution of (HJ). It is easy to prove (using for instance the explicit solution formulas for (HJ)) that every nonnegative viscosity solution of (HJ) with initial data having compact support converges as $t \to \infty$ to (2.11) with constant $K = \|u_0\|_\infty$. On the other hand the Barenblatt solutions $\overline{v}(x,t)$ also represent the asymptotic profiles for every nonnegative solution of (2.1), (2.2) with $u_0 \in L^1(\mathbb{R})$, see Section 5.

3. Properties of the Interface

Suppose now that u is a solution of (2.1), (2.2) whose initial data u_0 is compactly supported. Let $s_1(t)$, $s_2(t)$ be the left and right interfaces to this solution as defined by (1.7). Since their properties are similar we shall concentrate on $s_2(t)$, which we shall call simply $s(t)$.

The first property of the interface is monotonicity: for $t_1 > t_0$ we have $s(t_1) \geq s(t_0)$ (and $s_1(t_1) \leq s_1(t_0)$). This can be obtained as follows: From (2.3) and (2.5) we get

$$v_t \geq -\frac{(m-1)v}{(m+1)t} \tag{3.1}$$

This inequality implies that whenever $v(x_0,t_0) > 0$ and $t_1 > t_0$, then $v(x_0,t_1) > 0$ (retention property). The monotonicity of the positive set follows and it implies the monotoncity of the interface.

Aronson [A1] and Knerr [Kn] studied the equation that governs the interface and proved that for every $t>0$

$$D^+s(t) = v_x(s(t)-, t). \tag{3.2}$$

Here in the first member we have the right-hand derivative of s at t and the second member denotes the left-hand derivative of $v_x(\cdot,t)$ at $x = s(t)$. Since $v_x(s(t)+,t)$ is obviously 0 the second member equals the jump of $v_x(\cdot,t)$ at $x = s(t)$ and (3.2) is just a <u>Rankine-Hugoniot</u> condition. This is precisely the type of equation that controls the free-boundaries of conservation laws like (CL). Observe that the existence of lateral limits for $v_x(\cdot,t)$ is again a consequence of (2.5).

Aronson [A2] and Knerr [Kn] also studied the phenomenon of <u>metastability</u>. For certain initial configurations $v_0(x)$ the interface $s(t)$ is stationary for a certain time $0 < t < t^*$ and then begins to move. t^* is called the <u>waiting</u>

time. In particular, if $v_0(x) = 0$ for $x>0$ and

(3.3.a) $$v_0(x) = \alpha x^2 + 0(x^2) \quad \text{as} \quad x \uparrow 0,$$

(3.3.b) $$v_0(x) < \beta x^2 \qquad \text{for} \quad x < 0,$$

we have ([ACK]),

(3.4) $$\frac{1}{2(m+1)\beta} < t^* < \frac{1}{2(m+1)\alpha} .$$

Moreover the necessary and sufficient condition to have a positive waiting time is, according to [V3], if $s(0) = 0$,

(3.5) $$B \equiv \limsup_{x \to 0} \left(|x|^{-\frac{m+1}{m-1}} \int_x^0 u_0(x)dx\right) < \infty$$

and then we have

(3.6) $$\frac{T_m}{B^{m-1}} < t^* < \frac{\theta_m}{B^{m-1}}$$

for some constants T_m, $\theta_m > 0$.

In [CF1] Caffarelli and Friedman studied the behaviour of the interface for $t > t^*$. First by means of a delicate comparison with the Barenblatt solutions they proved that $s(t)$ is semiconvex. Their argument was subsequently made precise in [V1] as follows: if we observe that the interface $r(t)$ of the Barenblatt solution satisfies

(3.7) $$r''(t) + \frac{m}{(m-1)t} r'(t) = 0 ,$$

using their argument we can obtain

(3.8) $$s''(t) + \frac{m}{(m+1)t} s'(t) \geq 0 \quad \text{in} \quad \underline{D}'(\mathbb{R}^+) ,$$

which can also be written as

(3.9) $$\frac{d}{dt} (s'(t)/r'(t)) \geq 0 ,$$

so that the fastest rate of growth corresponds to the interface of the Barenblatt solution cf. (5.13). See also [ACV, pg. 383]. In [CF] the authors also prove the following regularity results: The functions v_t and v_x, which are smooth functions in $[u>0]$, can be continuously extended to the moving free-boundary $x = s(t)$, $t>t^*$, so that v becomes a C^1 function there. Also the derivative $D^+s(t)$ is a full derivative $s'(t)$ if $t>t^*$. Equation (3.2) implies that in fact $s \in C^1(t^*,\infty)$. Finally they prove that on the moving interface the equation (2.1) becomes the Hamilton-Jacobi equation

$$v_t = v_x^2 ,$$

i.e. $v(x,t)v_{xx}(x,t) \to 0$ as $(x,t) \to (x_0,t_0)$ with $t_0 > t^*$, $x_0 = s(t_0)$ and $v(x,t) > 0$.

This result leaves open the question: Is $s \in C^1(0,\infty)$? Since $s(t) = s(0)$ for $0 < t < t^*$, the question amounts to knowing if $s'(t^*)$ exists when $t^* > 0$. We recall that for conservation laws of the form $w_t + (w^m)_x = 0$ the interfaces are in general Lipschitz-continuous functions and there can be a countably infinite number of times at which the one sided-derivatives $D^+s(t_i)$, $D^-s(t_i)$ exist but $D^-s(t_i) < D^+s(t_i)$. In [ACV] Aronson, Caffarelli and Vazquez showed that this phenomenon happens at the waiting time to the solutions of (2.1), (2.2) for a large class of initial data with compact support. Suppose for instance that $u_0(x) > 0$ in $(a,0)$, $a<0$, and u_0 vanishes otherwise, that the waiting time t^* is positive, that $v_0(x) < \alpha|x|^2 + O(x^2)$ as $x \uparrow 0$ and that

$$t^* < \frac{1}{2(m+1)\alpha} \quad . \tag{3.10}$$

Then $D^+s(t^*) > 0 = D^-s(t^*)$. Observe that this occurs, in particular, if $v_0(x) = o(x^2)$ as $x \uparrow 0$ (by (3.5), (3.6) t^* is then finite and positive). Therefore Lipschitz continuity is the optimal global regularity for the interfaces in general. Nevertheless, the interface $s(t)$ may be C^1 when we have equality instead of inequality in (3.10), cf. [ACK]. See also [LOT]. And the interface is of course a C^1 function when $t^* = 0$.

Further details about the behaviour of the interfaces can be found in [V3].

4. The Limit $m \downarrow 1$

Bénilan and Crandall [BC] proved that the solution u to problem (2.1), (2.2) depends continuously in the $C(\mathbb{R}^+ : L^1(\mathbb{R}))$-norm on both the initial data $u(\cdot,0) = u_0 \in L^1(\mathbb{R})$ and on m. In particular, if u_0 is kept fixed and we let m tend to 1, then $u = u(x,t;m)$ converges to a solution of the heat equation

$$(4.1) \qquad u_t = u_{xx}$$

with initial datum u_0. Thus, for m near 1, the porous medium equation can be regarded as a perturbation of the heat equation. Since the heat equation has an infinite speed of propagation, if we assume that u_0 is nonnegative and has bounded suport and call $s_m(t)$ the right interface to the solution of (2.1), (2.2) with $m>1$ we have

$$(4.2) \qquad \lim_{m\to 1} s_m(t) = +\infty$$

for every $t>0$ unless $u_0 \equiv 0$. Here we want to show a different way of taking the limit $m \downarrow 1$ in (2.1) that preserves the finite propagation aspects of the solutions to the PME. It consists in taking the pressure v as the main variable and letting $m \to 1$ in (2.3) while keeping $v(\cdot,0)$ fixed. This approach has been pursued in [AV] by Aronson and Vazquez. If $v_m = v(x,t;m)$ is the solution of the problem

$$(4.3) \qquad \begin{aligned} &v_t = (m-1)vv_{xx} + v_x^2 && \text{in } Q, \\ &v(x,0) = v_0(x) && \text{for } x \in \mathbb{R} \end{aligned}$$

and v_0 is continuous, nonnegative and bounded, then as $m \to 1$ the sequence $\{v_m\}$ converges uniformly on compact subsets of $\overline{Q} = \mathbb{R} \times [0,\infty)$ to a solution of the problem

$$(4.4) \qquad \begin{aligned} &v_t = v_x^2 && \text{in } \underline{D}'(Q), \\ &v(x,0) = v_0(x) && \text{for } x \in \mathbb{R}. \end{aligned}$$

In fact, $v \in C(Q) \cap \mathrm{Lip}(\mathbb{R} \times (\tau,\infty))$ for every $\tau > 0$. Moreover, (2.5.a) gives in the limit

$$v_{xx} \geq -\frac{1}{2t}. \tag{4.5}$$

These properties uniquely characterize v as a solution of (4.4); in fact, v is the viscosity solution of (4.4).

Moreover, we have convergence of the velocities $w_m = -v_{mx}$ in $L^p_{loc}(Q)$ for every $1 \leq p < \infty$, to a solution of the conservation law (CL) in Q. In [AV] the condition on v_0 is relaxed to allow a sequence v_{0m} of nonnegative, uniformly bounded and continuous initial functions such that v_{0m} converges to v_0 uniformly on compact subsets of R.

Let us assume again that $v_{0m} \equiv v_0$ and moreover that v_0 vanishes for all large positive x so that the solutions of the problems (4.3) exhibit a right interface $s_m(t)$. Then as $m \to 1$ $s_m(t)$ converges uniformly on compact subsets of $[0,\infty)$ to the right interface $s(t)$ of the solution to problem (4.4) and $s_m'(t) \to s'(t)$ a.e. and in $L^p_{loc}(\mathbb{R}^+)$ for every $p \in [1,\infty)$, cf. [AV, Theorem 2]. These convergence results cannot be substantially improved in general because of the lack of regularity of the solution to (4.4), but in particular cases better results can be obtained. Thus if v_0 is concave in its support, then so are $v_m(\cdot,t)$ and $v(\cdot,t)$ for every $t>0$, v is C^1 in its support and the v_m converge monotonically to v in the C^1 - topology inside the support of v. We refer for a proof of all these results to [AV].

Let us finally remark that the convergence to the Hamilton-Jacobi equation and the results of [BC] can be reconciled because they occur at different scales of magnitude. It is clear that if $v_m \to v$ uniformly in a subset of Q and we define u_m from v_m through the relation (1.2), then $u_m \to 0$ uniformly on the same set. Therefore in passing to the limit $m \to 1$ with v_0 fixed we lose all the information at the u-level.

5. The Asymptotic Behaviour

Now let $u(x,t)$ be a solution of problem (2.1), (2.2) and let

$$M = \int u_0(x)dx > 0 \tag{5.1}$$

be its <u>mass</u>. It is an invariant of the dynamics, i.e. for all $t>0$

$$M = \int u(x,t)dx.$$

Using rescaling arguments Kamin [K] proved that as $t \to \infty$, u approaches the Barenblatt solution $\overline{u}$ with the same mass, i.e. the solution u given by (1.4), (1.5) with a constant C such that

$$\overline{u}(x,0) = M\delta(x),$$

δ being the Dirac mass at the origin. Then we have, [K] ,

$$\lim_{t \to \infty} t^{1/(m+1)} | u(x,t) - \overline{u}(x,t)| = 0 \tag{5.2}$$

uniformly in $x \in \mathbb{R}$. In trying to relate the solutions of (2.1), (2.2) to the equations (HJ) or (CL) we are therefore interested in looking at the v- and w-profiles of the Barenblatt solutions. In particular, the velocity of <u>all</u> the Barenblatt solutions is given by the formula

$$\overline{w}(x,t) = \frac{x}{(m+1)t} \quad \text{whenever} \quad \overline{v}>0, \tag{5.3}$$

i.e. whenever $|x| < r(t) = Ct^{1/(m+1)}$. Profiles like this are called N-<u>waves</u> and appear as the typical asymptotic profile of solutions to conservation laws of the form

$$u_t + (|u|^{m+1})_x = 0, \tag{5.4}$$

where the velocity is given by $w = |u|^{m+1}u$, cf. [L], [D], [DP], [LP]. Let us remark in passing that the N-waves for (5.4) can be nonsymmetric, i.e. (5-3) holds for

$$-pt^{1/(m+1)} < x < qt^{1/(m+1)}, \tag{5.5}$$

where p and q depend on <u>two</u> invariants of the problem. These results show that asymptotically the velocity distribution is limited by two shock fronts and behaves linearly between them both in the case of the porous medium equation and in the case of scalar conservation laws with a power-like nonlinearity. Moreover the magnitude and position of the shocks are determined by the invariants of the motion and the shocks can be said to carry the first-order asympotic information about the solution.

In [V4] I studied how a general solution of (2.1), (2.2) approaches this N-wave behaviour as $t \to \infty$. In fact, if u is a solution with mass M and interface $r(t)$ we have: For every $\varepsilon > 0$ there exists $t_\varepsilon > 0$ such that if $t > t_\varepsilon$

(5.6.a) $$|w(x,t) + \frac{x}{(m+1)t}|\, t^{\frac{m}{m+1}} < \varepsilon \quad \text{if} \quad |x| < (1-\varepsilon)r(t),$$

(5.6.b) $$|w(x,t)|\,t^{\frac{m}{m+1}} < \varepsilon \quad \text{if} \quad |x| > (1+\varepsilon)r(t)$$

and

(5.6.c) $$-\varepsilon < w(x,t)t^{\frac{m}{m+1}}\,\text{sign}(x) < \frac{r(t)}{(m+1)t} \quad \text{if} \quad ||x|-r(t)| < \varepsilon r(t)$$

This can be visualized as follows. If we introduce the rescaled variables

(5.7) $$y = \frac{x}{r(t)}, \quad W(y,t) = \frac{(m+1)t}{r(t)}\, w(x,t),$$

then (5.6) says that, as $t \to \infty$, $W(\cdot,t)$ converges to the piecewise linear profile

(5.8) $$\overline{W}(y) = \begin{cases} y & \text{if} \quad |y| < 1, \\ 0 & \text{if} \quad |y| > 1, \end{cases}$$

uniformly in the graph topology in $\mathbb{R}^2$. Convergence in the graph-topology had been used by Lax in the study of conservation laws of the type $u_t + f(u)_x = 0$, cf. [L].

In case u_0 has compact support, the above convergence rates can be improved by introducing a second invariant of the problem, namely the <u>center of mass</u>

(5.9) $$x_c = M^{-1} \int u_0(x)dx,$$

and considering the Barenblatt solution with same mass and centered at x_c, i.e.

$$\tilde{u} = u(x-x_c, t; M). \tag{5.10}$$

It was proved in [V1] (see also [V2]) that in that case we have, with the usual notations, as $t \to \infty$

$$t^{m/m+1} \ |v(x,t) - \tilde{v}(x,t)| \to 0 \ \text{ uniformly in } \ x \in R, \tag{5.11.a}$$

$$t\left|w(x,t) + \frac{x-x_c}{(m+1)t}\right| \to 0 \ \text{ uniformly in } \ \Omega(t) \tag{5.11.b}$$

where $\Omega(t) = \{x \in \mathbb{R}: u(x,t) > 0\}$,

$$s(t) - (r(t) + x_c) \downarrow 0, \tag{5.11.c}$$

$$s'(t)/r'(t) \to 1 \ \text{ and } \ s'(t) = r'(t) + o(1/t). \tag{5.11.d}$$

The main techniques used in the proof of the above results are i) a principle of comparison by shifting which allows us to prove that

$$s(t) - r(t) = 0(1) \tag{5.12}$$

by comparison with Barenblatt solutions conveniently shifted to the left or to the right and ii) the convexity estimate (3.9). In fact, it is immediate from (5.12) and (3.9) that

$$s'(t)/r'(t) \uparrow 1 \tag{5.13}$$

so that $s(t) - r(t)$ is nonincreasing and has a limit. A technical computation using the invariance of M shows that this limit is necessarily x_c.

An important geometrical aspect of the Barenblatt solutions is the fact that they are concave inside their support. More precisely they satisfy the basic relation (2.5.a) with equality on the set where they are positive:

$$\bar{v}_{xx} = -\frac{1}{(m+1)t} \quad \text{if} \quad \bar{v} > 0.$$

It is proved in [BV] that if u is a solution of problem (2.1), (2.2) whose initial pressure is concave in its support (so that there exist $a,b \in \mathbb{R}$ and $C>0$ such that $v_0(x) = 0$ if $x>b$ or $x<a$ and $v_0'' < -C$ in $D'(a,b)$), then for every $t>0$ $v(\cdot,t)$ is concave in its support $(s_1(t),s_2(t))$ and

$$(5.14) \qquad v_{xx} < -\frac{C}{1+(m+1)Ct} \quad \text{in } \underline{D}'(P),$$

P being the positivity set of v. From this we can derive sharp estimates for the above asymptotic convergence. In particular, we have as $t \to \infty$

$$(5.15.a) \qquad v_{xx} = -\frac{1}{(m+1)t} + O(\frac{1}{t^2}) \quad \text{in } P,$$

$$(5.15.b) \qquad s''(t) = r''(t)\ (1 + O(\frac{1}{t})).$$

By integration we obtain expressions for v, v_x, s, s'. The rates of convergence are the best possible since they are satisfied by the time-delayed Barenblatt solutions $\overline{v}(x,t) = v(x,t+\tau)$ with $\tau > 0$.

Optimal rates can also be obtained for symmetric solutions without the assumption of concavity, cf. [V1], pg. 521.

6. The Problem in Several Space Dimensions

Both from the mathematical point of view and for the sake of the applications mentioned in the Introduction we are interested in discussing the Cauchy problem

$$(6.1.a) \qquad u_t = \Delta(u^m) \qquad \text{in } Q = \mathbb{R}^n \times (0,\infty),$$

$$(6.1.b) \qquad u(x,0) = u_0(x) \qquad \text{for } x \in \mathbb{R}^n,$$

when $m,n>1$. While the one-dimensional case is by now well understood as we have shown above, the theory for $n>1$ is considerably more difficult and much work is being done at the present moment. There exists a satisfactory theory of existence and uniqueness of weak solutions (6.1), in particular for nonnegative solutions, cf. [BCP], [AC], [DK]. The solutions are Hölder continuous in Q and C^∞ in their positivity set P, [CF2].

From what is known it follows that the phenomena discussed in previous sections also occur to a large extent if $n>1$. Thus, [CF] proves that they are Hölder continuous surfaces and [CVW] proves that they are Lipschitz-continuous surfaces for all large times and that ∇v is not continuous in general across the interface. The asymptotic behaviour has been studied in [FK] where the authors prove convergence of u towards a Barenblatt solution. The limit $m\downarrow 1$ has been recently studied in [LSV], who obtain convergence of the pressures and the interfaces to the first-order equation

$$(6.2) \qquad v_t = |\nabla v|^2 \quad .$$

In view of these results we conjecture that the whole picture described in Sections 2,3,4,5 for n=1 holds, appropriately translated, if n>1.

Acknowledgements

The author is indebted to the Institute for Mathematics and its Applications and the University of Minnesota for its hospitality while this and other papers were written and for giving him the opportunity of discussing these ideas with many experts. He would especially like to thank Donald Aronson and Luc Tartar for numerous conversations held during the spring of 1985.

References

[A1] D.G. Aronson, Regularity properties of flows through porous media: the interface, Arch. Rational Mech. Anal. 37 (1970), 1-10.

[A2] D.G. Aronson, Regularity properties of flows through porous media: a counterexample, SIAM J. Appl. Math. 19 (1970), 299-307.

[AB] D.G. Aronson-Ph. Bénilan, Régularité des solutions de l'équation des milieux poreux dans $\mathbb{R}^n$, C. Rendus Acad. Sci. Paris Sér. A-B 288 (1979), 103-105.

[AC] D.G. Aronson-L.A. Caffarelli, The initial trace of a solution of the porous medium equation, Trans. Amer. Math. Soc. 280 (1963), 351-366.

[ACK] D.G. Aronson-L.A.Caffarelli-S. Kamin, How an initially stationary interface begins to move in porous medium flow, SIAM J. Math. Anal. 14 (1983), 639-658.

[ACV] D.G. Aronson-L.A. Caffarelli-J.L. Vazquez, Interfaces with a corner-point in one-dimensional porous medium flow, Comm. Pure Applied Math. 38 (1985), 375-404.

[AV] D.G. Aronson-J.L. Vazquez, The porous medium equation as finite-speed approximation to a Hamilton-Jacobi equation, J. d'Analyse Nonlinéaire, to appear.

[B] G.I. Barenblatt, On some unsteady motions of a liquid or a gas in a porous medium, Prikl. Mat. Mekh. 16 (1952), 67-78 (in Russian).

[BC] Ph. Bénilan-M.G. Crandall, The continuous dependence on ϕ of the solutions of $u_t - \Delta\phi(u) = 0$, Indiana Univ. Math. J. 30 (1971), 161-177.

[BCP] Ph. Bénilan - M.G. Crandall - M. Pierre, Solutions of the porous medium equation in $\mathbb{R}^n$ under optimal conditions on initial values, Indiana Univ. Math. J. 33 (1984), 51-87.

[BV] Ph. Bénilan, - J.L. Vazquez, Concavity of solutions of the porous medium equation, MRC TS Report #2851, to appear in Trans. Amer. Math. Soc.

[Bu] J. Buckmaster, Viscous sheets advancing over dry beds, J. Fluid Mech. 81 (1977), 735-756.

[CF1] L.A. Caffarelli, - A. Friedman, Regularity of the free-boundary for the one-dimensional flow of gas in a porous medium, Amer. J. Math. 101 (1979), 1193-1218.

[CF2] L.A. Caffarelli - A. Friedman, Regularity of the free boundary of a gas in an n-dimensional porous medium, Indiana Univ. Math. J. 29 (1980), 361-391.

[CVW] L.A. Caffarelli - J.L. Vazquez - N.I. Wolanski, Lipschitz continuity of solutions and interfaces of the N-dimensional porous medium equation, IMA Preprint Series # 191.

[CEL] M.G. Crandall - L.C. Evans - P.L. Lions, Some properties of viscosity solutions of Hamilton-Jacobi equations, Trans. Amer. Math. Soc. 282 (1984), 487-502.

[CL] M.G. Crandall - P.L. Lions, Viscosity solutions of Hamilton-Jacobi equations, Trans Amer. Math. Soc. 277 (1983), 1-42.

[D] C.M. Dafermos - Asymptotic behaviour of solutions of hyperbolic laws, in Bifurcation Phenomena in Mathematical Physics and related Topics, Proceedings NATO Adv. Study Inst. in Cargese, Corsica, 1979, 521-533.

[DK] B.E.J. Dahlberg - C.E. Kenig, Nonnegative solutions of the porous medium equation, Comm. P.D.E. 9 (1984), 409-437.

[DP] R.T. Di Perna, Decay and asymptotic behaviour of solutions to nonlinear hyperbolic systems of conservation laws, Indiana Univ. Math. J. 24 (1975), 1047-71.

[FK] A. Friedman - S. Kamin, The asymptotic behaviour of a gas in an n-dimensional porous medium, Trans. Amer. Math. Soc. 262 (1980), 551-563.

[GM] M. Gurtin - R.C. Mac Camy, On the diffusion of biological populations, Mathematical Biosciences 33 (1977), 35-49.

[K] S. Kamin, The asymptotic behaviour of the solution of the filtration equation, Israel J. Math. 14 (1973), 76-78.

[Kn] B. Knerr, The porous medium equation in one dimension, Trans. Amer. Math. Soc. 234 (1977), 381-415.

[Kr] S.N. Kruzhkov, Generalized solutions of first order nonlinear equations in several independent variables, I, Mat. Sb. 70(112)(1966), 394-415; II. Mat. Sb. 72(114)(1967), 93-116 (in Russian).

[LOT] A.A. Lacey - J.R. Ockendon - A.B. Tayler, 'Waiting-time' solutions of a nonlinear diffusion equation, SIAM J. Appl. Math. 42 (1982), 1252-1264.

[L] P. Lax, Hyperbolic systems of conservation laws II, Comm. Pure Applied Math. 10 (1957), 537-566.

[Li] P.L. Lions, Generalized solutions of Hamilton-Jacobi equations, Research Notes in Math. #69, Pitman, Boston, 1982.

[LSV] P.L. Lions, - P.E. Souganidis, - J.L. Vazquez, A degenerate viscosity approximation to the eikonal equation in several space dimensions, to appear.

[LP] T.-P. Liu, - M. Pierre, Source-solutions and asymptotic behavior in conservation laws, J. Diff. Equations 51 (1984), 419-441.

[M] M. Muskat, The Flow of Homogeneous Fluids through Porous Media, Mc Graw-Hill, New York, 1937.

[O] O.A. Oleinik, Discontinuous solutions of nonlinear differential equations, Amer. Math. Soc. Transl. (2) 26 (1983), 95-172.

[OKC] O.A. Oleinik, - A.S. Kalashnikov, Czhou Y.L., The Cauchy problem and boundary problems for equations of the type of nonstationary filtration, lzv. Akad. Nauk SSSR Ser. Mat. 22 (1958), 667-704 (in Russian).

[S] E.S. Sabinina, On the Cauchy problem for the equation of nonstationary gas filtration in several space dimensions, Dokl. Akad. Nauk. SSSR 136 (1961), 1034-1037.

[V1] J.L. Vazquez, Asymptotic behaviour and propagation properties of the one-dimensional flow of a gas in a porous medium, Trans. Amer. Math. Soc. 277 (1983), 507-527.

[V2] J.L. Vazquez, Large-time behaviour of the solutions of the one-dimensional porous media equation, in Free-Boundary Problems: Theory and Applications, vol. I, A. Fasano and M. Primicerio eds., Pitman's Research Notes in Math. # 78, 1983, 167-177.

[V3] J.L. Vazquez, The interfaces of one-dimensional flows in porous media, Trans. Amer. Math. Soc. 285 (1984), 717-737.

[V4] J.L. Vazquez, Behaviour of the velocity of one-dimensional flows in porous media, Trans. Amer. Math. Soc. 286 (1984), 787-802.

[ZR] Ya. B. Zeldovich - Yu. P. Raizer, Physics of shock-waves and high-temperature hydrodynamic phenomena, Vol. II, Academic Press, New York, 1966.

GREEN'S FORMULAS FOR LINEARIZED PROBLEMS WITH LIVE LOADS

Giorgio Vergara Caffarelli

Dipartimento de Matematica
Universita di Pisa
Via F. Buonarroti 2
56100 Pisa, ITALY

Let $\mathcal{B}$ be an elastic body, that is, a pair (Ω,T) where $\Omega \subset R^3$, the reference placement, is a bounded set with smooth or piecewise smooth boundary $\partial\Omega$ and exterior normal n, and T is a response function:

$$T(x,F) : \Omega \times \mathrm{Lin}^+ \to \mathrm{Lin} .$$

By Lin (respectively, Lin^+) we denote, as usual, the space of second-order tensors (with positive determinant).

We may specify the choice of T in order to satisfy the axioms of frame-indifference and balance of angular momentum and, possibly, to reflect certain material symmetries, but we don't need to consider such issues.

If we denote by (b^f, s^f) the loading pair, where b^f is the body force in $x \in \Omega$ and s^f is the surface traction at $x \in \partial\Omega$ exerted by the environment on $\mathcal{B}$ in the deformation f, the traction problem in finite elastostatics can be stated as follows:

Find a smooth, orientation-preserving $(\det \nabla f > 0)$ diffeomorphism $f : \overline{\Omega} \to f(\overline{\Omega})$, such that

$$-\partial_j T_{ij} (x, \nabla f) - b_i^f = 0 \quad \text{in } \Omega .$$

$$T_{ij} (x, \nabla f) n_j - s_i^f = 0 \quad \text{in } \partial\Omega .$$

If the loads b^f and s^f don't depend on the deformation f, they are said to be dead and we have the classical traction problem; otherwise, the loading is live.

An interesting class of live loadings has been considered by Spector [6,7]. These are the simple loadings, defined by constitutive equations of the form:

$$b^f(x) = b(x, f(x), \nabla f(x)), \in x \in \Omega ,$$

$$s^f(x) = s(x, f(x), D_t f(x)), \quad x \in \partial\Omega ,$$

where D_t denotes the tangential gradient operator. For example, a hydrostatic environment, practically the only well understood example of live loading, is a simple loading in the sense of Spector.

We consider more general live loadings, replacing the condition on surface traction by the condition

$$s^f(x) = s(x, f(x), \nabla f(x)) \quad x \in \partial\Omega ,$$

that is, allowing the surface traction to depend on the whole gradient and not only on the tangential derivatives [3].

With this choice of loading, the traction problem linearized about a reference deformation f_0 with gradient F_0 is the following boundary-value problem:

Find a vector field $u(x)$ such that

$$L[u]_i = -(S_{ijhk}u_{h,k})_j \; -b_{ihk}u_{h,k} \; -B_{ih}u_h = g_i \quad \text{in } \Omega ,$$

$$M[u]_i = S_{ijhk}n_j u_{h,k} \; -s_{ihk}u_{h,k} \; -S_{ih}u_h = d_i \qquad \text{on } \partial\Omega$$

where the comma indicates differentiation with respect to the corresponding space variable,

$$S_{ijhk} = \frac{\partial T_{ij}(x,F_0)}{\partial F_{hk}} \quad \text{is the linearized elasticity tensor} ,$$

$$s_{ihk} = \frac{\partial s_i(x,f_0,F_0)}{\partial F_{hk}} \quad \text{is the linearized environment tensor,}$$

and

$$b_{ihk} = \frac{\partial b_i}{\partial F_{hk}} , \quad B_{ih} = \frac{\partial b_i}{\partial f_h} , \quad S_{ih} = \frac{\partial s_i}{\partial f_h} .$$

The study of the linearized problem may be a useful tool to attempt to solve the nonlinear problem, for example, by the inverse function theorem and the implicit function theorem in suitable Banach spaces. This has been done by Valent [8] for the nonlinear displacement problem, following an idea of Gurtin and Spector.

The linearized problem is also useful when we try to study, in finite elasticity, the stability of a solution by Signorini's perturbation method (see, for example, [1]).

Moreover, the linearized problem coincides formally with the corresponding problem in the theory of small deformations superimposed on large. In this context, vector $b_{ihk}u_{h,k} + B_{ih}u_h$ specifies the body forces exerted by the environment in the small deformation u, $s_{ihk}u_{h,k} + S_{ih}u_h$ is the surface traction exerted by the environment at $x \in \partial\Omega$, and g,d are small incremental loadings, independent of u.

We do not attempt justify the linearization of the traction problem; in fact, Stoppelli has shown that the solutions of the linearized theory do not always approximate the solutions of the nonlinear problem even when the applied traction is small.

Now, for the pair of operators (L,M) we look for a Green's formula generalizing Betti's reciprocity theorem. Finding a Green's formula is a crucial point in order to apply the theory of existence of linear boundary value problems for elliptic oerators. More precisely, if we denote by L^* the operator formally adjoint to L, we wish to find an operator M^*, depending on L,M such that

$$\text{(G)} \qquad \int_\Omega L[u] \cdot v \, dx + \int_{\partial\Omega} M[u] \cdot v \, d\sigma = \int_\Omega L^*[v] \cdot u \, dx + \int_{\partial\Omega} M^*[v] \cdot u \, d\sigma$$

$$\text{for every } u,v \in C^2(\bar{\Omega}).$$

Capriz and Podio Guidugli [2] demonstrate this Green's formula in the particular case of Spector's simple loadings, supposing $\partial\Omega$ smooth. Their method of proof is inspired by that used by Fichera to deal with the problem of regular oblique derivatives for a scalar equation. They construct a bilinear form

$$a(u,v) = \int_\Omega (\alpha_{ijhk}\, u_{h,k}\, v_{i,j} + \beta_{ihk}\, u_{h,k}\, v_i + \gamma_{ih}\, u_h\, v_i)dx - \int_{\partial\Omega} S_{ih}\, u_h\, v_i\, d\sigma$$

such that

$$a(u, v) = \int_\Omega L[u] \cdot v\, dx + \int_{\partial\Omega} M[u] \cdot v\, dx \qquad \text{for every} \quad u, v \in C^2(\overline{\Omega}) \quad ,$$

and then, by an integration by parts, they obtain (G).

In order to construct the bilinear form, they make use of smooth extensions in Ω of both the vector field n and the tensor field s defined on $\partial\Omega$, and hence they need the regularity of the boundary (for example, $\partial\Omega \in C^3$). Moreover, they show that a Green's formula like (G) can be obtained if the linearized environment tensor is tangential, that is, if $s_{ihk}n_k = 0$. To see that the last condition is also necessary, we begin by setting

$$t_{ihk} = S_{ijhk}n_j - s_{ihk} , \qquad E_{ih}(n) = S_{ijhk}n_jn_k ,$$

$$A_{ih}(n) = t_{ihk}n_k = E_{ih}(n) - s_{ihk}n_k .$$

Then we consider the classical Green-Betti formula

(G.B.)
$$\int_\Omega L[u] \cdot v\, dx + \int_{\partial\Omega} S_{ijhk}n_ju_{h,k}v_i\, d\sigma =$$
$$= \int_\Omega L^*[v] \cdot u\, dx + \int_{\partial\Omega} (S_{ijhk}n_kv_{i,j} - b_{ihk}n_kv_i)u_h\, d\sigma .$$

Since we can split the gradient of a function u into its normal and tangential parts and since the value on $\partial\Omega$ of a function u and its normal derivatives are independent, we can choose $u = 0$ and $\partial u/\partial n$ arbitrary on $\partial\Omega$ in formulas (G) and (G.B.). Subtracting the formulas so obtained from each other, we get the condition

$$A(n) = E(n) ,$$

that is,

$$s_{ihk}n_k = 0 .$$

It follows that in order to deal with nontangential loadings, one must first generalize the Green's formula (G), for example, in this way:

Find a matrix V and an operator M^*, depending on L and M such that

$$(\underline{G})\quad \int_\Omega L[u]\cdot v\,dx + \int_{\partial\Omega} M[u]\cdot Vv\,d\sigma = \int_\Omega L^*[v]\cdot u\,dx + \int_{\partial\Omega} M^*[v]\cdot u\,d\sigma\,.$$

For the matrix V the previous argument immediately gives us the condition

$$V^T A(n) = E(n)\,.$$

Hence a sufficient condition in order to find V is that $\det A(n) \neq 0$: we may call this the normality condition [5].

If n is not characteristic for L on $\partial\Omega$, that is, if $\det E(n) \neq 0$, then the normality condition becomes a necessary condition also and, moreover, $\det V \neq 0$. This situation occurs when L is elliptic up to the boundary.

To construct M^* under the normality condition, we introduce the tensor

$$c_{ihk} = S_{ijhk}n_j - V_{pi}t_{phk}\,.$$

We call this tensor the tangential correction, since $c_{ihk}n_k = 0$ by definition of V. When the loading is simple,

$$c_{ihk} = s_{ihk}\,.$$

Looking for M^*, we multiply the tangential correction by $u_{h,k}v_k$ and substitute into (G.B.) to get

$$\int_\Omega L[u]\cdot v\,dx + \int_{\partial\Omega} M[u]\cdot Vv\,d\sigma + \int_{\partial\Omega} c_{ihk}u_{h,k}v_i\,d\sigma =$$

$$= \int_\Omega L^*[v]\cdot u\,dx + \int_{\partial\Omega} (S_{ijhk}n_kv_{ij} - b_{ihk}n_kv_i - S_{jh}V_{ji})u_h\,d\sigma\,.$$

The integral $\int_{\partial\Omega} c_{ihk}u_{h,k}v_i\,d\sigma$ can be evaluated by Stokes' formula in the case of $\partial\Omega$ a single regular surface or else, for example, in the case $\partial\Omega = \Sigma_1 \cup \Sigma_2$ with $\partial\Sigma_1 = \partial\Sigma_2 = \Gamma$ and Σ_1, Σ_2, Γ regular.

Indeed, set $w_k = c_{ihk}v_i u_h$ and consider $z_\ell = N_{\ell k}w_k$, where N is the skew tensor associated with n and $P - N_2 = I$, with $P = n \quad n$. For any smooth extension of $z = Nw$ to a neighborhood of Σ_1, we have

$$n \cdot \text{rot}\, z = N_{\ell s} z_{\ell,s} = N_{\ell s}N_{\ell k,s}w_k + N_{\ell s}N_{\ell k}w_{k,s} = -N^2_{ks}w_{k,s} \quad \text{on } \Sigma_1 ,$$

since $w \cdot n = 0$ and $n \cdot n = 1$ on Σ_1. Then by applying the Stokes formula, we obtain

$$\int_{\Sigma_1} -N^2_{sk}(c_{ihs}u_h v_i)_k \, d\sigma = \int_\Gamma c_{ihk}u_h v_i N_{\ell k} t_\ell^{(1)} \, ds$$

where $t^{(1)}$ is the tangential vector to Γ oriented coherently with the normal n on Σ_1. The same argument on Σ_2 leads to an analogous formula with $t^{(2)} = -t^{(1)}$. If the surfaces Σ_1 and Σ_2 have the same normal on Γ, by adding these two formulas we find that the line integral vanishes and, since $c_{ihk} = -c_{ihs}N^2_{sk}$, we get

$$-\int_{\partial\Omega} c_{ihk}u_{h,k}v_i d\sigma = \int_{\partial\Omega} (c_{ihk}v_{i,k} - N^2_{ks}c_{ihk,s}v_i)u_h d\sigma .$$

From this, we can easily derive that

$$M^*[v]_h = (S_{ijhk}n_k + c_{ihj})v_{i,j} - (b_{ihk}n_k + S_{\ell h}V_{\ell i} + N^2_{ks}c_{ihk,s})v_i ,$$

hence the Green's formula is constructed in the case that Ω is regular.

If n is discontinuous across Γ, that is, if on Γ the normal $n^{(1)}$ of Σ_1 is different from the normal $n^{(2)}$ of Σ_2, then the line integral

$$\int_\Gamma u_h v_i [c_{ihk}N_{k\ell}]_2^1 \, t_\ell^{(1)} \, ds$$

does not vanish, and we don't have a Green's formula.

However, consider for simplicity the case of a hydrostatic environment in which $s_{ihk} = c_{ihk} = -\pi(n_i \delta_{hk} - n_h \delta_{ik})$; in this case the environment tensor depends on the normal and the line integral disappears. More generally, for any tangential correction such that $c_{ihk} = c_{ihk}(x,n)$, the condition

$$c_{ihk}(x,n^{(1)})n_k^{(2)} + c_{ihk}(x,n^{(2)})n_k^{(1)} = 0 \quad \text{on} \quad \Gamma$$

guarantees the vanishing of the line integral.

In summary, if $\partial\Omega$ is smooth, or if $\partial\Omega$ is piecewise smooth but the environment tensor satisfies suitable conditions, we can obtain the Green's formula (G) under the hypothesis that the boundary operator M is normal.

From the Green's formula we can easily derive a necessary condition for the existence of solutions of the linearized traction problem. In fact, if we consider (L^*,M^*) as the adjoint of (L,M), we can state the following compatibility condition for the data (g,d):

$$\int_\Omega g \cdot v \, dx + \int_{\partial\Omega} d \cdot Vv d\sigma = 0$$

for every solution v of the homogeneous adjoint problem

$$L^*[v] = 0 \quad \text{in} \quad \Omega ,$$

$$M^*[v] = 0 \quad \text{on} \quad \partial\Omega .$$

This condition is analogous to the compatibility condition well known in linear elasticity except for the presence of the matrix V.

Moreover, if we suppose that L is elliptic in the sense of Petrowskii, M is normal, L and M satisfy the complementing condition, $\partial\Omega$ and the coefficients are smooth, then, by the usual techniques for elliptic boundary value problems, we can prove the existence and the regularity of the solution under the compatibility condition for the data.

To conclude, we wish now to give an example of the ambiguities that may accompany the failure of the normality condition [4]. Let S_{ijhk} be the classical elasticity tensor with L the associated differential operator, and let $P = n \otimes n$. If we choose $s_{ihk} = P_{i\ell} S_{\ell jhk} n_j$ and $S_{ih} = -P_{ih}$ we obtain a boundary operator which is not normal. Precisely, if we denote by $t_i[u] = S_{ijhk} n_j u_{h,k}$ the traction vector on $\partial\Omega$, the boundary operator becomes

$$M[u] \equiv (1 - P)t[u] + Pu$$

and the resulting boundary condition can be equally well classified as a live-boundary condition of traction or as a dead-boundary condition of contact.

In addition, a Green's formula for this problem follows directly from the Betti's formula

$$\text{(B)} \qquad \int_\Omega L[u] \cdot v \, dx + \int_{\partial\Omega} t[u] \cdot v \, d\sigma = \int_\Omega L^*[v] \cdot u \, dx + \int_{\partial\Omega} t^*[v] \cdot u \, d\sigma$$

by virtue of the identity

$$t[u] \cdot v - t^*[v] \cdot u = ((1 - P)t[u] + Pu) \cdot (-Pt^*[v] + (1 - P)v) -$$

$$-((1 - P)t^*[v] - Pv) \cdot (Pt[u] + (1 - P)u) .$$

In fact, if we set

$$M[u] = (1 - P)t[u] + Pu , \qquad M^*[v] = (1 - P)t^*[v] - Pv ,$$

$$V[v] = -Pt^*[v] + (1 - P)v , \qquad V^*[u] = Pt[u] = (1 - P)u ,$$

we can rewrite (B) as

$$\int_\Omega L[u] \cdot v \, dx + \int_{\partial\Omega} M[u] \cdot V[v] d\sigma = \int_\Omega L^*[v] \cdot u \, dx + \int_{\partial\Omega} M^*[v] \cdot V^*[u] d\sigma .$$

But this Green's formula differs from the one we have obtained under the normality assumption, since V is a differential operator of order 1 and not a matrix.

References

1. Capriz, G. and P. Podio Guidugli: The role of Fredholm conditions in Signorini's perturbation method, ARMA 70 (1979), 261-288.

2. Capriz, G. and P. Podio Guidugli: Duality and stability questions for the linerized traction problem with live loads in elasticity, in Stability in the Mechanics of Continua, F.H. Schroeder, Ed., Springer Verlag, Berlin (1982).

3. Podio Guidugli, P. and G. Vergara Caffarelli: On a class of live traction problems in elasticity, in Trends and Applications of Pure Mathematics to Mechanics, P.G. Ciarlet and M. Roseau, Eds., Springer Verlag, Berlin (1984).

4. Podio Guidugli, P., G. Vergara Caffarelli and E.G. Virga: The role of ellipticity and normality assumptions in formulating live boundary conditions in elasticity. M.R.C. Research Reports, University of Wisconsin - Madison (1984).

5. Podio Guidugli, P., G. Vergara Caffarelli and E.G. Virga: Una formula di Green per il problema di carichi vivi dell'elastotatica linearizzata, in Atti VII Congresso AIMETA, Trieste, Ottobre 1984.

6. Spector, S.J.: On uniqueness in finite elasticity with general loading, J. Elasticity 10 (1980), 145-161.

7. Spector, S.J.: On uniqueness for the traction problem in finite elasticity, J. Elasticity 12 (1982), 367-383.

8. Valent, T.: Local theorems of existence and uniqueness in finite elastostatics, in Proc. UTAM Symposium in Finite Elasticity, D.E. Carlson and R.T. Shield, Eds. (1980), 401-421.

SOME ASPECTS OF ADIABATIC SHEAR BANDS

T.W. Wright

Ballistic Research Laboratory
Aberdeen Proving Ground, MD 21005

Abstract

A one dimensional continuum formulation for the initiation and growth of adiabatic shear bands is reviewed, including some remarks on the underlying assumptions. A short description of some perturbation calculations introduces the idea of stress collapse and band formation as a bifurcation from homogeneous deformation. An approach for estimating the critical time of collapse for infinitesimal perturbations is introduced. Steady solutions are exhibited and their interpretation as central boundary layers is suggested. Finally it is shown that in a certain limit shear bands occur following a hyperbolic to elliptic transition in the governing equation. In a concluding remark dipolar plasticity is introduced as a possible alternative continuum description.

Introduction

Adiabatic shear banding is the name given to a localization phenomenon that occurs during rapid plastic deformation of many materials, including both metals and plastics. The process is usually thought of as coming about in the following manner. Plastic deformation heats the material locally, and if the flow stress decreases with increasing temperature, there is the possibility that thermal softening will overcome strain and strain-rate hardening and that stress will subsequently decrease with further straining. When net softening occurs, the material becomes extremely sensitive to inhomogeneities, and the deformation tends to accumulate in narrow regions where the plastic strain-rate is enormous, whereas in neighboring regions the material may even return to a purely elastic state. As localization occurs and the deformation becomes extremely inhomogeneous, heat conduction tends to damp the process.

Shear banding of this type is a major damage mechanism that can occur in machining and forming processes as well as during impact and penetration. Bands may form as macroscopic failure surfaces or as distributed microscopic damage. Many examples of naturally occurring shear bands are given in references [1-4], and in references [5-7] controlled shear experiments, designed specifically to produce isolated shear bands, are reported. Photomicrographs in those articles give a good idea of the morphology and scale of individual bands, which vary in thickness from less than 1 micron to more than 100 or 150 microns, with lateral extent in the plane of the band being many, many times its thickness. In general it appears that when shear bands form, they cut through the underlying metallurgical structures indiscriminately so that a continuum model, based on gross physical properties, can be expected to give an accurate description.

In references [5-7] the experimental data is resolved either in space or in time, but not in both simultaneously. Data exhibiting both space and time resolution during the formation and development of a single band would be of considerable importance for further theoretical advances.

Continuum Formulation

According to the general description given above, a proper continuum setting for the phenomenon should treat finite deformations of a thermo-visco-plastic material, and therein lie several substantial problems. Namely, there is no generally agreed upon version of plasticity for finite deformations in existence today; the same is true for viscoplasticity even for small deformations; and heat conduction and heat generation from plastic dissipation only make the situation more unsatisfactory. In this paper, as in all the other recent literature on adiabatic shear banding, these problems are largely bypassed or ignored by restricting consideration to a one-dimensional initial-boundary value problem for a very highly idealized material. The way in which a theoretical description of shear band formation should be embedded in a three-dimensional theory, with proper account being taken of finite deformations, invariance requirements, finite band size in lateral directions, etc., remains as a major problem for the future.

Accordingly, consider a block of incompressible material with upper and lower boundaries at $\overline{Y} = \pm H$ and undergoing only a simple shearing motion in the $\overline{X}$ direction

$$\overline{x} = \overline{X} + \overline{u}(\overline{Y},\overline{t}) \ , \ \overline{y} = \overline{Y} \ , \ \overline{z} = \overline{Z} \ ; \tag{1}$$

all dependent variables depend only on the space coordinate $\overline{Y}$ and time $\overline{t}$. The overbar signifies a dimensional quantity. Since shear bands are observed to be extremely thin with respect to their lateral extent, this motion is intended to simulate the actual motion far from the edges of the band. In nondimensional terms the governing equations may be written as follows.

$$\text{Momentum:} \quad \rho \, v_{,t} = s_{,Y} \ , \tag{2}$$

$$\text{Energy:} \quad \theta_{,t} = k\,\theta_{,YY} + s\dot{\gamma}_{\rho} \ , \tag{3}$$

$$\text{Constitutive:} \quad s_{,t} = \mu(v_{,Y} - \dot{\gamma}_{\rho}) \ , \tag{4}$$

$$K_{,t} = \frac{sh(K)}{K}\,\dot{\gamma}_{\rho} \ , \tag{5}$$

$$\text{Yield:} \quad f(s,\theta,\dot{\gamma}_{\rho}) = K \ . \tag{6}$$

Boundary conditions are taken to be

$$v(\pm 1,t) = \pm 1, \ \theta_{,Y}(\pm 1,t) = 0 \tag{7}$$

and only solutions with θ and $v_{,Y}$ symmetric about $Y = 0$ are considered. In these equations $v = u_{,t}$ is particle velocity, s is shear stress, θ is temperature change, K is a work hardening parameter, and $\dot{\gamma}_{\rho}$ is the plastic strain rate. These variables are related to the dimensional (barred) quantitities as follows.

$$Y = \overline{Y}/H, \ \ t = \dot{\gamma}_0\overline{t}, \ \ v = \overline{v}/H\dot{\gamma}_0 \ , \ \dot{\gamma}_p = \dot{\overline{\gamma}}_p/\dot{\gamma}_0 \ , \tag{8}$$

$$s = \overline{s}/K_0 \ , \ K = \overline{K}/K_0 \ , \ \theta = \overline{p}\, c\, \overline{\theta}/K_0 \ .$$

In addition nondimensional constants have been introduced for density, thermal conductivity, and elastic modulus. Respectively these are

$$\rho = \overline{\rho}H^2\,\dot{\gamma}_0^2/K_0 \ , \ k = \overline{k}/\overline{\rho}\ \ H^2\, c\, \dot{\gamma}_0 \ , \ \overline{\mu} = \mu/K_0 \ . \tag{9}$$

In (7) and (8), $\dot{\gamma}_0 = v(H,t)/H$ is the characteristic strain rate, K_0 is the initial yield stress, and c is the specific heat of the material.

In equations (2) - (6), a number of implicit simplifying assumptions have been made. In (3) it has been assumed that the elastic and thermal parts of the internal energy are completely decoupled so that there is no thermoelastic effect. In particular there is no thermal expansion, c does not depend on strain, and the elastic modulus does not depend on temperature. The term $\dot{\gamma}_p$ in (3) is the rate of plastic work, and it is assumed to be converted completely into heat. In (4) it has been assumed that the total strain rate may be decomposed into the sum of elastic and plastic parts, and that the stress is a linear function of elastic strain. In equation (5) the function $h(K)$ is the slope of a reference isothermal stress strain curve. Implicit in this equation is the assumption that the work hardening parameter, K, evolves according to the plastic work done, so that its evolution is the same in a slow isothermal test as in a rapid adiabatic test. Equation (6) states that the yield surface depends on the plastic strain rate, as well as the stress and the temperature. If $\dot{\gamma}_p = 0$, then the equation simply gives the static yield surface in stress temperature space. It is natural to assume that the function f has the properties $f_s > 0$, $f_\theta > 0$, and $f_{\dot{\gamma}p} < 0$. Thus if $f(s,\theta,0)$ is greater than K, plastic flow occurs, and if it is less than or equal to K, the plastic strain rate is zero, and the material deforms only elastically.

Alternatively, by inverting (6) it is possible to write a rate equation for γ_p :

$$\dot{\gamma}_p = g(s,\theta,K), \tag{10}$$

where to be consistent the function g must have the properties $g_s > 0$, $g_\theta > 0$, and $g_K < 0$. In this form it is clear that K and $\dot{\gamma}_p$ have the status of internal variables, which are controlled by rate equations.

In the preceding discussion it has been assumed that there is a definite yield surface where the plastic strain rate vanishes. Many authors prefer to use (10), but without a definite yield surface, for ease in computations. Several examples of this approach are to be found in reference [2], but in every case it turns out that there is a transition region where the plastic strain rate becomes extremely small and the deformation is essentially elastic.

In all the calculations described in this paper, the functions g and h have been taken as follows.

$$g(s,\theta,K) = \frac{\text{sgn } s}{b} \{[\frac{|s|}{K(1-a\theta)}]^{1/m} - 1\}, \; h(K) = \frac{n}{\psi_0} K^{\frac{n-1}{n}} \; . \quad (11)$$

These are empirical functions, which are intended to be fit to representative data for particular materials. They will be more readily recognized in the forms

$$|s| = K(1-a\theta)(1+b|\gamma_p|)^m \; , \; K = (1 + \frac{\psi}{\psi_0})^n$$

where ψ is plastic strain in a reference test. Equation (11) has introduced five more nondimensional constants, namely

$$n \; , \; \psi_0 \; , \; b = \bar{b}\,\dot{\gamma}_0, \; m \; , \; a = \frac{\bar{a}k_0}{\bar{\rho}c} \; . \quad (12)$$

The first one is known as the work hardening exponent, the second can be used to adjust the slope at first yield for the reference curve, $\bar{b}$ is a characteristic time, m is the rate hardening exponent, and $\bar{a} = d\bar{s}/d\bar{\theta}$ is the slope of the thermal softening curve, which has been assumed to be linear. In equations (9) and (12) there are eight independent nondimensional parameters in all that control the behavior of the response. Of the eight, only three contain the externally imposed quantities H and γ_0, and five are determined entirely by intrinsic physical properties of the material. In any case it is to be expected that the overall response may be quite different in different regions of parameter space, and because there are so many independent parameters, it may be some time before the full range of effects can be known.

Results of Finite Element Calculations

In two other papers [16,17] the results of finite element calculations for the system of equations (2) - (5) and (11) have been described in some detail. Some of the principal features of those calculations are summarized here.

First note that the equations have solutions with $v = Y$ and all other response functions independent of Y. Examples of these so-called homogeneous solutions for particular choices of the eight parameters have been given in [17] and are shown in Figure 1. The choice a = b = 0 generates the reference stress

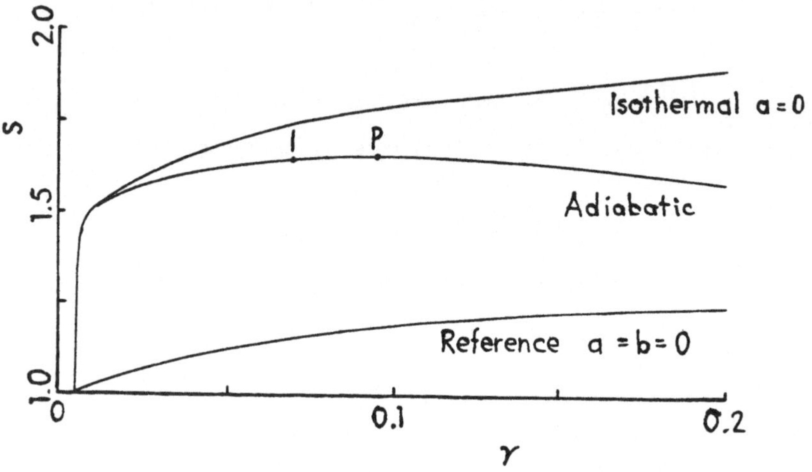

Figure 1

Sketch of Typical Reference, Isothermal, and Adiabatic Stress Strain Response for Homogeneous Deformation.

strain curve, and the choice a = 0 generates an isothermal response curve for each applied strain rate. When both a and b have finite values, the response curve starts out like an isothermal curve, but as the plastic work accumulates, the temperature rises and softening begins. Finally at a critical strain the stress response reaches a maximum, denoted P in the figure, and decreases thereafter. This type of behavior for thermal softening materials is well known (eg., [5,7,8,11-14] and others), although the details of the thermo-visco-plasticity models vary among the various authors. In fact the softening response is typical for many cases of materials with damage, with the damage often being represented by internal variables. Softening may also be induced by geometric changes, the most familiar example being change of cross sectional area in a tension test.

No matter what its origin, once softening begins there is opportunity for localization to occur. With the system of equations described above, the onset of localization has been observed by adding a temperature perturbation to the homogeneous solution and restarting the finite element calculation as a new initial-boundary value problem. The perturbation was positive and symmetric about the center of the interval, and it was added at the point marked I in Figure 1, just before peak homogeneous stress. Complete details have been given in reference [16].

Two cases were considered: one with central height of the temperature bump $\Delta\theta = 0.1$ and width 0.5, and one with 1/5th the height and width of the first. The stress response, which is nearly constant in y, is shown in Figure 2. In the case

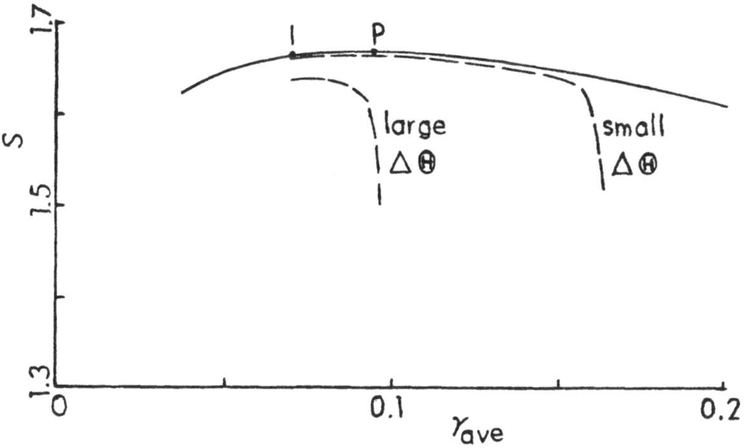

Figure 2

Response to Temperature Perturbations.

of the smaller perturbation the stress follows the homogeneous response rather closely at first, until well past the peak homogeneous stress, but eventually the response deviates abruptly, and the stress appears to collapse. For the larger perturbation the pattern is similar except that stress collapse sets in even before peak homogeneous stress. In both cases cross sectional plots of temperature or plastic strain rate at fixed times show large peaks developing in the center of the interval as the stress collapses. At the end of the calculation the central value for $\dot{\gamma}_p = \dot{\bar{\gamma}}_p/\dot{\gamma}_0 > 60$. As the central peaks develop, the plastic strain rate drops to zero at the edges of the interval, and the temperature becomes constant there since plastic working vanishes with plastic strain rate. Calculations for finite deformations have also been reported in [7,13-15] where somewhat different versions of visco-plasticity or just nonlinear elasticity were used. Delayed stress collapse was noted in [13], as well, so that the phenomenon appears to be generic for softening behavior and not specific for the particular constitutive description used.

Estimate for the Maximum Critical Time of Stress Collapse

Figure 2 is reminiscent of curves for imperfection sensitivity in a bifurcation problem. All that is missing is the bifurcation branch itself. In the usual buckling problem some measure of deflection is plotted against an external loading parameter. As the load increases, no change occurs in the perfect system until a critical load occurs where the possible solution paths in load-deflection space split into multiple branches. These principal bifurcation branches represent limiting behavior for an imperfect system so they give useful information about the response, although the imperfection sensitivity may be such that, as a practical matter, it is extremely difficult to approach the ideal response. From this point of view there should be a curve (or curves) in Figure 2 that breaks off from the homogeneous response, plunging steeply and lying somewhat to the right of both perturbation curves. Such a curve would indicate the largest strain that could ordinarily be achieved without localization. References [22-24] discuss bifurcation and stability in the context of elastic, plastic, and general systems (mostly in steady motion) respectively.

In the present case, time plays the role of the external load, but it is much more than just a passive parameter, as in a static buckling problem, since here it is also an independent variable in the differential equations. In an attempt to examine the stability of the homogeneous response, several authors have considered the equations of first variation with respect to the homogeneous response [8-12,14], and that will be the approach taken here. With the variations expressed as $v = v - v_H$, etc., where the subscript H indicates the homogeneous response, the linearized variational equations have the form

$$u_{,t} = u + A(t)u \tag{13}$$

where $u = (\tilde{v}, \tilde{s}, \tilde{\theta}, \tilde{K})$, is a linear matrix differential operator that contains first and second derivatives on Y, and A(t) is a time dependent matrix with entries determined by the homogeneous motion. Homogeneous boundary conditions (13) are $\tilde{v}(\pm 1,t) = \theta_{,Y}(\pm 1,t) = 0$.

In previous work the variational equations have usually been treated either as an initial-boundary value problem or as a system with constant coefficients. Since the coefficients are not constant, the latter approach tacitly assumes without further justification that in some sense the coefficients are slowly varying in comparison with the solutions to be found. When treated as an initial-boundary value problem, it is always found that solutions can begin to grow exponentially once the peak in the homogeneous response has been passed. However, it is also always found that the rate of growth is far too small to explain the most important aspects of observed behavior. In fact the growth of the perturbation in Figure 2 for the smaller disturbance is approximately exponential at first, but at a very low rate, and the early behavior gives no hint about the timing for stress collapse. In view of the results shown in Figure 2 it seems likely that the treatment as an initial-boundary value problem can only give limited information about the sensitivity to infinitesimal imperfections, and no information at all about either the point of bifurcation or the bifurcation branch. In [11] the use of a Lyapunov function provides a third approach to examine stability and is used to analyze the unsteady shearing of a thermoviscous fluid, that is, a Newtonian fluid with a temperature dependent viscosity. However, a complete bifurcation analysis is not actually carried through, the difficulty being that solutions of the perturbation equations with the time-dependent coefficients are not known.

When the left-hand side of (13) is multiplied through by the row matrix $(\tilde{v},\tilde{s}\beta\tilde{\theta},\alpha\tilde{\kappa})$, where β and α are arbitrary (at this stage) but positive weights, it becomes the exact derivative of a sum of squares. After integration over the interval followed by use of the boundary conditions and division of both sides by the integral of the weighted sum of squares, (13) becomes

$$\frac{1}{E}\frac{dE}{dt} = \frac{1}{E}\left\{-\beta k \int_I \tilde{\theta}_Y^2\, dY + A_{ij}(t) \int_I u_i\, u_j dY\right\} \tag{14}$$

$$= \Lambda\, (\tilde{v},\tilde{s},\tilde{\theta},\tilde{K})$$

$$\text{where}\quad E = \frac{1}{2} \int_I \left(\rho\tilde{v}^2 + \frac{1}{\mu}\, \tilde{s}^2 + \alpha\tilde{\kappa}^2 + \beta\tilde{\theta}^2\right) dY \; .$$

For all possible solutions of (13) we would like to know when the right-hand side of (14) can first become positive. Thus for each value of time it is necessary to look for the maximum over solutions of (13). Only the symmetric part of A_{ij} enters into (14), and the most advantageous choices for α and β will be left for later.

Rather than maximizing over solutions, which are unknown, it is easier to regard the right hand side as a Raleigh quotient for which stationary values are sought among all smooth functions which satisfy the boundary conditions. Since this set of functions contains the solutions, the maximum for the Rayleigh problem will be a lower bound for the critical time in the original problem. This approach for estimating the bifurcation point when the coefficients are time varying was introduced by Serrin [25] and has been used to advantage in unsteady fluids problems by Neitzel and Davis [26]. The addition of the free constants in the present case should allow some optimization of the estimate.

To illustrate this idea consider the simple case where $K = 1$ and $\mu = \infty$. Then the homogeneous problem has the solution

$$v_H = Y,\ \dot{\gamma}_{pH} = 1,\ s_H = (1+b\dot{\gamma}_{pH})^m(1-a\theta_H)\ ,$$
$$(1-a\theta_H) = \exp\{-a(1+b)^m t\} \tag{15}$$

and the Rayleigh problem becomes

$$\Lambda = \frac{1}{E} \{-\beta k \int_I \theta,_Y^2 \, dY + A^S_{\alpha\beta}(t) \int_I u_\alpha u_\beta \, dY\} \tag{16}$$

where now $E = \frac{1}{2} \int_I (\rho\tilde{v}^2 + \beta\tilde{\theta}^2)dY$, $u = (\tilde{s},\tilde{\theta})$ and the matrix A is given by

$$\{A\} = \{\begin{matrix} -g_s & -g_\theta \\ \beta(sg)_s & \beta(sg)_\theta \end{matrix}\} . \tag{17}$$

The plastic strain rate $\dot{\gamma}_{pH} = g(s_H,\theta)$ is found by inverting $(15)_3$. Next s is eliminated in favor of $\tilde{v}$ and $\tilde{\theta}$ by using the exact perturbation equation

$$\tilde{s} = -\frac{1}{A_{11}} (\tilde{v},_Y + A_{12}\tilde{\theta}). \tag{18}$$

Now the Rayleigh quotient, considered as a variational problem, leads to two coupled, second order, ordinary differential equations in $\tilde{v}$ and $\tilde{\theta}$ with Λ as an eigenvalue. The choice

$$\tilde{v} = \tilde{v}_0 \sin \lambda_j Y, \quad \tilde{\theta} = \tilde{\theta}_0 \cos \lambda_j Y ,$$

$$\lambda_j = j\pi , \quad j = 1,2,3.....$$

has the desired symmetry about $Y = 0$ and satisfies the necessary boundary conditions. The characteristic matrix is symmetric so the eigenvalues are real, and the characteristic equation for Λ turns out to be

$$\Lambda^2 - (-2k\lambda_j^2 - \frac{2\lambda_j^2}{\rho g_s} - 2\frac{g g_\theta}{g_s})\Lambda + \{\frac{4k\lambda_j^2}{\rho g_s} - \frac{[\beta(sg)s - g_\theta]^2}{\beta\rho g_s^2} - \frac{4(sg)_\theta}{\rho g_s}\}\lambda_j^2 = 0 . \tag{19}$$

From the sign of the coefficients in (19) it is clear that

$$\Lambda^{(1)} + \Lambda^{(2)} < 0, \text{ always}$$

and

$$\Lambda^{(1)}\Lambda^{(2)} > 0, \text{ provided that} \tag{20}$$

$$(1+m)^2x^2 - [2a(-1 + m) + 4ks_j^2\, m(1+b)^{-m}]x + a^2 < 0$$

where $x = \beta(1-a\theta)$. When (20) and (21) hold, both eigenvalues are negative, and therefore, the left side of (16) is guaranteed to be negative for all solutions of the perturbation equations. The left-hand side of (21) is a parabola in x that opens upwards and has two real positive roots, $r_1 > r_2 > 0$. The constant β has not yet been chosen, but since x decreases with increasing t, and since $\theta(0) = 0$, the best choice is $\beta = r_1$. Then the left hand side of (21) will be less than or equal to zero from $t = 0$ until the time when $x = r_2$. That is to say, for the perturbation problem

$$\frac{1}{E}\frac{dE}{dt} < 0 \quad \text{if} \quad t < \frac{1}{a(1+b)^m} \ln \frac{r_1}{r_2} \tag{22}$$

and equality in $(22)_2$ gives a lower bound estimate for the critical bifurcation time. Note that since the two roots r_1 and r_2 depend on the bifurcation mode j, there is actually a sequence of increasing, critical bifurcation times that corresponds to increasing mode numbers $j = 1,2,3...$.

It turns out that if either the (nondimensional) thermal conducitivity k or the rate sensitivity m vanishes, then $r_1 = r_2$ for all modes, so that the best estimates for the critical bifurcation times are all $t = 0$. As one final comment, note that having an estimate for the critical bifurcation time only provides an estimate for the time of stress collapse in the case of infinitesimal perturbations; it does not provide an estimate for any case of finite perturbation.

The success of this approach for the simple example given above shows that it holds promise for the more general case, which will be reported in a future paper.

Steady Solutions

Equations (2) - (6) also have steady solutions, where the four variables $(v,s,K\theta)$ depend only on the spatial coordinate Y, but not on time t. For this to be strictly true, equation (5) implies that $h(K) = 0$ throughout the whole interval, which in turn implies that K is constant. That is to say, the material has saturated with respect to work hardening. Since s = const by (2), and (6) can be solved to get $\dot{\gamma}_p = \Gamma(\theta; s, K)$, equation (3) reduces to an ordinary differential equation in θ with a first integral

$$\frac{1}{2}\ \theta_Y^2 = \frac{s}{k} \int_\theta^{\theta_c} \Gamma \, d\theta \tag{23}$$

where θ_c is the temperature at the center of the band. Equation (23) may be solved by quadrature to obtain a solution of the form

$$Y = F(\theta; \theta_c, s, K)\ . \tag{24}$$

Since the saturation flow stress K may be regarded as a material property, equation (24) shows that in general there is a two-parameter family of steady solutions for a given material. Once (23) has been obtained, the velocity field may be found by integrating the steady version of (4). Typical profiles are shown in Figure 3. To give a better idea of the physical scaling in a shear band, for

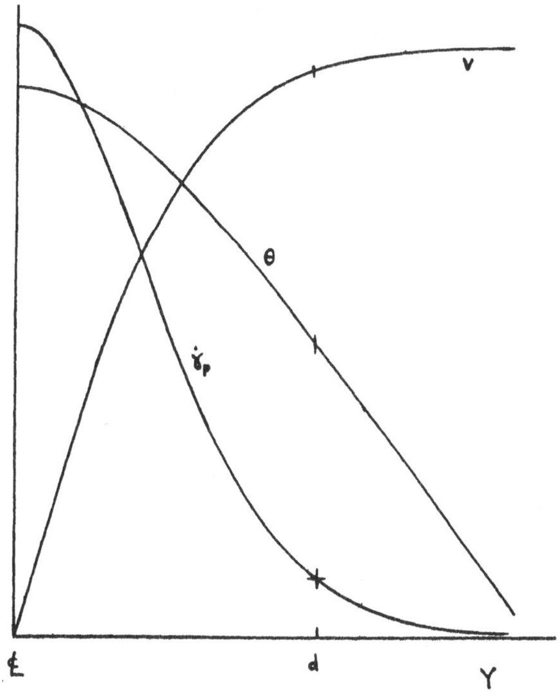

Figure 3

Typical Configuration for Steady Response.

the Litonski law in (11) equation (24) may be written as

$$\bar{Y} = \frac{1}{\sqrt{2}} \sqrt{\frac{bk}{aK}} \left(\frac{1-a\theta_c}{1-a\theta_a} \right)^{\frac{1+m}{2m}} \int_0^{\frac{\bar{a}(\bar{\theta}_c - \bar{\theta})}{1-a\theta_c}} \frac{dt}{t^{1/2} \{\frac{m}{1-m} [\frac{1 - (\frac{1}{1+t})^{\frac{1-m}{m}}}{t}] - (\frac{1-a\theta_c}{1-a\theta a})^{\frac{1}{m}}\}^{\frac{1}{2}}} \quad (25)$$

where $Y = 0$ locates the center of the band. In this form all quantities are dimensional, and the residual integral is $O(1)$. θ_a is the temperature where $\dot{\bar{\gamma}}_p = 0$, and there $\bar{s} = \bar{K}(1-\bar{a}\bar{\theta}_a)$ holds. Note the square root scaling on the basic physical quantities, but note especially the extremely strong dependence on the temperature contrast between the center and the edge of the band when the strain rate sensitivity is small. Typical values for m in many metals are around 0.02. The temperature term may also be written

$$\frac{1 - \bar{a}\,\bar{\theta}_c}{1 - \bar{a}\,\bar{\theta}_a} = (1 + \bar{b}\,\dot{\bar{\gamma}}_p^c)^{-m}$$

so for m small and $\bar{b}\,\dot{\bar{\gamma}}^c$ large (the usual case) the length scale varies nearly as the inverse square root of the central plastic strain rate and only weakly with $\bar{b}$.

In Figure 3, d is defined arbitrarily as the distance from the center of the band where the plastic strain rate has fallen to 1/10th its central value. At that distance the velocity has nearly reached its extreme value and the temperature gradient has become nearly constant. Thus, it gives a convenient point for measuring the size and strength of the shear band. The following table shows typical results for a realistic material.

Shear Band Values for $\bar{\theta}_c$ = 400°K and Various Strain Rates

$\dot{\gamma}_p^c$, s^{-1}	d, μm	$\bar{\theta}(d)$, °K	$\bar{v}$, m/s	$\bar{s}$, GPa
10^3	138	334	0.076	0.467
10^4	43	334	0.234	0.489
5×10^4	19	334	0.514	0.505

Note that the temperature difference in the band is not large, but the plastic strain rate varies by a factor of 10 over distances measured in 10's or at most 100's of microns. These characteristic values have been computed using realistic numbers for the material constants of a strong steel, so it is encouraging that the numbers come out in the right general range for known experimental values.

The steady solutions by themselves are of limited utility because they tend to constant, and rather large, temperature gradients away from the center, as can be seem by rough estimation from values in the table. However, note that when the time in (2) - (5) is scaled according to the rule $t = \tau/\delta$, where δ is a small parameter, then the steady equations (vanishing left hand sides) result in the limit as $\delta \to 0$. With τ being held fixed, $t \to \infty$, so the steady equations apparently have the interpretation of holding asymptotically at large times. but since they cannot meet arbitrary boundary conditions away from the center, they must hold only in a central boundary layer. With that interpretation the outer configuration of the steady solutions defines inner boundary values for an exterior solution. The steady solutions and their probable interpretation as boundary layers will be explored in a future paper.

The Essential Embedded Problem

Another point of view has been suggested by Varley [27], namely that the essential features of the localization type of material response can be captured by some simpler problem, which is embedded within the full set of equations. From that point of view, many of the mathematical complications stem from "small higher order terms" that modulate the basic response. To look for a simpler embedded problem, consider the special case of a rigid/plastic, non-heat-conducting, and non-rate-sensitive material. With $\mu = \infty$, and $k = f_{\gamma_p} = 0$, equations (2) - (6) become

$$s_{,y} = \rho v_{,t} \;,\; v_{,y} = p \;,$$
$$K_{,t} = \frac{s}{K} H(K)p,\; \theta_{,t} = sp \;, \qquad (26)$$
$$s = K_g(\theta) .$$

where $p = \dot{\gamma}_p > 0$ and $g(\theta)$ carries the temperature sensitivity of the material. When $p = 0$, unloading occurs, and in an unloaded region the stress is constant. With v eliminated from the first two equations and sp eliminated from the next two, (26) becomes

$$s_{,YY} = \rho p_{,t} \ , \ M(K) = \theta + A(Y), \ s = S(\theta) \ . \tag{27}$$

In (27) M(K) is a monotonically increasing function since h(K) is positive, A(Y) is an arbitrary function, and in $S(\theta)$ the dependence on Y has been suppressed. Since $g(\theta)$ must be a decreasing function to represent thermal softening, it is not difficult to find combinations of $g(\theta)$ and h(K) such that the function $S(\theta)$ has a single maximum. For example for linear work hardening with constant modulus G_p and linear thermal softening, the stress is given by $s = S(\theta) = \sqrt{2G_p(\theta+A)(1-a\theta)}$, which has a maximum at $\theta_m = \frac{1}{3a}[1-2aA(Y)]$. It may be assumed that θ_m is positive as shown in Figure 4a. Now with the aid of $(26)_4$ and $(27)_3$ the momentum equation (27) becomes

$$[s(\theta)]_{,YY} = \rho\left[\frac{\theta_{,t}}{S(\theta)}\right]_{,t} \ . \tag{28}$$

This equation is hyperbolic when $S_\theta > 0$ and elliptic when $S_\theta < 0$, and since θ can only increase, each point either evolves toward the hyperbolic/elliptic transition and beyond or else unloading occurs. Figure 4b shows a sketch of the conjectured domains for a typical initial/boundary value problem. Alternatively the H-E boundary would extend to $Y = \pm 1$, and the E-U boundary would come in from the sides at a later time. In either case both θ and $\theta_{,t}$ would be specified along part of the boundary of the E-region, which will generally tend to produce singular behavior at later times in accordance with Varley's conjecture. Without heat conduction or a rate effect there is no mechanism to limit the singularity.

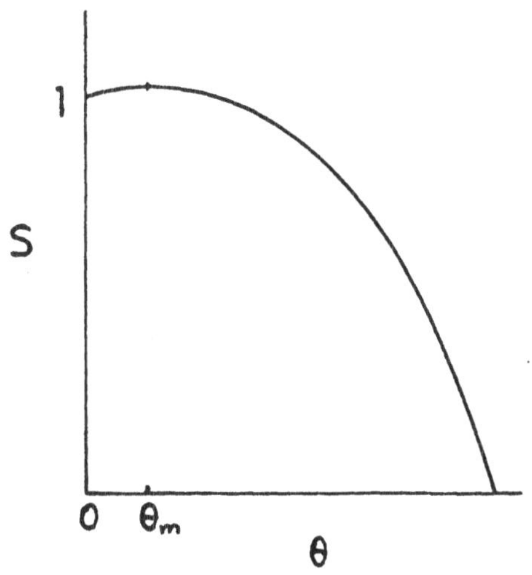

Figure 4a

Sketch of Stress-Temperature Response in the Absence of Heat Conduction and Viscous Effects for a Rigid-Plastic Material

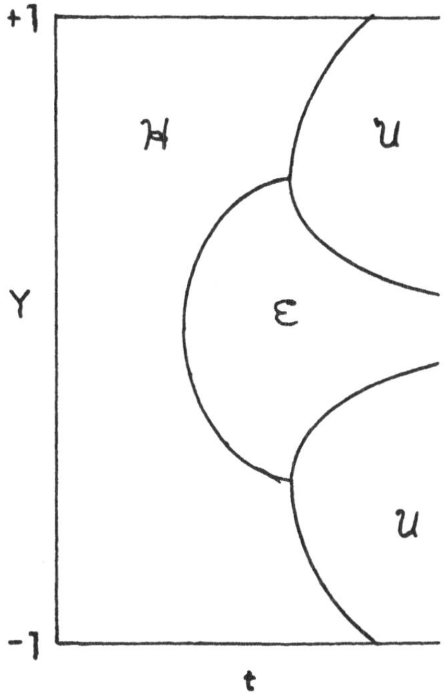

Figure 4b

Conjectured Hyperbolic, Elliptic, and Unloaded Regions for an Initially Inhomogeneous Temperature Distribution.

Dipolar Plasticity

Since large temperature and strain gradients occur during localization, it seems reasonable to examine the consequences of using a dipolar type theory of plasticity, as formulated by Green, McInnis, and Naghdi [28]. A rate sensitive version of the dipolar theory has been set down by Wright and Batra [17]. The equations have homogeneous solutions, which are the same as for the simple case, and preliminary perturbation calculations of the same type as discussed above indicate that the dipolar effect is stabilizing as the homogeneous response changes from increasing stress to decreasing stress, much in the same way that bending stiffness in a thin wire is stabilizing as the axial force changes from tension to compression. Further calculations are in progress.

References

Material Aspects, Occurance and Qualitative Description

1. H.C. Rogers, Adiabatic Shearing - General Nature and Material Aspects, in Material Behavior Under High Stress and Ultrahigh Loading Rates Sagamore Army Materials Research Conference Proceedings 29, eds. J. Mescall and V. Weiss, Plenum Press, New York (1983) 101-118 (see also Rogers, Ann. Rev. Mat. Sci. 9 (1979) 283).

2. S.L. Semiatin, G.D. Lahoti and S.I. Oh, The Occurance of Shear Bands in Metalworking, Sagamore 29, 119-160.

3. I.M. Hutchings, The Behavior of Metals Under Ballistic Impact at Sub-Ordnance Velocities, Sagamore 29, 161-198.

4. D.E. Grady, J.R. Asay, R.W. Rohde, and J.L. Wise, Microstructure and Mechanical Properties of Precipitation Hardened Aluminum Under High Rate Deformation, Sagamore 29, 81-100.

Experimental Observations on Isolated Bands

5. L.S. Costin, E.E. Crisman, R.H. Hawley and J. Duffy, On the Localization of Plastic Flow in Mild Steel Tubes Under Dynamic Torsional Loading, in Mechanical Properties at High Rates of Strain, Proc. 2nd Oxford Conf. 1979, ed. J. Harding, Inst. Phys., London (1980) 90-100

6. G.L. Moss, Shear Strains, Strain Rates, and Temperature Changes in Adiabatic Shear Bands, in Shock Waves and High-Strain Rate Phenomena in Metals, eds. M.A. Meyers and L.E. Murr, Plenum Press, New York (1981) 299-312 (See also Technical Report ARBRL-TR-02242 Ballistic Research Laboratory, Aberdeen Proving Ground, MD 21005, May 1980)

7. U.S. Lindholm and G.R. Johnson, Strain-Rate Effects in Metals at Large Shear Strains, Sagamore 29, 61-79. (See also Johnson, Hoegfeldt, and Lindholm, J. Eng. Mat. and Tech., Trans. ASME 105 (1983) 42-47 and 48-53).

Linear Stability Analyses

8. R.J. Clifton, Adiabatic Shear Banding, Chap. 8 in Materials Response to Ultra-High Loading Rates, NRC Rept. NMAB-356 (1980).

9. Y.L. Bai, Thermo-Plastic Instability in Simple Shear, J. Mech. Phys. Sol. 30 (1982) 195-207.

10. J. Pan, Perturbation Analysis of Shear Strain Localization in Rate Sensitive Materials, Int. J. Solids and Structures 19 (1983) 153-164.

11. T.J. Burns and T.G. Trucano, Instability in Simple Shear Deformations of Strain Softening Materials, Mech. Mat. 1 (1982) 313-324.

12. T.J. Burns, Approximate Linear Stability Analysis of a Model of Adiabatic Shear Band Formation, SAND83-1907 (Oct. 1983) Sandia National Laboratories, Albuquerque, NM 87185; Quart. Appl. Math. 43 (1985) 65-84.

Nonlinear Calculations of Shear Band Formation

13. A.M. Merzer, Modelling of Adiabatic Shear Band Development from Small Imperfections, J. Mech. Phys. Sol. 30 (1983) 323-338.

14. R.J. Clifton, J. Duffy, K.A. Hartley and T.G. Shawki, On Critical Condtions for Shear Band Formation at High Strain Rates, Scripta Met. 18 (1984) 443-448 (See also Shawki, Clifton, Majda, Analysis of Shear Strain Localization in Thermal Visco-Plastic Materials, Brown Univ. Report ARO DAAG29-81-K-0121/3 Oct 1983)

15. F.H. Wu and L.B. Freund, Deformation Trapping due to Thermo-plastic Instability in One-Dimensional Wave Propagation, J. Mech. Phys. Sol. 32 (1984) 119-132.

16. T.W. Wright and R.C. Batra, The Initiation and Growth of Adiabatic Shear Bands, Int. J. Plasticity 1 (1985) 205-212.

17. T.W. Wright and R.C. Batra, Further Results on the Initiation and Growth of Adiabatic Shear Bands at High Strain Rates, Int. Symp. DYMAT 85, Paris, J. de Physique 46 (1985), Colloque C5, 323-330.

Continuum Model with Distributed Shear Bands

18. L. Seaman, D.R. Curran and D.H. Shokey, Scaling of Shear Band Fracture Processes, Sagamore 29, 295-307 (see also Erlich, Seaman, Shocky and Curran, Development and Application of a Computational Shear Band Model, Stanford Research Institute DAAD05-76-C-0762 (1977)).

Viscoplasticity

19. N. Cristescu, Dynamic Plasticity, North Holland, Amsterdam (1967).

20. P. Perzyna, Thermodynamic Theory of Viscoplasticity, Advances in Applied Mechanics 11 (1971) 313-354.

21. L.E. Malvern, Experimental and Theoretical Approaches to Characterization of Material Behavior at High Rates of Deformation, in Mechanical Properties at High Rates of Strain, Proc. 3rd Oxford Conf. 1984, ed. J. Harding, Inst. Phys., London (1984) 1-20.

Bifurcation Theory

22. B. Budiansky, Theory of Buckling and Post-Buckling Behavior of Elastic Structures, Adv. Appl. Mech. 14, ed. C.-S. Yih, Academic Press, New York (1974) 2-66.

23. J.W. Hutchinson, Plastic Buckling, Adv. Appl. Mech. 14, ed. C.-S. Yih, Academic Press, New York (1974) 57-145.

24. G. Iooss adn D.D. Joseph, Elementary Stability and Bifurcation Theory, Springer-Verlag, New York (1980).

25. J. Serrin, On the Stability of Viscous Fluid Motions, Arch. Rat. Mech. Anal. 3 (1959) 1-13.

26. G.P. Neitzel and S.H. Davis, Energy Stability Theory of Decelerating Swirl Flows, Phys. Fluids 23 (1980) 432-437.

27. E. Varley, private communication (1985).

28. A.E. Green, B.C. McInnis, and P.M. Naghdi, Elastic-Plastic Continua with Simple Force Dipole, Int. J. Eng. Sci. 6 (1968) 373-394.

Table of Contents from Other Volumes from the Program in Continuum Physics and Partial Differential Equations

Homogenization and effective moduli of materials

October 22 – October 26, 1984

J. L. Ericksen
D. Kinderlehrer
R. Kohn
J.-L. Lions
Conference Committee

Theory and applications of liquid crystals

January 21 - January 25, 1985

J. L. Ericksen
D. Kinderlehrer
Conference Committee

Tentative contributors: Berry, G., Brezis, H., Capriz, G., Choi, H. I., Cladis, P., Di Benedetto, E., Hardt. R. and Kinderlehrer, D., Leslie, F., Luskin, M., Miranda, M., Ryskin, G., Sethna, J., and Spruck, J.

Amorphous polymers and non-newtonian fluids

March 4 - March 8, 1985

J. L. Ericksen
D. Kinderlehrer
M. Tirrell
S. Prager
Conference Committee

Tentative contributors: Bird, R., Caswell, B., Dafermos, C., Hrusa, W. and Renardy, M., Joseph, D. D., Kearsley, E., Marcus, M. and Mizel, V., Nohel, J. and Renardy, M., Rabin, M., Wool, R. P.

Oscillation theory, computation, and methods of compensated compactness

April 1 - April 4, 1985

C. Dafermos
J. L. Ericksen
D. Kinderlehrer
M. Slemrod
Conference Committee

Chacon, T. and Pironneau, O.	Convection of microstructures by incompressible and slightly compressible flows
DiPerna, R.	Oscillations in solutions to nonlinear differential equations
Forest, M.G. and Lee, J.-L.	Geometry and modulation theory for the periodic nonlinear Schrödinger equation
Harten, A.	On high-order accurate interpolation for non-oscillatory shock capturing schemes
Lax, P.	On the weak convergence of dispersive difference schemes
Majda, A.	Nonlinear geometric optics for hyperbolic systems of conservation laws
McLaughlin, D.	On the construction of a modulating multiphase wavetrain for a perturbed KdV equation
Nunziato, J., Gartling, D., and Kipp, M.	Evidence of nonuniqueness and oscillatory solutions in in computational fluid mechanics
Osher, S. and Chakravarthy, S	Very high order accurate T V D schemes
Rascle, M.	Convergence of approximate solutions to some systems of conservation laws: a conjecture on the product of the Riemann invariants

Schonbek, M.	Applications of the theory of compensated compactness
Serre, D.	A general study of the commutation relation given by L. Tartar
Slemrod, M.	Interrelationships among mechanics, numerical analysis, compensated compactness, and oscillation theory
Venakides, S.	The solution of completely integrable systems in the continuum limit of the spectral data
Warming, R. and Beam, R.	Stability of finite-difference approximations for hyperbolic initial boundary value problems
Yee, H.	Construction of a class of symmetric T V D schemes

Dynamical problems in continuum physics

June 3 - June 7, 1985

J. Bona
C. Dafermos
J. L. Ericksen
D. Kinderlehrer
Conference Committee

Amick, C. and Kirchgässner, K.	Solitary water-waves in the presence of surface tension
Beals, M.	Presence and absence of weak singularities in nonlinear waves
Beatty, M.	Some dynamical problems in continuum physics
Beirao da Veiga, H.	Existence and asymptotic behavior for strong solutions of the Navier Stokes equations in the whole space
Bell, J.	A confluence of experiment and theory for waves of finite strain in the solid continuum
Bona, J.	Shallow water waves and sediment transport
Chen, P.	Classical piezoelectricity: is the theory complete?
Keller, J.	Acoustoelasticity
McCarthy, M.	One dimensional finite amplitude pulse propagation in electroelastic semiconductors
Morawetz, C.	Weak solutions of transonic flow by compensated compactness
Müller, I.	Extended thermodynamics of ideal gases

Pego, R.	Phase transitions in one dimensional nonlinear viscoelasticity: admissibility and stability
Shatah, J.	Recent advances in nonlinear wave equations
Slemrod, M.	Dynamic phase transitions and compensated compactness
Spagnolo, S.	Some existence, uniqueness, and non-uniqueness results for weakly hyperbolic equations in Gevrey classes
Strauss, W.	On the dynamics of a collisionless plasma